This text introduces the theoretical framework for describing the quark–gluon plasma, an important new state of matter.

The first part of the book is a self-contained introduction to relativistic thermal field theory. Topics include the path integral approach, real- and imaginary-time formalisms, fermion fields and gauge fields at finite temperature. Useful techniques such as the evaluation of frequency sums and the use of cutting rules are illustrated by various examples. The second part of the book is devoted to recent developments, giving a detailed account of collective excitations (bosonic and fermionic), and showing how they give rise to energy scales which imply a reorganization of perturbation theory. The relationship with kinetic theory is also explained. Applications to processes which occur in heavy ion collisions and in astrophysics are worked out in detail. Each chapter ends with exercises and a guide to the literature.

T0181922

CAMBRIDGE MONOGRAPHS ON
MATHEMATICAL PHYSICS

General Editors: P. V. Landshoff, D. R. Nelson, D. W. Sciama, S. Weinberg

THERMAL FIELD THEORY

Cambridge Monographs on Mathematical Physics

[†] Issued as paperback

THERMAL FIELD THEORY

MICHEL LE BELLAC
University of Nice

CAMBRIDGE
UNIVERSITY PRESS

PUBLISHED BY THE PRESS SYNDICATE OF THE UNIVERSITY OF CAMBRIDGE
The Pitt Building, Trumpington Street, Cambridge, United Kingdom

CAMBRIDGE UNIVERSITY PRESS
The Edinburgh Building, Cambridge CB2 2RU, UK   http://www.cup.cam.ac.uk
40 West 20th Street, New York, NY 10011–4211, USA   http://www.cup.org
10 Stamford Road, Oakleigh, Melbourne 3166, Australia
Ruiz de Alarcón 13, 28014 Madrid, Spain

First published 1996
First paperback edition (with corrections) 2000

Typeset in 11/13pt Times

A catalogue record for this book is available from the British Library

ISBN 0 521 46040 9 hardback
ISBN 0 521 65477 7 paperback

Transferred to digital printing 2004

TAG

Contents

Contents

Preface

Non-relativistic field theory at finite temperature and finite density was invented in the late 1950s for the theoretical description of condensed matter and nuclear matter under standard laboratory conditions. Although it does not involve concepts beyond the Schrödinger equation and statistical mechanics, it offers a most convenient theoretical framework with which to deal with a large number of particles. It is often referred to as the 'Many-Body Problem' or as the 'N-Body Problem'. Relativistic field theory at finite temperature was first studied by Fradkin (1965) and rediscovered ten years later, the main motivation at the time being a description of the phase transition which occurs in the electroweak theory, at a temperature of the order of 200 MeV. This transition is of course of great interest for the history of the early Universe.

In the early 1980s, lattice gauge theories suggested the existence of a deconfined phase of quarks and gluons, which has been called the 'quark–gluon plasma phase', above a temperature which is estimated nowadays to lie around 150 MeV. The possibility of observing this new state of matter in ultrarelativistic heavy ion collisions gave a further boost to the study of finite temperature field theory. Furthermore, Braaten and Pisarski (1990a,b) made a theoretical breakthrough, which enabled them to reorganize perturbation theory, whose original formulation had led to nonsensical results in some circumstances.

It seems appropriate at present to write a review which gives a self-contained introduction to the subject and, at the same time, an account of the recent developments. Needless to say, all the problems are far from being solved, and much remains to be done, for example, in our understanding of some delicate infrared problems and of non-equilibrium phenomena.

After an introductory chapter which is aimed to outline the physical motivations (and which can be skipped at first reading), the book is divided into two parts. The first part, chapters 2–5, is devoted to setting up the formalism. Chapter 2 examines the one-dimensional motion of a quantum particle in a potential, in thermal equilibrium: since many important concepts can be understood on this elementary example, it is used in order to introduce path integrals and thermal Green's functions. Chapter 3 describes perturbation theory and Feynman rules in the case of scalar fields, while chapter 4 deals with some simple applications of these rules. Chapter 5 generalizes the results to the case of Dirac and gauge fields. The second part, chapters 6–10, is devoted to the theoretical developments which have occurred in recent years (although some results were known long ago). Chapter 6 introduces hard thermal loops which are used to describe collective excitations of gauge bosons and fermions. The full perturbative scheme relying on the hard thermal loop concept is developed in chapter 7, which also gives some applications to physical processes. Further applications are found in chapter 8 (dynamical screening) and in chapter 9 (neutrino emission from stars), while chapter 10 concludes with some open problems.

The reader is assumed to have a working knowledge of elementary quantum statistical mechanics, for example at the level of Mandl (1988), and to be familiar with the basic methods of quantum field theory. However, no advanced methods will be needed, other than in chapter 5, where some use is made of the technical details of the quantization of non-Abelian gauge fields in a section which can be skipped at first reading; in particular, renormalization problems will play only a minor role.

In order to keep the book to a reasonable size, and also because of my limited competence, a number of interesting topics had to be omitted. Fortunately I can refer the interested reader to excellent reviews: thermofield dynamics is described in Umezawa, Matsumoto and Tachiki (1982), Landsman and van Weert (1987), van Weert (1994) or Umezawa (1994). Higher order calculations of the partition function in QED and QCD may be found in Kapusta (1989); see Corianò and Parwani (1994) for recent results. References to the literature on symmetry restoration in the electroweak theory can be found in Kapusta (1989) and in Arnold and Espinosa (1993). I have tried to avoid, as far as is possible, overlap with the material already covered by Kapusta (1989), and I decided to leave aside for the time being the whole field of non-equilibrium phenomena in the quark–gluon plasma, which has not yet reached a sufficiently mature stage such that it can be reasonably included in a textbook.

There is an abundant literature on thermal field theory, and I had to make a choice from available references. As a general rule, and except for some historical landmarks, I have not quoted articles prior to 1988, since a complete list of references can be found in Landsman and van Weert (1987) and in Kapusta (1989).

While writing this book, I was assisted by many people who I would like to thank warmly. First of all the book would not have been written without the constant encouragement of Peter Landshoff. Jean-Louis Meunier helped me to find my way through the many versions of TEX, Frithjof Karsch provided me with his latest results, given in figures 1.2 and 1.3. Jean-Paul Blaizot, Thierry Grandou, Edmond Iancu, Peter Landshoff and Art Weldon made useful comments on specific points. I am especially indebted to Patrick Reynaud, who took care of the diagrams, and to Rolf Baier, Françoise Guérin, Hubert Mabilat and Toni Rebhan who read a first version of the manuscript thoroughly and made a number of observations and suggestions. The responsibility for the final text is of course entirely mine. Last but not least, Joanna provided me with her moral support.

I cannot conclude this preface without a sad thought for my friend and colleague Tanguy Altherr, whose life ended in a tragic climbing accident at the age of 31. The numerous references to his work bear witness to the importance of his contribution to finite temperature field theory.

Nice, 1996

1
Introduction

1.1 Quantum chromodynamics

Most high-energy physicists will readily agree that quantum chromodynamics (QCD) is today the well-established theory of strong interactions; at least it has no serious competitor. Quantum chromodynamics is a non-Abelian gauge field theory whose gauge group is the colour group $SU(3)$; it will sometimes be convenient to let the number of colours vary, and to take $SU(N)$ as the gauge group: then the number of colours is N. There are $(N^2 - 1) = 8$ gauge bosons, called gluons, and the matter particles are spin 1/2 quarks. There are six families of quarks, or six different flavours: up, down, strange, charm, beauty and top. The number of flavours will be denoted by N_f. The last three quarks are heavy and will not play any role at all in the development of this book because their mass is much larger than the characteristic energy scale of a few hundred MeV that we are interested in, while the role of the strange quark will be intermediate. The up and down quarks will often be taken as massless since their mass, of the order of a few MeV, is much smaller than our characteristic energy scale. Note that we use a system of units where $\hbar = c = k_B = 1$, where $\hbar$, c and k_B are, respectively, the Planck constant, the speed of light and the Boltzmann constant. Masses and temperatures will be often measured in MeV or GeV, lengths and times in MeV^{-1} or GeV^{-1}.

Quantum chromodynamics is a renormalizable field theory, and, as in all field theories of this kind, the coupling constant g is a function of the energy scale Q: $g = g(Q)$. The remarkable property of QCD, characteristic of non-Abelian gauge theories, is that the coupling constant tends to zero as the inverse of the logarithm of the energy scale, when this energy scale tends to infinity: this is the famous property of asymptotic freedom, which implies that perturbation theory may be used at high energies, or more precisely in processes with large-momentum transfers, the so-called

1

'hard processes'. This aspect of the theory has been tested in a variety of experiments. The cleanest test is deep inelastic scattering of leptons on nucleons, and there is little doubt that QCD reproduces the pattern of scaling violations observed experimentally. Other processes (lepton pair production, jets, prompt (real) photon production, large-angle scattering, τ-decay,...) have also been studied theoretically and experimentally. No discrepancy with QCD has been found, but the accuracy of the comparison is no better than a few percent, due to theoretical and experimental uncertainties.

When the energy scale decreases, the counterpart of asymptotic freedom is that the coupling constant increases, which leads to a breakdown of perturbation theory as a series expansion in powers of $g^2(Q)$. Thus 'soft' processes, namely, processes with small-momentum transfers, or low-energy properties of hadrons (mass spectrum, resonance widths, scattering lengths,...) cannot be studied in perturbation theory. Actually, this is a matter of principle and not simply a question of the coupling constant being large: assuming massless quarks in order to simplify the discussion, all hadron masses, for example, must be proportional to the QCD scale Λ:

$$\Lambda = \lim_{Q \to \infty} Q \ e^{-1/2\beta_0 g^2(Q)} [\beta_0 g^2(Q)]^{-\beta_1/2\beta_0^2} \qquad (1.1)$$

where g is the (running) gauge coupling constant, β_0 and β_1 are the first two coefficients of the renormalization group β-function: this expression is clearly non-perturbative. Since perturbation theory is irrelevant in low-energy hadron physics, the situation of QCD is much less bright than for hard processes. No good analytic guess has been made, and non-perturbative approaches have been explored following two main avenues:

(i) low-energy models which attempt to incorporate some of the features of the QCD Lagrangian;
(ii) numerical simulations of a discretized version of QCD, that is, lattice QCD.

The most interesting model of type (i) is the effective chiral Lagrangian: it lies on the observation that one of the most conspicuous features of low-energy hadronic physics is the existence of a light particle, that is, the pion. This particle is presumably a Goldstone boson, which would be massless in the case of massless up and down quarks. The corresponding broken symmetry is the chiral $SU(2) \times SU(2)$ symmetry, and the chiral Lagrangian model uses this property as a basic starting point. However, in spite of its usefulness and its phenomenological success, this model, as well as other models, cannot be used in any serious quantitative comparison between QCD and experiment.

The other road, that of numerical simulation of lattice QCD, is in principle limited only by computing power. However, in spite of the ingenuity which has been displayed by lattice practitioners in order to keep track of finite-size effects, it will still take some time before anyone is able to reproduce, for example, the hadron spectrum with an accuracy of a few percent. On the other hand, lattice QCD has been decisive in convincing physicists that a phase transition should occur for sufficiently high temperatures and/or densities. At zero temperature, for densities of the order of the nuclear density, quarks and gluons are confined in hadrons: there is no such thing as a free quark or a free gluon propagating in the vacuum. The situation may change when the temperature rises above a critical temperature T_c of the order of 150 MeV; then one may obtain a gas of almost free quarks and gluons, which has been dubbed the quark–gluon plasma. The same situation may occur at zero temperature, for sufficiently large densities (or, more correctly, for sufficiently large chemical potentials), or when one increases both the temperature and the chemical potential. The occurrence of this phase transition is important from a conceptual point of view, as it implies the existence of a new state of matter. In the standard cosmological model, this state of matter prevailed until a time $\sim 10^{-6} - 10^{-5}$ s after the Big-Bang. It could also occur in the core of heavy neutron stars, and it could be produced in collisions of ultrarelativistic heavy ions. Actually, the possibility of producing a quark–gluon plasma in such collisions is the main motivation for interest in the details of this new phase, as one may hope to detect signals of its existence experimentally and to measure some of its properties.

This book will be mainly concerned with the perturbative study of the quark–gluon plasma in an idealized situation where the plasma is at equilibrium, or only slightly out of equilibrium. That a perturbative approach is possible is a consequence of asymptotic freedom: at high temperatures the only scale which is available is the temperature T itself, and the coupling constant $g(T)$ is small at large T: $g(T) \sim 1/\ln(T/\Lambda)$. For large T (very large indeed!), the quark–gluon plasma is close to an ideal gas. However this *a priori* simple situation is spoiled by infrared divergences, which lead to a hierarchy of energy, or length, scales. In addition to the perturbative scale $l_P = 1/T$, there is an electric, or Debye, length scale $l_{el} = 1/gT$, a magnetic length scale $l_{mag} = 1/g^2T$, and perhaps other scales which limit the applicability of naïve perturbation theory.

The same hierarchy of scales is also found in QED plasmas, where the particles are electrons, positrons and ions, and which are relevant, for example, in the core of some stars (see chapter 9). The QCD coupling constant g is replaced by the electron charge e: $l_{el} = 1/eT$, $l_{mag} = 1/e^2T$.

Over the past few years important progress has been made in our understanding of the role of these scales in perturbation theory, and a large part of this book will be devoted to explaining this progress and the present limitations of the theory. Unfortunately, there will be little contact with phenomenology: it is likely that one will have to await the results of the future colliders before a real interaction between theory and experiment is possible. These colliders are the Relativistic Heavy Ion Collider (RHIC) located at Brookhaven, which will provide Au–Au collisions at energies of 200 MeV per nucleon around 1997, and the Large Hadron Collider (LHC), which will be running at CERN at the beginning of the next century and will provide Pb–Pb collisions at energies of a few TeV per nucleon.

1.2   The deconfinement/chiral transition

1.2.1 · The bag model

We first examine the possibility of a phase transition in a very simple model, the bag model, which attempts to incorporate, in a crude way, the two main features of QCD: asymptotic freedom and confinement. Hadrons are represented as bubbles of (perturbative) vacuum, or 'bags', in a confining medium. Inside the bag, quarks move freely to a first approximation (in principle one could try to improve the model by taking interactions with gluons in perturbation theory into account, but it is not clear that this strategy is viable, due to the crudeness of the model). Outside the bag, quarks and gluons cannot appear as free particles. This feature is achieved by assuming a constant energy density B, the bag constant, for the vacuum, which keeps quarks and gluons confined. The energy of a hadron is thus composed of the energy associated with the volume of the bag due to the finite energy density of the vacuum, and the kinetic energy of the quarks inside the bag. Assuming, for simplicity, that the bag is spherical with radius R, the kinetic energy is of the form C/R, from the uncertainty principle. Thus the hadron energy E_H is

$$E_H = \frac{4\pi}{3}R^3B + \frac{C}{R} \qquad (1.2)$$

By minimizing E_H with respect to R, one finds

$$R = \left(\frac{C}{4\pi B}\right)^{1/4} \qquad (1.3)$$

The hadron mass M is given by

$$M = \frac{16\pi}{3}R^3B \qquad (1.4)$$

and the pressure P by

$$P = -\frac{\partial E_H}{\partial V} = -B + \frac{C}{4\pi R^4} \qquad (1.5)$$

The contribution $-B$ due to the vacuum is balanced by the contribution from the quark kinetic energy, so that the total pressure vanishes at equilibrium.

In order to derive the condition for a (deconfinement) phase transition between a hadron phase and a quark–gluon plasma phase, we compare their thermodynamical functions. The grand potential $\Omega = -PV$ in a volume V for an ideal gas of bosons with one degree of freedom is

$$\Omega = VT \int \frac{d^3k}{(2\pi)^3} \ln\left[1 - e^{-\beta(\omega_k - \mu)}\right] \tag{1.6}$$

where μ is the chemical potential, V is the volume, $T = \beta^{-1}$ is the absolute temperature and $\omega_k = (k^2 + m^2)^{1/2}$ is the boson energy. For massless bosons and $\mu = 0$, (1.6) yields, after an integration by parts,

$$\Omega = -\frac{VT^4}{6\pi^2} \int_0^\infty \frac{x^3 dx}{e^x - 1} = -\frac{\pi^2 VT^4}{90} \tag{1.7}$$

a result that is familiar from blackbody radiation. We write the general result for a gas of massless bosons with zero chemical potential as

$$\Omega = -v_b \frac{\pi^2 VT^4}{90} \tag{1.8}$$

where v_b is the number of degrees of freedom. From equation (1.8), $P = -\partial\Omega/\partial V$ and the relation $E = 3PV$ between the energy and the pressure which is valid for an ultrarelativistic ideal gas, one derives for the pressure and the energy density ε

$$P = v_b \frac{\pi^2 T^4}{90} \qquad \varepsilon = v_b \frac{\pi^2 T^4}{30} \tag{1.9}$$

Since the chemical potential vanishes by assumption, the entropy S is immediately derived from $\Omega = E - TS$, which leads to an entropy density $s = 4\varepsilon/3T$. We shall also need the particle density deduced from the average number of particles $\overline{N}$ (for one degree of freedom):

$$\overline{N} = V \int \frac{d^3k}{(2\pi)^3} \frac{1}{e^{\beta k} - 1} = \frac{VT^3}{2\pi^2} \int_0^\infty \frac{x^2 dx}{e^x - 1} = \frac{VT^3}{\pi^2} \zeta(3) \tag{1.10}$$

where $\zeta(p)$ is the Riemann function: $\zeta(3) \simeq 1.202$. Equation (1.10) yields an interesting relation between the particle density $n = \overline{N}/V$ and the entropy density:

$$s = \frac{2\pi^4}{45\zeta(3)} n \simeq 3.6n \tag{1.11}$$

Let us now turn to fermions: the grand potential is

$$\Omega = -VT \int \frac{d^3p}{(2\pi)^3} \ln\left[1 + e^{-\beta(\varepsilon_p - \mu)}\right] \tag{1.12}$$

with $\varepsilon_p = (p^2 + m^2)^{1/2}$. For massless fermions, taking particles and antiparticles into account leads to the following grand potential, for one degree of freedom:

$$\begin{aligned}
\Omega &= -\frac{VT^4}{6\pi^2}\int_0^\infty k^3 dk \left[\frac{1}{e^{\beta(\varepsilon_p - \mu)} + 1} + \frac{1}{e^{\beta(\varepsilon_p + \mu)} + 1} \right] \\
&= -\frac{V}{6\pi^2}\left[\frac{7\pi^4 T^4}{60} + \frac{\mu^2 \pi^2 T^2}{2} + \frac{\mu^4}{4} \right]
\end{aligned} \tag{1.13}$$

Let us first examine the possibility of a phase transition with zero chemical potential; in this case the hadronic system will mainly consist of pions. In order to work with simple analytical formulae we neglect the pion mass; the pressure of the hadronic gas is then given by (1.9) with $v_b = 3$, corresponding to the three charge states of the pion, while the pressure of the plasma phase is, from (1.9) and (1.13),

$$P = \left(v_b + \frac{7}{4}v_f \right)\frac{\pi^2 T^4}{90} - B \tag{1.14}$$

In writing (1.14) we have generalized relation (1.5) to a gas of quarks and gluons by adding to the free energy a factor of BV representing the energy of the perturbative vacuum inside the confining medium. The number of bosonic degrees of freedom is $2(N^2 - 1)$, where N is the number of colours, while $v_f = 2NN_f$, N_f being the number of flavours. Inserting the physical values $N = 3$ and $N_f = 2$ corresponding to up and down massless quarks in (1.14) leads to

$$P = \frac{37}{90}\pi^2 T^4 - B \tag{1.15}$$

The equality of the pressure in the plasma and hadron phases gives the critical temperature T_c:

$$T_c = \left(\frac{45B}{17\pi^2} \right)^{1/4} \simeq 144\,\mathrm{MeV} \tag{1.16}$$

with a bag constant $B^{1/4} \simeq 200\,\mathrm{MeV}$ from (1.4). The transition is first-order, and it is a simple exercise to compute the latent heat; note that we use the terminology 'critical temperature' even for first-order phase transitions, although it is often reserved for second-order phase transitions.

In principle the bag model could be used to determine the transition line between the hadronic and plasma phases in the whole (μ, T) plane. Unfortunately, the model is much less convincing for non-zero μ, as is clear

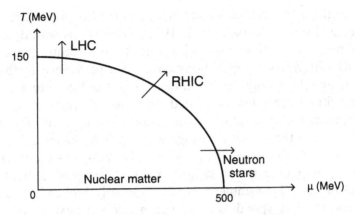

Fig. 1.1   The phase diagram in the μ–T plane.

from considering the $T = 0$ situation. The baryonic density n_B is related to the baryonic chemical potential μ_B and to the nucleon mass m_N through

$$n_B = \frac{2}{3\pi^2}(\mu_B^2 - m_N^2)^{3/2} \qquad (1.17)$$

The critical chemical potential can be determined by writing the equality of the pressures together with $\mu_B = 3\mu$, where μ is the quark chemical potential. One finds that the transition is possible only for a narrow range of values of the bag constant (exercise 1.1), which seems difficult to understand physically. Thus the bag model does not seem to give much information on the $T = 0$, $\mu \neq 0$ case. Nevertheless, it is useful to have in mind the following orders of magnitude: for the nuclear density $n_B^{(0)} \simeq 0.15 \text{ fm}^{-3}$, $\mu \simeq 0.26$ GeV, while for a density equal to five times the nuclear density $\mu \simeq 0.45$ GeV.

We shall see in the next section that lattice QCD confirms the value $T_c \simeq 150$ MeV for the critical temperature with zero chemical potential. Unfortunately, there is no reliable estimate of the critical chemical potential at zero temperature. It is generally assumed that the transition should occur for a density of three to five times the nuclear density, so that we may guess the phase diagram of fig. 1.1 in the (μ, T) plane; the regions in this plane which are reached in various physical situations are also indicated on the plot.

1.2.2 Lattice quantum chromodynamics

The discretized version of QCD can be studied on computers using Monte-Carlo simulations. The details of 'how one puts QCD on a lattice' will not be given here, as a detailed account of lattice QCD is outside the scope of this book; the interested reader is referred to the literature: see, for example, Creutz (1983) or Karsch (1990, 1992). Simulations with dynamical quarks are much more time consuming than those without quarks, and this explains why many computations have been performed in the pure gauge sector by taking only the gluon degrees of freedom into account.

Phase transitions are often associated with broken symmetries, and, of course, symmetries of the lattice action play a crucial role. The action S_G of a pure gauge theory with a gauge group $SU(N)$ has a global $Z(N)$ symmetry, where $Z(N)$ (the Abelian group of the Nth-roots of unity) is the centre of the group, namely, the subgroup whose elements commute with all group elements. This symmetry is spontaneously broken at the phase transition and the corresponding order parameter is the expectation value of the Polyakov loop L (see, for example, Karsch (1990) for a definition). This expectation value $\langle L \rangle$ is related to the excess of free energy $F_q(T)$ induced by a static quark source in the gluonic heat bath:

$$\exp(-F_q(T)/T) \simeq \langle L \rangle \tag{1.18}$$

In the confined phase $F_q(T)$ is infinite since a coloured quark cannot be screened, and $\langle L \rangle = 0$. On the other hand, in the deconfined case $F_q(T)$ is finite and $\langle L \rangle \neq 0$. Note that the order parameter is zero in the low-temperature phase and non-zero in the high-temperature phase.

When quarks are added to the Lagrangian, global $Z(N)$ symmetry is lost. However, in the case of massless quarks, there is another global symmetry of the action, that is, chiral symmetry, with symmetry group $SU(N_f) \times SU(N_f)$, which is spontaneously broken at low temperature. The (local) order parameter in this case is the quark condensate $\langle \bar{\psi}\psi \rangle$, where ψ is the quark field. The quark condensate is non-zero at low temperature, where chiral symmetry is broken, and zero above the phase transition temperature, where chiral symmetry is restored.

With massive quarks, no global symmetry of the action survives. Nevertheless, numerical simulations indicate a sharp rise in entropy, energy density and pressure around a temperature T_c, which is thought to be the critical temperature for the deconfinement/chiral symmetry restoration phase transition. There is no indication of two different transitions, one corresponding to deconfinement and the other to chiral symmetry restoration, at least in the case of zero chemical potential. Let us give a brief summary of what has been learned in recent years on the following topics:

  (i) the nature of the phase transition;
 (ii) the value of the critical temperature;
(iii) the equation of state above the critical temperature;
(iv) the Debye screening length.

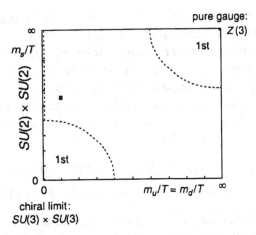

Fig. 1.2.   Character of the transition as a function of the light quark masses $m_u = m_d$ and the strange quark mass m_s. The circle with a cross represents the physical values of the parameters. Private communication from F. Karsch.

In the case of a pure gauge theory, there are convincing universality arguments which allow one to relate the $SU(N)$ deconfinement transition in dimension four to the ferromagnetic–paramagnetic transition of a $Z(N)$ spin model with ferromagnetic interactions in dimension three. In the $SU(2)$ case, the spin model is the Ising model and the phase transition is well-known to be second order. There is remarkable agreement between the numerical values of the critical exponents of the $SU(2)$ deconfinement transition and those of the three-dimensional Ising model, which gives a beautiful confirmation of universality. In the $SU(3)$ case, the spin model is the three-state Potts model, whose transition is first order. A careful analysis of the deconfinement transition, using finite-size scaling, does confirm the first-order character of the deconfinement transition.

In the case of chiral symmetry restoration, it is also possible to use universality arguments in order to study the character of the transition, by relating the gauge theory to simpler models. The transition can be shown to be first order for $N_f \geq 3$, while for $N_f = 2$ the transition is probably second order, although the situation is not entirely clear.

Our present knowledge of the character of the transition, which relies on the preceding universality arguments and on numerical simulations, is summarized in fig. 1.2. At present, with two massless up and down quarks and a strange quark with a mass around 150 MeV, the best guess seems to be that the deconfinement/chiral transition is second order. However, one must remember that, without symmetry breaking, it is always possible that the transition between the hadronic and plasma phases occurs gradually and a critical temperature may not be precisely defined.

Let us now turn to the critical temperature. With present numerical simulations, asymptotic scaling is still not established for $SU(2)$ and $SU(3)$ gauge theories; however scaling violations drop out in the ratio of physical quantities, so that one can compute $T_c/\Lambda_{\overline{MS}}$, or $T_c/\sqrt{\sigma}$, or T_c/m_ρ, where $\Lambda_{\overline{MS}}$, σ and m_ρ are the QCD mass scale (see (1.1)) in the so-called $\overline{MS}$ renormalization scheme, the string tension and the ρ-meson mass respectively. It seems that the best estimates are deduced from the T_c/m_ρ ratio and one finds, in MeV,

$$-\text{pure gauge } SU(3) : \quad T_c = 240 \pm 20 \text{ MeV}$$
$$-N_f = 2 \qquad\qquad\quad : \quad T_c = 150 \pm 10 \text{ MeV}$$

In numerical simulations of the equation of state, it is very important to proceed in a fully non-perturbative manner. Up to now, this has been possible only in the pure gauge sector. From the expectation value of the plaquette action and the non-perturbative determination of the renormalization-group β-function, one computes the pressure and the quantity $\Delta = (\varepsilon - 3P)/T^4$ which measures the deviation with respect to the ideal gas behaviour $\varepsilon = 3P$. The results are first obtained as functions of the coupling $2N/g^2$, which may then be converted into T/T_c thanks to the non-perturbative β-function. Figure 1.3 shows the behaviour of the ratios ε/T^4 and $3P/T^4$, compared with the value $\varepsilon_{SB}/T^4 = 8\pi^2/15$ of an ideal gluon gas, and the value $\simeq 6.2$ on an $N_\tau = 6$ lattice. Even at $T = 2T_c$, the energy density only reaches about 70% of the ideal gas value. However, at least for $T \gtrsim 2T_c$, one finds $\varepsilon \simeq 3P$.

Let us finally examine screening lengths. In QCD a naïve perturbative calculation with resummed self-energy bubbles for the gluon propagator yields the following behaviour of the potential $V(r)$ between gauge invariant sources:

$$V(r) \propto \frac{e^{-2m_{\mathrm{D}}r}}{r^2} \tag{1.19}$$

where the Debye mass $m_{\mathrm{D}} = l_{\mathrm{el}}^{-1}$ is given by (see chapter 6)

$$m_{\mathrm{D}}^2 \simeq \frac{1}{3}\left(N + \frac{1}{2}N_f\right)g^2(T)T^2 \tag{1.20}$$

The behaviour (1.19) is controlled by $2m_{\mathrm{D}}$ because at least two gluons must be exchanged between two gauge-invariant sources. The Debye mass can be estimated in numerical simulations by computing the large-distance behaviour of the Polyakov loop correlation function. For reasons of computing time, this can be done in practice only in the case of pure gauge theories. One finds for $T \geq T_c$, both in $SU(2)$ and $SU(3)$, that $m_{\mathrm{D}} \simeq T$ within a 10% accuracy, which confirms the proportionality with

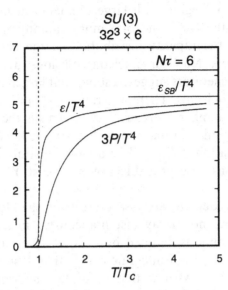

Fig. 1.3. Ratios ε/T^4 and $3P/T^4$ as functions of T/T_c on the lattice. Private communication from F. Karsch.

T of equation (1.20) and is even in quantitative agreement with (1.20) if one chooses $g(T) \simeq 1$.

1.3 Heavy ion collisions

As has already been mentioned, the main reason for being interested in 'hot QCD' is that one may hope to produce a quark–gluon plasma in collisions of ultrarelativistic heavy ions. In this section, we wish to show that such collisions are indeed able to produce energy densities that are sufficiently large for the phase transition to occur. In order to estimate the available energy density, we shall give a short account of the space-time and thermal evolution of the collision. In what follows, we always view the collision in the centre of mass frame; E_{cm} denotes the energy per nucleon in this frame, the z-axis is chosen along the collision axis and we neglect transverse motion. Finally, for simplicity, we restrict ourselves to collisions of identical nuclei with atomic number A.

The generally accepted picture for the collision of two ultrarelativistic heavy nuclei rests on two concepts which are consistent with the present experimental data: nuclear transparency and the inside–outside cascade.

Nuclear transparency, or the leading particle effect, was first observed in nucleon–nucleon collisions: a large fraction (about a half) of the available energy is carried away by two secondary nucleons, which are called the leading baryons. Similarly, in the collision of two heavy nuclei, the baryonic number is carried away essentially by two systems of nucleons leaving the interaction region with large momenta in the centre of mass frame. More quantitatively, the maximum rapidity in the collision is $Y = \ln(2E_{cm}/m_N)$; the rapidity distribution of the baryonic number is centred at $\pm(Y - \delta y)$. Nucleon–nucleon collisions give $\delta y \simeq 2$ units of rapidity, but event generators suggest a somewhat larger value, of the order of 3–3.5, for A–A collisions. There is a range $\simeq [-(Y - 2\delta y), (Y - 2\delta y)]$ where the baryonic number averages to zero. This is the so-called central region, while the regions in rapidity which carry the baryonic number are called the fragmentation regions. For a clear separation of central and fragmentation regions, one probably needs at least $Y \simeq 5 - 6$, namely, $E_{cm} \geq 100$ GeV.

The concept of inside–outside cascade rests on the idea that an excited hadronic system does not decay instantaneously: in its rest frame, the characteristic decay time is denoted by τ_0, with $\tau_0 \simeq 1$ fm/c. Because of time dilation, slow systems disintegrate earlier than faster ones: this is the inside–outside cascade. Assume that an excited hadronic system leaves the interaction point $z = 0$ at time $t = 0$ and moves freely with velocity v along the collision axis. Its proper time is $\tau = (t^2 - z^2)^{1/2} = t(1-v^2)^{1/2}$. A collection of hadronic systems created at $t = z = 0$ will thus disintegrate on average on a hyperbola of constant proper time $(t^2 - z^2)^{1/2} = \tau_0$; in a hydrodynamical description, the initial conditions for the fluid will be given on this hyperbola. Note that the initial conditions are not given at an initial time, but, rather, at an initial proper time.

We may now summarize our picture of the collision: two Lorentz contracted nuclei, with transverse radius R and longitudinal extension $d_L \sim 2Rm_N/E_{cm}$ collide during a time $\sim d_L/c$. The two nuclei cross each other and two Lorentz contracted hadronic systems, carrying the initial baryonic number, recede from each other at high velocity in opposite directions from the interaction region. They leave behind them excited hadronic matter with zero baryonic number. It is interesting to observe that for Pb–Pb collisions at LHC energies, with $E_{cm} = 2$ TeV, $d_L/c \sim 7 \times 10^{-3}$ fm/c $\ll \tau_0$, so that the duration of the collision may be neglected. For lower energies, it may be necessary to take the finite time of crossing into account.

We shall now limit ourselves to the study of the central hadronic system, which carries zero baryon number. Recall that in our picture a collection of excited hadronic systems leaves the origin at $t = 0$ and disintegrates on a hyperbola $(t^2 - z^2)^{1/2} = \tau_0$; the coordinate which parametrizes a point on this hyperbola is denoted by α:

$$\alpha = \frac{1}{2} \ln \frac{t+z}{t-z} \qquad (1.21)$$

and it can be identified with the rapidity y, since our hadronic systems are assumed to move freely. Experimental data indicate that the particle density is approximately independent of the rapidity in the central region. We may thus assume that the initial conditions on the hyperbola $\tau = \tau_0$ are invariant under Lorentz boosts of moderate velocity (in order not to leave the central region), that is, they do not depend on y: these are Bjorken's initial conditions.

In order to predict the subsequent evolution of the collision, we assume the validity of a hydrodynamical picture, with a relativistic perfect fluid; these assumptions will be discussed at the end of the section. Note that, for the time being, we do not specify the character of the fluid, which may be hadronic-like or plasma-like. The rest frame of the fluid is defined as the frame in which the energy flux is zero. We denote the four-velocity by u^μ, $u^\mu = (1 - v^2)^{-1/2}(1, \mathbf{v})$, where $\mathbf{v}$ is the fluid velocity in the centre of mass frame of the collision. Let us consider a locally conserved scalar density N, and its volume density n measured in the rest frame of the fluid. In the case of a perfect fluid, there are no transport phenomena, so that there is no energy flux without scalar flux. Thus, in the rest frame of the fluid, the flux of any conserved quantity is zero. This leads to the local conservation law:

$$\partial_\mu(nu^\mu) = 0 \tag{1.22}$$

Similarly, we write the conservation of the energy-momentum tensor $T^{\mu\nu}$ as

$$\partial_\mu T^{\mu\nu} = 0 \tag{1.23}$$

In the fluid rest frame, we have $T^{00} = \varepsilon$, $T^{ij} = P\delta_{ij}$, where, as before, ε is the energy density and P is the pressure. From Lorentz covariance one deduces the well-known expression that is valid in any frame:

$$T^{\mu\nu} = (\varepsilon + P)u^\mu u^\nu - Pg^{\mu\nu} \tag{1.24}$$

Equations (1.22)–(1.24), together with the equation of state, form a closed system. Let us now derive from this formalism an equation for entropy conservation. Contracting (1.23) with u^ν yields

$$u^\mu(\partial_\mu \varepsilon) + (\varepsilon + P)\partial_\mu u^\mu = 0 \tag{1.25}$$

On the other hand, if there are conserved scalar quantities N_i, with associated densities n_i and chemical potentials μ_i, we have thermodynamical identities:

$$d\varepsilon = T\,ds + \sum_i \mu_i dn_i \tag{1.26a}$$

$$\varepsilon + P = Ts + \sum_i \mu_i n_i \tag{1.26b}$$

where s is the entropy density. Using (1.22) and (1.26) and assuming all densities to be continuous functions of space-time allows us to derive the equation for entropy conservation:

$$\partial_\mu(su^\mu) = 0 \tag{1.27}$$

We are now in a position to describe the longitudinal expansion of the hadronic matter produced in the central region, ignoring transverse expansion and making use of Bjorken's initial conditions. From the Lorentz invariance of our hydrodynamical equations and of our initial conditions, the properties of the fluid in a slice around the longitudinal coordinate z can be deduced from those of the fluid in a slice around $z = 0$ by a longitudinal Lorentz boost. It is convenient to use the coordinates (τ, α), with $t = \tau \cosh \alpha$, $z = \tau \sinh \alpha$. In this coordinate system, the components (V^τ, V^α) of a four vector V^μ read

$$
\begin{aligned}
V^\tau &= V^0 \cosh \alpha - V^z \sinh \alpha \\
V^\alpha &= -V^0 \sinh \alpha + V^z \cosh \alpha
\end{aligned}
\tag{1.28}
$$

from which follows

$$\partial_\mu V^\mu = \frac{1}{\tau}\left(\frac{\partial(\tau V^\tau)}{\partial \tau} + \frac{\partial V^\alpha}{\partial \alpha}\right) \tag{1.29}$$

The invariance of the initial conditions under Lorentz boosts, together with the Lorentz invariance of the hydrodynamical equations, requires that no quantity depends on α: see exercise 1.2 for further details. A particular solution of the hydrodynamical equations is given by $u^\alpha = 0$, $u^\tau = 1$; in this case the fluid rapidity y coincides with the space-time rapidity α while equation (1.27) for entropy conservation becomes

$$\frac{\mathrm{d}}{\mathrm{d}\tau}(\tau s) = 0 \tag{1.30}$$

This implies that the entropy density decreases as the inverse of the proper time τ. Since the system is expanding in the z-direction, the volume of a slab of fluid in the rapidity range $[y, y + \mathrm{d}y]$ grows as τ : $\mathrm{d}^3x = \tau \mathrm{d}y\, \mathrm{d}^2x_\perp$ in a frame where the fluid is at rest. Equation (1.30) means that the entropy contained in a given rapidity interval is conserved.

If we assume that the system thermalizes at a proper time $\tau_0 \simeq 1$ fm/c, with an initial entropy density s_0, we shall have for $\tau \geq \tau_0$

$$s(\tau) = \frac{s_0 \tau_0}{\tau} \tag{1.31}$$

We now turn to numerical estimates. The total entropy contained in a slice $[y, y + \Delta y]$ at a proper time τ_0 is

$$\Delta S = \pi R^2 s_0 \tau_0 \Delta y \tag{1.32}$$

where the nuclear radius $R \simeq 1.2 \times A^{1/3}$. Equation (1.11) allows one to relate s_0, or more properly $s_0 \tau_0$, to an observable, namely, the rapidity density dN/dy: as the baryonic number vanishes in the central region, entropy is mainly carried by pions and we have

$$s_0 \tau_0 \simeq \frac{3.6}{\pi R^2} \frac{dN}{dy} \qquad (1.33)$$

The rapidity density dN/dy in A–A collisions can be estimated from

$$\frac{dN}{dy} \simeq A^{1.1} \frac{dN_0}{dy} \qquad \frac{dN_0}{dy} \simeq 0.9 \ln \frac{E_{cm}}{m_N} \qquad (1.34)$$

where dN_0/dy is the rapidity density in nucleon–nucleon collisions. The exponent of A in (1.34) reflects multiple scattering, and could be slightly higher than the conservative estimate of 1.1. With $\tau_0 \simeq 1$ fm/c, we obtain for Pb–Pb collisions

$$s_0 \simeq 7.5 \ln \frac{E_{cm}}{m_N} \text{ fm}^{-3} \qquad (1.35)$$

For $E_{cm} = 2$ TeV, in the case of an ideal quark–gluon plasma with $N_f = 2$, this value corresponds to a temperature $T_0 \simeq 300$ MeV, about twice the critical temperature, and to an energy density $\varepsilon \simeq 12.8$ GeV/fm^3.

It is also interesting to follow the evolution of the temperature; in the case of zero chemical potential we have from (1.26)

$$\frac{s}{T} \frac{\partial T}{\partial s} = \frac{\partial P}{\partial \varepsilon} = c_s^2 \left(= \frac{1}{3} \right) \qquad (1.36)$$

where c_s is the velocity of sound ($c_s = 1/\sqrt{3}$ for an ideal ultra-relativistic gas). From entropy conservation (1.30) we obtain

$$T = T_0 \left(\frac{\tau_0}{\tau} \right)^{c_s^2} \qquad (1.37)$$

which shows that the temperature drops more slowly than the entropy density.

From the preceding discussion we may now give the following picture of the central region. At a proper time $\tau_0 \simeq 1$ fm/c after the collision, thermal equilibrium is reached and the temperature is large enough for the system to be in the quark–gluon plasma phase. The assumption of thermal equilibrium is difficult to justify in an entirely convincing way, but it is not unreasonable: for a quark–gluon plasma at $T \simeq 200$ MeV, one has approximately a particle density n of four particles per fm^3; if the cross-section σ is estimated from the additive quark model to be around 10 mb, the mean free path $\lambda = 1/n\sigma \sim 0.25$ fm, so that the size of the interaction region is 15–20 times λ for Pb–Pb collisions. However, a recent discussion suggests that the situation could be more complicated:

the gluons might reach equilibrium much faster than the quarks, because the gluon—gluon cross-section, as computed in perturbative QCD, is much larger than the gluon—quark and quark—quark cross-sections. The gluon mean free path is estimated to be around 0.3 fm, while the quark mean free path is around 2 fm. If only the gluons come to thermal equilibrium, the initial temperature is estimated to be around 400 MeV, instead of the 300 MeV found earlier.

Thermalization justifies the subsequent use of hydrodynamics, and the evolution is obtained by assuming the perfect fluid approximation, where entropy is conserved since there is no dissipation. The system expands mainly in the longitudinal direction, and the temperature drops until it reaches T_c. Then a mixed plasma—hadronic phase occurs which lasts for some time until a pure hadronic phase is obtained. Finally, when the mean free path is of the order of the size of the system, free particles are produced and may be observed in detectors; this is the so-called freeze-out transition. The assumptions which underly this hydrodynamical description seem less crucial than that of thermalization. Entropy may be produced by viscous effects, shock waves which develop from the phase transition and the freeze-out transition. This entropy production has been estimated to be small and, in any case, the main virtue of the hydrodynamical approach is that it includes the fundamental conservation laws: most of the results can be viewed as consequences of the basic conservation laws, and one probably need not assume the detailed validity of perfect fluid dynamics.

References and further reading

Standard references to statistical mechanics are Huang (1962), Reif (1965), Ma (1985) or Mandl (1988). A good introduction to QCD may be found in the book by Muta (1987) and recent results are reviewed in the articles collected in the proceedings of the conference *QCD: 20 Years Later*; see in particular the contributions by Wilczek (1992) and Altarelli (1992). Lattice QCD is explained in the book by Creutz (1983). More recent reviews, with emphasis on finite temperature, are Karsch (1990, 1992). The universality arguments relating lattice gauge theory to $Z(N)$ models are given by Svetitsky and Yaffe (1982), while analogous arguments in the case of chiral transition are due to Pisarski and Wilczek (1984). The general space-time picture of heavy ion collisions was first developed by Bjorken (1983). See also the review article by Cleymans, Gavai and Suhonen (1986), the book by Shuryak (1988) and the articles by Satz (1990, 1992). A very clear account of the hydrodynamical picture is to be found in Blaizot and Ollitrault (1990); see also the original article by Baym, Friman, Blaizot, Soyeur and Czysz (1983). The discussion of equilibration times is due to Shuryak (1992).

Exercises

1.1 Discuss the transition between the hadronic phase and the plasma phase at zero temperature in the framework of the bag model. First compute the pressure and the energy density of the hadronic phase as functions of the chemical potential μ and the nucleon mass m_N. Show that the transition is possible only if

$$\frac{1}{2\pi^2}\left(\frac{m_N}{3}\right)^4 \le B \le \left(\frac{3}{4} - \ln 2\right)\frac{m_N^4}{8\pi^2}$$

where B is the bag constant.

1.2 Assuming that the fluid four-velocity may be written as $u^\mu = (\cosh\theta, \sinh\theta)$, where we neglect transverse dependence and where θ is *a priori* a function of (t, z), show that the hydrodynamical equations read

$$\tau\frac{\partial\varepsilon}{\partial\tau} + \tanh(\theta - y)\frac{\partial\varepsilon}{\partial y}$$

$$+ (\varepsilon + P)\left(\frac{\partial\theta}{\partial y} + \tanh(\theta - y)\tau\frac{\partial\theta}{\partial\tau}\right) = 0$$

$$\frac{\partial P}{\partial y} + \tanh(\theta - y)\tau\frac{\partial P}{\partial\tau}$$

$$+ (\varepsilon + P)\left(\tau\frac{\partial\theta}{\partial\tau} + \tanh(\theta - y)\frac{\partial\theta}{\partial y}\right) = 0$$

with $t = \tau\cosh y$, $z = \tau\sinh y$. Also write the equation for baryon number conservation. Show that the above equations lead to (1.30) when one uses Bjorken's initial conditions $\theta(y, \tau_0) = y$.

2
Quantum statistical mechanics

We now start what will be the subject of this book: quantum field theory at non-zero temperature and/or non-zero chemical potential. Two competing formalisms have been used at zero temperature in order to study field theory: the operator formalism, which is the older one, and the path integral formalism, which represents the more modern approach. Actually, it may often be illuminating to look at a given problem from both points of view, although in some cases one of the formalisms may prove to be definitely superior to the other: for example, path integrals are much simpler when quantizing gauge theories. At finite temperature, both formalisms are useful, and it is instructive to be able to switch from one approach to the other. Since the simplest case of field theory is field theory with zero space dimension, or, in other words, quantum mechanics, we shall begin with a short description of quantum mechanics at finite temperature, or, equivalently, quantum statistical mechanics. We first deal with the path integral approach, and later on revert to the more conventional operator formalism. We shall be particularly interested in time-ordered products of position operators, which will generalize to time-ordered products of field operators in field theory.

At zero temperature, it is often convenient to perform an analytical continuation from real to imaginary time: $t \rightarrow -i\tau$, or $x^0 \rightarrow -ix_4$, where τ (x_4) is real. This also means going from Minkowski to Euclidean space, since the Minkowski metric transforms into a Euclidean metric (with a change of sign):

$$t^2 - \mathbf{x}^2 \rightarrow -(\tau^2 + \mathbf{x}^2) \tag{2.1}$$

In momentum space the corresponding operation is $k^0 \rightarrow -ik_4$. For reasons which will become clear in the following chapters, the use of Euclidean space plays a much more important role at finite temperature than at zero temperature. Thus the first section of the present chapter will

18

be devoted to a short review of the path integral formalism in imaginary time.

2.1 Path integral formalism and imaginary time

In conventional quantum mechanics, the motion of a particle in a time-independent potential $V(q)$ can be described by the probability amplitude $F(q',t';q,t)$ of finding the particle at position q' at time t', when one knows that it was located at point q at time t:

$$F(q',t';q,t) = < q'|e^{-i\hat{H}(t'-t)}|q > \qquad (2.2)$$

$\hat{H}$ is the full, time-independent Hamiltonian; for simplicity, we restrict ourselves to a one-dimensional motion, and we set $\hbar = 1$. Whenever there is a risk of confusion, circumflexes will be used to denote operators acting in the Hilbert space of states. There exists a path integral representation of F, which can be found in textbooks and will not be described here, as we shall be interested in the analytical continuation of F to imaginary time:

$$t \rightarrow -i\tau \quad t' \rightarrow -i\tau'$$
$$F(q',-i\tau';q,-i\tau) = < q'|e^{-\hat{H}(\tau'-\tau)}|q > \qquad (2.3)$$

Let us derive a path integral representation for F in (2.3). We proceed in a heuristic way and ignore the subtleties linked to the definition of the integration measure. We divide the interval $[\tau,\tau']$ into $(n+1)$ subintervals, each of length $\varepsilon = (\tau - \tau')/(n+1)$, with $n \rightarrow \infty$, and write

$$\exp[-\hat{H}(\tau'-\tau)] = \exp\left[-(\tau'-\tau)\left(\frac{\hat{p}^2}{2m} + V(\hat{q})\right)\right] \qquad (2.4)$$

where $\hat{q}$ and $\hat{p}$ are the position and momentum operators. We then use Trotter's (or Lie product) formula:

$$\lim_{n\to\infty} \left(e^{A/n}e^{B/n}\right)^n = e^{A+B} \qquad (2.5)$$

Equation (2.5) is easily proved when A and B are bounded operators, and the necessary generalization to unbounded operators can be found in the literature: see, for example, Glimm and Jaffe (1987). We insert complete sets of eigenstates of the position operator $\hat{q}$ at 'times' $\tau_1,...,\tau_n$

$$F(q',-i\tau';q,-i\tau) = \lim_{\varepsilon\to 0} \int \prod_{l=1}^{n} dq_l < q_{l+1}|\exp\left(-\frac{\varepsilon}{2}V(\hat{q})\right)$$

$$\times \exp\left(-\varepsilon\frac{\hat{p}^2}{2m}\right) \exp\left(-\frac{\varepsilon}{2}V(\hat{q})\right)|q_l > \qquad (2.6)$$

and evaluate the matrix elements in (2.6). The action of $V(\hat{q})$ on the ket $|q_l>$ is of course trivial, while for the matrix element of the momentum operator we insert a complete set of states $|p_l>$ of the momentum operator $\hat{p}$:

$$< q_{l+1}| \exp\left(-\varepsilon\frac{\hat{p}^2}{2m}\right)|q_l >$$

$$= \int dp_l < q_{l+1}| \exp\left(-\varepsilon\frac{\hat{p}^2}{2m}\right)|p_l >< p_l|q_l >$$

$$= \int \frac{dp_l}{2\pi} e^{i(q_{l+1}-q_l)p_l} e^{-\varepsilon p_l^2/2m} = \left(\frac{m}{2\pi\varepsilon}\right)^{1/2} e^{-m(q_{l+1}-q_l)^2/2\varepsilon} \quad (2.7)$$

Equation (2.7) allows us to rewrite F as

$$F(q',-i\tau';q,-i\tau) = \lim_{\varepsilon\to 0} \left(\frac{m}{2\pi\varepsilon}\right)^{1/2} \int \prod_{l=1}^{n}\left[\left(\frac{m}{2\pi\varepsilon}\right)^{1/2} dq_l\right]$$

$$\times \exp\left[-\varepsilon\left(\sum_{l=0}^{n}\frac{m(q_{l+1}-q_l)^2}{2\varepsilon^2}\right.\right.$$

$$\left.\left.+\sum_{l=0}^{n}V\left(\frac{q_l+q_{l+1}}{2}\right)\right)\right] \quad (2.8)$$

We have written the argument of V as $(q_l + q_{l+1})/2$ for aesthetic reasons and have neglected some terms on the boundary, which is possible in the limit $\varepsilon \to 0$. Proceeding in a formal way, we call $\mathscr{D}q(\tau'')$ the limit $\varepsilon \to 0$ of the integration measure in (2.8) and recognize in the argument of the exponential a Riemann sum for the integral:

$$S_E(\tau' - \tau) = \int_{\tau}^{\tau'} d\tau'' \left[\frac{1}{2}m\dot{q}^2(\tau'') + V(q(\tau''))\right] \quad (2.9)$$

where the dot indicates a derivative with respect to τ; S_E in (2.9) is often called the Euclidean action. We may finally write F in the form of a path integral:

$$F(q',-i\tau';q,-i\tau) = \int \mathscr{D}q(\tau'')$$

$$\times \exp\left[-\int_{\tau}^{\tau'} d\tau'' \left(\frac{1}{2}m\dot{q}^2(\tau'') + V(q(\tau''))\right)\right] \quad (2.10)$$

The boundary conditions on the paths $q(\tau'')$ are $q(\tau) = q$, $q(\tau') = q'$. Although the above 'derivation' is purely formal, it should be emphasized that a rigorous meaning can be given to the integration measure in (2.10). Standard derivations of (2.10) use the Hamiltonian form of the path integral as an intermediate step (see exercise 2.1). Here we have chosen a more direct derivation, which is valid because the Hamiltonian has the form given in (2.4). In more complicated cases, it may be necessary to go through the Hamiltonian form of the path integral.

At this point we make the connection with quantum statistical mechanics by using (2.10) to express the partition function $Z(\beta)$ as a path integral; as usual $\beta = 1/T$ is the inverse of the temperature in a system of units where the Boltzmann constant $k_B = 1$. We start from the definition of the partition function:

$$Z(\beta) = \mathrm{Tr}\ e^{-\beta\hat{H}} = \sum_n e^{-\beta E_n} \tag{2.11}$$

where the trace has been evaluated by using a complete set of eigenvectors of $\hat{H}$:

$$\hat{H}|n> = E_n|n> \tag{2.12}$$

However, it is also possible to write the trace by using a complete set of eigenvectors of the position operator:

$$Z(\beta) = \int dq < q|e^{-\beta\hat{H}}|q> \tag{2.13}$$

It is clear that $e^{-\beta\hat{H}}$ can be interpreted formally as an evolution operator in imaginary time. Comparing with (2.3) we see that

$$Z(\beta) = \int dq F(q, -i\beta; q, 0) \tag{2.14}$$

so that $Z(\beta)$ can be expressed as a path integral:

$$Z(\beta) = \int \mathscr{D}q(\tau)\exp\left[-\int_0^\beta d\tau\left(\frac{1}{2}m\dot{q}^2(\tau) + V(q(\tau))\right)\right]$$
$$= \int \mathscr{D}q(\tau)\exp\left[-S_E(\beta)\right] \tag{2.15}$$

where one integrates over paths subject to the boundary condition

$$q(\beta) = q(0) \tag{2.16}$$

namely, over paths with period β in imaginary time.

As usual in quantum field theory, we define a generating functional $Z(\beta; j)$, with $Z(\beta) = Z(\beta; j = 0)$, through

$$Z(\beta; j) = \int \mathscr{D}q(\tau)\exp\left[-S_E(\beta) + \int_0^\beta j(\tau)q(\tau)d\tau\right] \tag{2.17}$$

Functional differentiation of (2.17) gives the propagator in imaginary time:

$$\frac{1}{Z(\beta)}\frac{\delta^2 Z(\beta; j)}{\delta j(\tau_1)\delta j(\tau_2)}\bigg|_{j=0} = \frac{1}{Z(\beta)}\int \mathscr{D}q(\tau)q(\tau_1)q(\tau_2)e^{-S_E(\beta)} \tag{2.18}$$

Indeed, we now show that the RHS of (2.18) can be identified as the thermal average of a time-ordered-, or T-product of position operators:

$$\langle T(\hat{q}(-i\tau_1)\hat{q}(-i\tau_2))\rangle_\beta = \frac{1}{Z(\beta)}\text{Tr}\left[e^{-\beta\hat{H}}\,T(\hat{q}(-i\tau_1)\hat{q}(-i\tau_2))\right] \qquad (2.19)$$

where $\hat{q}(-i\tau)$ is the position operator in the Heisenberg picture:

$$\hat{q}(t) = e^{i\hat{H}t}\hat{q}\,e^{-i\hat{H}t}$$
$$\hat{q}(-i\tau) = e^{\hat{H}\tau}\hat{q}\,e^{-\hat{H}\tau} \qquad (2.20)$$

We recall that the thermal average $\langle\hat{A}\rangle_\beta$ of an operator $\hat{A}$ is defined by

$$\langle\hat{A}\rangle_\beta = \frac{1}{Z(\beta)}\text{Tr}\left(\hat{A}e^{-\beta\hat{H}}\right) \qquad (2.21)$$

The T-product in (2.19) operates in imaginary time:

$$T(\hat{q}(-i\tau_1)\hat{q}(-i\tau_2)) = \hat{q}(-i\tau_1)\hat{q}(-i\tau_2) \text{ if } \tau_1 > \tau_2$$
$$= \hat{q}(-i\tau_2)\hat{q}(-i\tau_1) \text{ if } \tau_2 > \tau_1 \qquad (2.22)$$

That the functional differentiation of $Z(\beta;j)$ does give the thermal average of a T-product of position operators is easily proved: one writes, for example, for $\tau_1 > \tau_2$

$$Z(\beta)\langle T(\hat{q}(-i\tau_1)\hat{q}(-i\tau_2))\rangle_\beta$$
$$= \int dq < q|e^{-(\beta-\tau_1)\hat{H}}\hat{q}e^{-(\tau_1-\tau_2)\hat{H}}\hat{q}e^{-\tau_2\hat{H}}|q > \qquad (2.23)$$

Then one inserts a complete set of states of the position operator at 'times' τ_1 and τ_2 and repeats the procedure leading to the path integral (2.15). It is also straightforward to convince oneself that $Z(\beta;j)$ can be written in operator form:

$$Z(\beta;j) = \text{Tr}\left[e^{-\beta\hat{H}}\,T\left(e^{\int_0^\beta d\tau j(\tau)\hat{q}(-i\tau)}\right)\right] \qquad (2.24)$$

From the cyclicity of the trace in the operator form of the thermal average, or from the periodicity (2.16) of the paths in the path integral formalism, we deduce the following property of the propagator:

$$\langle T(\hat{q}(-i\beta)\,\hat{q}(-i\tau))\rangle_\beta = \langle T(\hat{q}(0)\hat{q}(-i\tau))\rangle_\beta \qquad (2.25)$$

Until now we have restricted τ to lie in the interval $[0,\beta]$; we shall see in the next section that $\Delta(\tau)$ is defined *a priori* in the interval $[-\beta,\beta]$. Then the cyclicity of the trace allows us to define a function $\Delta(\tau)$ which is periodic in imaginary time:

$$\Delta(\tau) = \langle T(\hat{q}(-i\tau)\hat{q}(0))\rangle_\beta \qquad (2.26)$$

and which obeys

$$\Delta(\tau - \beta) = \Delta(\tau) \tag{2.27}$$

for any value of τ in the interval $[0, \beta]$.

Up to now we have not specified the potential $V(q)$; we now specialize to the quantum mechanical analogue of the free field, namely, the harmonic oscillator with potential

$$V(q) = \frac{1}{2}\omega^2 q^2 \tag{2.28}$$

In order to simplify the formulae and to make a smooth connection with the following chapters, we set $m = 1$. We wish to compute the time-ordered product of position operators (2.19), which will generalize to the free field propagator. We use the path integral approach: the results will be rederived later on using the operator formalism. The generating functional $Z(\beta;j)$ is easily computed, since the integration is now Gaussian after an integration by parts in $S_E(\beta)$:

$$
\begin{aligned}
Z(\beta;j) &= \int_{q(0)=q(\beta)} \mathcal{D}q(\tau) \exp\left[-\int_0^\beta d\tau \left(\frac{1}{2}q(\tau)\right.\right. \\
&\quad \left.\left. \times \left(-\frac{d^2}{d\tau^2} + \omega^2\right)q(\tau) - j(\tau)q(\tau)\right)\right] \\
&= Z(\beta)\exp\left[\frac{1}{2}\int d\tau d\tau'\, j(\tau)K(\tau,\tau')j(\tau')\right]
\end{aligned}
\tag{2.29}
$$

In (2.29), the Green function $K(\tau,\tau')$ is the inverse of $(-d^2/d\tau^2 + \omega^2)$:

$$\left(-\frac{d^2}{d\tau^2} + \omega^2\right)K(\tau,\tau') = \delta(\tau - \tau') \tag{2.30}$$

From (2.18), (2.29) and the definition (2.26) of $\Delta(\tau)$ we may identify K and Δ; in the case of the harmonic oscillator, we add a subscript F, where F stands for 'free', as the field theoretic generalization of the harmonic oscillator will be the free field:

$$K(\tau,\tau') = \Delta_F(\tau - \tau') \tag{2.31}$$

The solution of (2.30), given the periodicity condition (2.27), is unique. If τ lies in the interval $[0, \beta]$, we have

$$\Delta_F(\tau) = \frac{1}{2\omega}\left[(1 + n(\omega))e^{-\omega\tau} + n(\omega)e^{\omega\tau}\right] \tag{2.32}$$

where $n(\omega)$ is the Bose–Einstein distribution:

$$n(\omega) = \frac{1}{\exp(\beta|\omega|) - 1} \tag{2.33}$$

An easy calculation allows one to check that (2.32), continued to negative values of τ thanks to the periodicity condition (2.27), does obey the differential equation (2.30), and that it is the unique solution of this equation with property (2.27): see exercise 2.3. One may also note that it is precisely the periodicity property of $\Delta_F(\tau)$ which allows one to perform the Gaussian integration in (2.29), as the periodicity of $q(\tau)$ is preserved in the change of integration variable (exercise 2.3).

2.2 Operator formalism

In the preceding section we already appealed to the operator formalism; however time was restricted to be pure imaginary. We now want to lift this restriction, as we are ultimately interested in real values of time. It is convenient to define the two-point functions $D^>(t, t')$ and $D^<(t, t')$:

$$D^>(t, t') = \langle \hat{q}(t)\hat{q}(t') \rangle_\beta \tag{2.34a}$$

$$D^<(t, t') = \langle \hat{q}(t')\hat{q}(t) \rangle_\beta = D^>(t', t) \tag{2.34b}$$

By inserting a complete set of eigenvectors of $\hat{H}$ one can express $D^>(t, t')$ as

$$D^>(t, t') = \frac{1}{Z(\beta)} \sum_{n,m} e^{-\beta E_n} e^{iE_n(t-t')}$$

$$\times e^{-iE_m(t-t')} |<n|\hat{q}(0)|m>|^2 \tag{2.35}$$

If the convergence in (2.35) is controlled by the exponentials, one sees that $D^>(t, t')$ is defined for

$$-\beta \le \text{Im}(t - t') \le 0 \tag{2.36a}$$

while $D^<(t, t')$ is defined for

$$\beta \ge \text{Im}(t - t') \ge 0 \tag{2.36b}$$

The functions $D^>$ and $D^<$ can be defined as distributions on the boundary of their domain of definition. We now use the fact that $\exp(-\beta\hat{H})$ is an evolution operator in imaginary time:

$$e^{-\beta\hat{H}} \hat{q}(t) e^{\beta\hat{H}} = \hat{q}(t + i\beta) \tag{2.37}$$

and we use the cyclicity of the trace to derive the Kubo–Martin–Schwinger (KMS) relation:

$$D^>(t, t') = D^<(t + i\beta, t') \tag{2.38}$$

It is useful to note that for τ lying in the interval $[0, \beta]$

$$\Delta(\tau) = D^>(-i\tau, 0) \tag{2.39}$$

Then the periodicity condition (2.27) of $\Delta(\tau)$ follows from the KMS relation (2.38). Finally, for real values of t and t', we define the time-ordered propagator:

$$D(t, t') = \langle T(\hat{q}(t)\hat{q}(t')) \rangle$$
$$= \theta(t - t')D^>(t, t') + \theta(t' - t)D^<(t, t') \tag{2.40}$$

The generalization of (2.40) to complex values of time will be dealt with in the following chapter.

2.3 The spectral function $\rho(k_0)$

The two-point Green function will appear in a variety of versions: imaginary time, real time, advanced, retarded. All these versions depend on a single function, which we call the spectral function $\rho(k_0)$. Because of translation invariance, $D^>(t, t')$ and $D^<(t, t')$ depend solely on the difference $t - t'$, and we shall use the shorthand notations $D^>(t) = D^>(t, 0)$ and $D^<(t) = D^<(t, 0)$. Let us define the Fourier transforms of $D^>(t)$ and $D^<(t)$:

$$D^>(k_0) = \int_{-\infty}^{\infty} dt \, e^{ik_0 t} D^>(t) \tag{2.41}$$

and

$$D^<(k_0) = \int_{-\infty}^{\infty} dt \, e^{ik_0 t} D^<(t) = \int_{-\infty}^{\infty} dt \, e^{ik_0 t} D^>(t - i\beta) \tag{2.42}$$

where the second equality follows from the KMS condition (2.38). Comparing (2.41) and (2.42) one deduces

$$D^<(k_0) = D^>(-k_0) = e^{-\beta k_0} D^>(k_0) \tag{2.43}$$

The physical interpretation of (2.43), which is related to detailed balance, is examined in exercise 2.4. Moreover, one notes from the hermiticity of $\hat{q}(t)$ and translation invariance that $D^>(k_0)$ and $D^<(k_0)$ are real functions of k_0. The spectral function $\rho(k_0)$ is defined by

$$\rho(k_0) = D^>(k_0) - D^<(k_0) \tag{2.44}$$

and is thus the Fourier transform of the thermal average of the commutator $[\hat{q}(t), \hat{q}(0)]$. We can write from (2.43)

$$D^>(k_0) = (1 + f(k_0))\rho(k_0) \qquad D^<(k_0) = f(k_0)\rho(k_0) \tag{2.45}$$

with

$$f(k_0) = (e^{\beta k_0} - 1)^{-1} \tag{2.46}$$

An explicit expression for $\rho(k_0)$ can be obtained by using (2.35) and by computing $(D^>(k_0) - D^<(k_0))$, making use of (2.43):

$$\rho(k_0) = \frac{2\pi}{Z(\beta)} \sum_{n,m} e^{-\beta E_n}(1 - e^{-\beta k_0})$$

$$\times \delta(k_0 + E_n - E_m)| < n|\hat{q}(0)|m > |^2$$

$$= \frac{2\pi}{Z(\beta)} \sum_{n,m} e^{-\beta E_n} [\delta(k_0 + E_n - E_m)$$

$$- \delta(k_0 + E_m - E_n)]| < n|\hat{q}(0)|m > |^2 \qquad (2.47)$$

Equation (2.47) shows explicitly that

(i) $\rho(k_0)$ is an odd real function of k_0: $\rho(k_0) = -\rho(-k_0)$,
(ii) $\rho(k_0)$ obeys the positivity condition $\varepsilon(k_0)\rho(k_0) > 0$, where $\varepsilon(k_0)$ is the sign function.

Furthermore, $\rho(k_0)$ obeys a sum rule which can be deduced from the equal-time canonical commutation relation

$$\left[\hat{q}(t), \frac{d}{dt'}\hat{q}(t')\right]\Big|_{t'=t} = i \qquad (2.48)$$

Indeed, we have

$$\int_{-\infty}^{\infty} \frac{dk_0}{2\pi} k_0 e^{-ik_0 t} \left(D^>(k_0) - D^<(k_0)\right)$$

$$= i\frac{d}{dt} \left(D^>(t) - D^<(t)\right) \qquad (2.49)$$

and, from the very definition of $D^>$ and $D^<$:

$$D^>(t) - D^<(t) = \langle[\hat{q}(t), \hat{q}(0)]\rangle_\beta \qquad (2.50)$$

Taking the $t \to 0$ limit of the above equations yields the sum rule:

$$\int_{-\infty}^{+\infty} \frac{dk_0}{2\pi} k_0 \rho(k_0) = 1 \qquad (2.51)$$

This sum rule can also be obtained from a simple manipulation of (2.47).
    Finally, it will be interesting to write down the analogue of the propagator in quantum mechanics; it is easily found from the expression of the position operator in the case of the harmonic oscillator (with $m = 1$):

$$\hat{q}(t) = \frac{1}{\sqrt{2\omega}}\left(a e^{-i\omega t} + a^\dagger e^{i\omega t}\right) \qquad (2.52)$$

(where $a^\dagger$ is the hermitian conjugate of a), and equation (2.50) which leads to

$$D^>(t) - D^<(t) = \frac{1}{2\omega}\langle[a, a^\dagger]e^{-i\omega t} + [a^\dagger, a]e^{i\omega t}\rangle_\beta \qquad (2.53)$$

Since $[a, a^\dagger] = 1$, one finds immediately the free (F) spectral function $\rho_F(k_0)$ by Fourier transformation

$$\rho_F(k_0) = 2\pi\varepsilon(k_0)\delta(k_0^2 - \omega^2) \tag{2.54}$$

It can easily be checked that this free spectral function obeys the positivity properties and the sum rule (2.51). It may also be worth noting that ρ_F is temperature-independent.

2.4   The Matsubara (or imaginary-time) propagator

We define the Fourier transform of the imaginary-time propagator $\Delta(\tau)$ (2.26) through

$$\Delta(i\omega_n) = \int_0^\beta d\tau\, e^{i\omega_n\tau}\Delta(\tau) \tag{2.55a}$$

the inverse formula being

$$\Delta(\tau) = T\sum_n e^{-i\omega_n\tau}\Delta(i\omega_n) \tag{2.55b}$$

Since $\Delta(\tau)$ is periodic $(\Delta(\tau - \beta) = \Delta(\tau))$, the Fourier transform is taken over a finite interval $[0, \beta]$, so that frequencies ω_n take discrete values:

$$\omega_n = \frac{2\pi n}{\beta} \tag{2.56}$$

which are called 'Matsubara frequencies'. If we choose τ in the interval $[0, \beta]$ we have $\Delta(\tau) = D^>(-i\tau)$ and, from (2.41):

$$\Delta(\tau) = \int \frac{dk_0}{2\pi} e^{-k_0\tau}D^>(k_0) \tag{2.57}$$

Using (2.55a) and the representation (2.45) of $D^>(k_0)$ in order to introduce the spectral function $\rho(k_0)$ leads to

$$\Delta(i\omega_n) = -\int_{-\infty}^\infty \frac{dk_0}{2\pi}\frac{\rho(k_0)}{i\omega_n - k_0} \tag{2.58}$$

In the free-field case, we use expression (2.54) for $\rho_F(k_0)$ and we get:

$$\Delta_F(i\omega_n) = \frac{1}{\omega_n^2 + \omega^2} \tag{2.59}$$

The inverse Fourier transform of (2.59) is, of course, $\Delta_F(\tau)$ given in (2.32). The Matsubara (full) propagator (2.58) is defined for discrete values (2.56) of the frequencies only and its analytic continuation to arbitrary values of the frequencies is not unique. However, from (2.58) one can define a unique analytic continuation provided that one requires

(i) $|\Delta(z)| \to 0$ if $|z| \to \infty$;
(ii) $\Delta(z)$ is analytic outside the real axis.

Then the analytic continuation is provided by

$$\Delta(z) = -\int_{-\infty}^{\infty} \frac{dk_0}{2\pi} \frac{\rho(k_0)}{z - k_0} \qquad (2.60)$$

This is in fact the continuation we are interested in, since for real values of z, or, more exactly, for $z = k_0 \pm i\eta$, $\eta \to 0^+$, with k_0 real, we obtain the retarded and advanced propagators which occur in linear response theory (see chapter 6) $D_R(t)$ and $D_A(t)$:

$$D_R(t) = \langle \theta(t)[\hat{q}(t), \hat{q}(0)] \rangle_\beta \qquad (2.61)$$
$$D_A(t) = -\langle \theta(-t)[\hat{q}(t), \hat{q}(0)] \rangle_\beta \qquad (2.62)$$

Indeed, from the usual representation of the θ-function:

$$\theta(t) = i \int_{-\infty}^{\infty} \frac{dk_0'}{2\pi} \frac{e^{-ik_0' t}}{k_0' + i\eta} \qquad (2.63)$$

one finds, for example, the Fourier transform $D_R(k_0)$ of $D_R(t)$:

$$D_R(k_0) = i \int_{-\infty}^{\infty} \frac{dk_0'}{2\pi} \frac{\rho(k_0')}{k_0 - k_0' + i\eta} \qquad (2.64)$$

so that:

$$D_R(k_0) = -i\Delta(k_0 + i\eta) \qquad D_A(k_0) = i\Delta(k_0 - i\eta) \qquad (2.65)$$

The function $D_R(k_0)$ is analytic in the upper half complex k_0 plane. Moreover, we note that the *free* retarded and advanced Green functions are temperature-independent because the free spectral function (2.54) is itself temperature-independent. This is not surprising: as will be explained in chapter 6, the physical content of these Green functions is that they give the energies and the lifetimes of the elementary excitations. In the free case we simply have $k_0 = \pm\omega$ and the excitations are of course stable.

2.5 The time-ordered propagator

We now wish to establish the expression of the thermal propagator in real time. The Fourier transform $D(k_0)$ of the propagator (2.40) reads

$$D(k_0) = \int dt \, e^{ik_0 t} \left(\theta(t) D^>(t) + \theta(-t) D^<(t) \right) \qquad (2.66)$$

From the representation (2.63) of the θ-function one derives

$$D(k_0) = i \int \frac{dk_0'}{2\pi} \frac{\rho(k_0')}{k_0 - k_0' + i\eta} + f(k_0)\rho(k_0) \qquad (2.67)$$

From this expression it is clear that $D(k_0)$ is *not* the analytic continuation toward real frequencies of the Matsubara propagator. Indeed, we may define a continuation D' which depends on the sign of k_0:

$$D'(k_0) = -i\Delta(k_0 + i\eta k_0) \tag{2.68}$$

and we have

$$D(k_0) = (1 + n(k_0))D'(k_0) + n(k_0)D'^*(k_0) \tag{2.69}$$

where

$$n(k_0) = (e^{\beta|k_0|} - 1)^{-1} \tag{2.70}$$

It is now easy to find the free propagator by substituting the expression (2.54) of the free spectral function in (2.67):

$$D_F(k_0) = \frac{i}{k_0^2 - \omega^2 + i\eta} + 2\pi n(k_0)\delta(k_0^2 - \omega^2) \tag{2.71}$$

There are two important points to be noted in the preceding formula. First, one sees that (2.71) allows one to separate a zero-temperature part and a thermal part, which vanishes in the $T = 0$ limit. Second, one has to note that it is the absolute value $|k_0|$ which is featured in (2.70), while k_0 appears in (2.46). We could hope that (2.71), suitably generalized to take the spatial degrees of freedom into account, would give the time-ordered propagator at non-zero T. Unfortunately, as we shall see in the next chapter, the full story is more complicated (even in ordinary quantum statistical mechanics!), and one has to double the field degrees of freedom in order to obtain the correct perturbative expansion.

2.6 Frequency sums

In calculations using the imaginary-time formalism, one often encounters sums over Matsubara frequencies. There exist standard techniques to handle these frequency sums. In the present section, we shall explain one of these techniques, which relies on the representation (2.55) of the Matsubara propagator and the explicit expression (2.32) of $\Delta_F(\tau)$. Another useful technique, using contour integration, will be explained in the following chapter.

Our first example concerns the calculation of the partition function of the harmonic oscillator. Of course, this partition function is most easily obtained by using definition (2.11) and a complete set of states of the Hamiltonian. Here we just want to gain some familiarity with path integrals in thermal field theory, and we try to derive $Z(\beta)$ from (2.15). The functional integral is of the Gaussian type and we start from the basic formula:

$$\int_{-\infty}^{\infty} \prod_{i=1}^{i=N} d\varphi_i \exp\left[-\frac{1}{2}\sum_{i,j}\varphi_i A_{ij}^{-1}\varphi_j\right]$$

$$= (2\pi)^{N/2}(\det A)^{1/2} = (2\pi)^{N/2}\exp\left[\frac{1}{2}\operatorname{Tr}\ln A\right] \qquad (2.72)$$

Within an additive constant, which we can drop because it does not affect thermodynamics, we obtain

$$\ln Z(\beta) = \frac{1}{2}\operatorname{Tr}\ln K = \frac{1}{2}\operatorname{Tr}\ln \Delta_F \qquad (2.73)$$

where K is the operator defined in (2.30). The trace is computed by going to Fourier space, where Δ_F is diagonal, and by using representation (2.59)

$$\ln Z(\beta) = -\frac{1}{2}\sum_{n=-\infty}^{n=+\infty}\ln(\omega^2 + \omega_n^2) \qquad (2.74)$$

One notes that the sum over n is divergent: this unfortunate feature stems from our careless handling of the integration measure $\mathscr{D}q(\tau)$. A more rigorous treatment (see exercise 2.2), using the proper definition (2.8) of $\mathscr{D}q(\tau)$, gives a perfectly finite result. In order to handle (2.74), we differentiate with respect to ω:

$$\frac{1}{2\omega}\frac{d\ln Z(\beta)}{d\omega} = -\frac{1}{2}\sum_{n=-\infty}^{n=+\infty}\frac{1}{(\omega^2+\omega_n^2)}$$

$$= -\frac{\beta}{2}\Delta_F(\tau = 0) = -\frac{\beta}{4\omega}(1 + 2n(\omega)) \qquad (2.75)$$

Then, by integration with respect to ω, we derive the free energy:

$$-\frac{1}{\beta}\ln Z(\beta) = \frac{1}{2}\omega + \frac{1}{\beta}\ln\left(1 - e^{-\beta\omega}\right) + \text{constant} \qquad (2.76)$$

where the constant is divergent, but independent of β and ω; (2.76), without the additive constant, is of course the standard result which can be found in any textbook on statistical mechanics in the part which is devoted to blackbody radiation.

Another typical frequency sum, which we shall encounter in the calculation of loop diagrams, is

$$S(i\omega_m) = T\sum_n \Delta(i(\omega_m - \omega_n))\Delta'(i\omega_n) \qquad (2.77)$$

where the oscillator frequencies of Δ and Δ' are ω and ω' respectively. Using the Fourier representation (2.55) of $\Delta(i\omega_n)$ and the relation

$$T\sum_n e^{i\omega_n\tau} = \sum_p \delta(\tau - p\beta) \qquad (2.78)$$

where $p = 0, \pm 1, \pm 2, ...$, yields

$$S(i\omega_n) = \int_0^\beta d\tau \, e^{i\omega_n \tau} \Delta(\tau) \Delta'(\tau) \qquad (2.79)$$

We now express $\Delta(\tau)$ and $\Delta'(\tau)$ in terms of their spectral functions $\rho(k_0)$ and $\rho'(k_0')$ (see (2.45) and (2.57)), integrate over τ remembering that $\exp(i\omega_m \beta) = 1$, and make use of simple identities involving the Bose–Einstein distribution to derive

$$S(i\omega_m) = - \int_{-\infty}^{+\infty} \frac{dk_0}{2\pi} \frac{dk_0'}{2\pi}$$
$$\times (1 + f(k_0) + f(k_0')) \frac{\rho(k_0)\rho'(k_0')}{i\omega_m - k_0 - k_0'} \qquad (2.80)$$

As explained in section 2.4, this expression can be analytically continued to values of ω_m which are not Matsubara frequencies, and in particular to real values of the energy: $i\omega_m \to q_0$, but of course only after the frequency sum over ω_n has been performed. Let us specialize this formula to the case of the 'free propagators' (2.59); we set $k_0 = s\omega$, $s = \pm 1$, and note that

$$f(\omega) = n(\omega) \qquad f(-\omega) = -(1 + n(\omega)) \qquad (2.81)$$

to write the final result as

$$S(q_0) = - \frac{1}{4\omega\omega'} \Bigg[(1 + n(\omega) + n(\omega'))$$
$$\times \left(\frac{1}{q_0 - \omega - \omega'} - \frac{1}{q_0 + \omega + \omega'} \right) + (n(\omega) - n(\omega'))$$
$$\times \left(\frac{1}{q_0 + \omega - \omega'} - \frac{1}{q_0 - \omega + \omega'} \right) \Bigg] \qquad (2.82)$$

This formula will lie at the basis of the one-loop calculations of the following chapters. We shall also need the imaginary part of $S(q_0 + i\eta)$, $\eta \to 0^+$; for arbitrary spectral functions we obtain from (2.80)

$$\text{Disc } S(q_0) = S(q_0 + i\eta) - S(q_0 - i\eta) = 2i \, \text{Im} \, S(q_0 + i\eta)$$
$$= 2i\pi \int_{-\infty}^{+\infty} \frac{dk_0}{2\pi} \frac{dk_0'}{2\pi} (1 + f(k_0) + f(k_0'))$$
$$\times \rho(k_0)\rho'(k_0')\delta(q_0 - k_0 - k_0') \qquad (2.83)$$

An equivalent formula is

$$\text{Disc } S(q_0) = 2i\pi(e^{\beta q_0} - 1) \int_{-\infty}^{+\infty} \frac{dk_0}{2\pi} \frac{dk_0'}{2\pi}$$
$$\times f(k_0)f(k_0')\rho(k_0)\rho'(k_0')\delta(q_0 - k_0 - k_0') \qquad (2.84)$$

In the case of free propagators, (2.83) becomes

$$\text{Disc } S(q_0) = \frac{2i\pi}{4\omega\omega'} \Big[(1 + n(\omega) + n(\omega'))(\delta(q_0 - \omega - \omega')$$
$$- \delta(q_0 + \omega + \omega')) + (n(\omega) - n(\omega'))$$
$$\times (\delta(q_0 + \omega - \omega') - \delta(q_0 - \omega + \omega')) \Big] \qquad (2.85)$$

References and further reading

There are many good accounts of path integrals in textbooks: see, for example, Feynman (1972), chap. 3; Schulman (1981); Rivers (1987); Negele and Orland (1988), chap. 1; Parisi (1988), chap. 13; Brown (1992), chap. 1. The integration measure in path integrals is examined from a mathematical point of view by Glimm and Jaffe (1987) and the extension of (2.5) to unbounded operators by Glimm and Jaffe (1987) and Schulman (1981). The analytical continuation from Matsubara frequencies to real values of the energy was first discussed by Baym and Mermin (1961); see also Fetter and Walecka (1971), chap. 9. Our presentation of sections 2.2–2.5 has been inspired by Dolan and Jackiw (1974), who were the first to derive the propagator (2.71), and Blaizot (1992).

Exercises

2.1 Obtain the Hamiltonian form of the path integral by writing $\exp(-\varepsilon\hat{H})$ $\simeq 1 - \varepsilon\hat{H}$ and leaving the p-integration in (2.7) undone:

$$F(q', -i\tau'; q, -i\tau) = \lim_{\varepsilon \to 0} \int \prod_{l=1}^{n} dq_l \prod_{l=0}^{n} \left[\frac{dp_l}{2\pi} \right.$$
$$\left. \times \exp(ip_l(q_{l+1} - q_l)) \exp(-\varepsilon H(p_l, q_l)) \right]$$

which is written formally as ($\dot{q} = dq/d\tau$)

$$\int \mathscr{D}p\mathscr{D}q \exp\left(\int_\tau^{\tau'} d\tau \left[ip\dot{q} - H(p, q) \right] \right)$$

2.2 We wish to compute the harmonic oscillator partition function from its path integral representation (2.15). We begin with the evaluation of the Euclidean action:

$$S[q(\tau)] = \frac{1}{2} \int_0^\beta d\tau \left(\dot{q}^2(\tau) + \omega^2 q^2(\tau) \right) d\tau$$

with $q(0) = q(\beta) = q$ (the calculation may be easily generalized to boundary conditions $q(0) = q'$, $q(\beta) = q''$). Let $q_{cl}(\tau)$ be a solution of the equation of motion $(d^2 q_{cl}/d\tau^2 - \omega^2 q_{cl}) = 0$ with $q_{cl}(0) = q_{cl}(\beta) = q$; we write

$$q(\tau) = q_{cl}(\tau) + h(\tau)$$

Use the stationarity of the action $([\delta S/\delta q]_{q=q_{cl}} = 0)$ and the equation of motion to obtain

$$S[q(\tau)] = \frac{1}{2} q[\dot{q}(\beta) - \dot{q}(0)] + S[h(\tau)]$$
$$= \frac{q^2 \omega (\cosh \beta \omega - 1)}{\sinh \beta \omega} + S[h(\tau)]$$

In order to evaluate

$$Z' = \int \mathcal{D}h \exp(-S[h(\tau)])$$

we discretize the h-integration. The determinant to be evaluated takes the form

$$D_n = \begin{pmatrix} 1 & -a & 0 & \cdots \\ -a & 1 & -a & 0 & \cdots \\ 0 & & \ddots & \\ & \vdots & & \end{pmatrix} \qquad |a| < \frac{1}{2}$$

Show that $D_n = D_{n-1} - a^2 D_{n-2}$ and solve this recursion relation. Taking the limit $n \to \infty$ yields

$$Z' = \left(\frac{\omega}{2\pi \sinh \beta \omega} \right)^{1/2}$$

Perform the q-integration and recover (2.76) with the constant $= 0$.

2.3 (a) Use the change of variable

$$q'(\tau) = q(\tau) - \int_0^\beta K(\tau, \tau') j(\tau') d\tau'$$

in the first two lines of (2.29) in order to obtain the last line of (2.29). Show that this change of variable preserves, as it should, the periodicity of $q(\tau)$.
(b) Show that (2.32) is the unique solution of (2.30) with the condition (2.27).

2.4 Assume that a harmonic oscillator in a thermal bath is submitted to a weak external force $j(t)$, so that the extra Hamiltonian is $j(t)\hat{q}(t)$. The transition probability $n \to m$ is given by Fermi's golden rule; it is proportional to

$$w_{n \to m} = 2\pi \delta(k_0 + E_n - E_m)| < n|\hat{q}|m > |^2$$

Show that (2.47) can be written as

$$\rho(k_0) = \sum_{n,m}(P_n - P_m)w_{n \to m}$$

where P_n is the probability of occupation of level n. Show that for $k_0 > 0$, $D^>$ gives the rate of energy absorption by the oscillator, while $D^<$ gives the rate of energy emission.

3

The scalar field at finite temperature

In the present chapter we study the quantization at finite temperature of the simplest case of field theory: the scalar field. The results of the preceding chapter will be generalized by including spatial degrees of freedom; however the time structure will closely parallel that already explained in the case of quantum statistical mechanics. We begin with the quantization of the neutral scalar field in the imaginary-time formalism and state the Feynman rules of the perturbative expansion. Section 3.2 is devoted to an extension to the case of the charged scalar field, and allows us to introduce the chemical potential associated with the conservation of a quantum number. Then we proceed to real-time formalism, where we shall discover some complications: the perturbative expansion leads to a doubling of the field degrees of freedom. In section 3.4 the two formalisms are compared in the case of the self-energy. The chapter concludes with some brief comments on renormalization.

3.1 The neutral scalar field

We begin with the neutral scalar field: in Minkowski space, the corresponding Lagrangian density reads

$$\mathcal{L} = \frac{1}{2}(\partial_\mu \varphi)(\partial^\mu \varphi) - \frac{1}{2}m^2\varphi^2 - \mathcal{V}(\varphi) \tag{3.1}$$

where $\varphi(x)$ is a real field, m is the mass of the corresponding particles and the interaction is contained in $\mathcal{V}(\varphi)$. This Lagrangian leads to the following Euclidean action (see (2.15)):

$$S_E(\beta) = \int_0^\beta d^4x \left(\frac{1}{2}(\partial_\mu \varphi)^2 + \frac{1}{2}m^2\varphi^2 + \mathcal{V}(\varphi)\right) \tag{3.2}$$

where $x = (\tau, \mathbf{x})$ and

$$\int_0^\beta \mathrm{d}^4 x = \int_0^\beta \mathrm{d}\tau \int \mathrm{d}^3 x$$
$$(\partial_\mu \varphi)^2 = (\partial_\tau \varphi)^2 + (\nabla \varphi)^2 \tag{3.3}$$

The interaction $\mathscr{V}(\varphi)$ is usually taken to be of the form $\lambda \varphi^4/4!$, but for the time being we shall not restrict the form of $\mathscr{V}(\varphi)$, except that it should not contain derivative interactions, that is, it should not depend on derivatives of φ; $\mathscr{V}(\varphi) = 0$ corresponds to the free-field case which generalizes the harmonic oscillator. From our experience in the preceding chapter, we may write the generating functional $Z(\beta; j)$ directly:

$$Z(\beta; j) = \int \mathscr{D}\varphi \exp\left(-S_E(\beta) + \int_0^\beta \mathrm{d}^4 x j(x)\varphi(x)\right) \tag{3.4}$$

Note that the generating functional of a $T = 0$ field theory in Euclidean space would also be given by (3.4), except for the fact that the τ-integration would run from $-\infty$ to $+\infty$. We could then read the Feynman rules from the zero-temperature Euclidean theory directly, but, in order to be self-contained, we shall give some hints on their derivation by using elementary examples. The derivation of Feynman rules relies on Wick's theorem, which we recall briefly.

3.1.1 Wick's theorem and the propagator

For pedagogical purposes, it is convenient to revert to discrete variables. Let us consider a (non-normalized) Gaussian probability distribution, defined by a positive symmetric matrix A^{-1}:

$$P(\varphi_1, \ldots, \varphi_N) = \exp\left(-\frac{1}{2}\sum_{i,j=1}^{i,j=N} \varphi_i A_{ij}^{-1} \varphi_j\right)$$
$$= \exp\left(-\frac{1}{2}\varphi^T A^{-1}\varphi\right) \tag{3.5}$$

where the superscript T stands for 'transpose', and let us write the generating function:

$$Z(j_1, \ldots, j_N) = \int \prod_{i=1}^{i=N} \mathrm{d}\varphi_i \exp\left(-\frac{1}{2}\varphi^T A^{-1}\varphi + j^T \varphi\right)$$
$$= Z(0)\exp\left(\frac{1}{2}j^T A j\right) \tag{3.6}$$

The moments of $P(\varphi)$ are obtained on differentiation of $Z(j)$:

$$\langle \varphi_{i_1} \cdots \varphi_{i_{2n}} \rangle = \frac{1}{Z(0)} \frac{\partial^{2n} Z(j_1, \ldots, j_N)}{\partial j_{i_1} \cdots \partial j_{i_{2n}}} \bigg|_{j=0}$$

$$= \frac{\partial^{2n}}{\partial j_{i_1} \cdots \partial j_{i_{2n}}} \frac{1}{n!} \frac{1}{2^n} \left(j^T A j \right)^n \qquad (3.7)$$

Let us start with the second moment:

$$\langle \varphi_i \varphi_k \rangle = \frac{\partial^2}{\partial j_i \partial j_k} \left(\frac{1}{2} \sum_{lm} j_l A_{lm} j_m \right) = A_{ik} \qquad (3.8)$$

The quantity $\langle \varphi_i \varphi_j \rangle$ is called the 'contraction' of φ_i and φ_j. The differentiations in (3.7) yield $(2n)!$ terms; we must, however, divide by $2^n n!$ hence the total number of terms on the right of (3.7) is $(2n-1)!!$. But this is simply the number of ways of choosing the pairs $\langle \varphi_{i_1} \varphi_{i_2} \rangle \langle \varphi_{i_3} \varphi_{i_4} \rangle \cdots \langle \varphi_{i_{2n-1}} \varphi_{i_{2n}} \rangle$. To see this, note that there are $(2n - 1)$ ways of forming the first pair $\langle \varphi_{i_1} \varphi_{i_2} \rangle$, $(2n - 3)$ ways of forming the second pair $\langle \varphi_{i_3} \varphi_{i_4} \rangle$, and so on. Wick's theorem follows (see Appendix B for further considerations on Wick's theorem):

$$\langle \varphi_{i_1} \cdots \varphi_{i_{2n}} \rangle = \langle \varphi_{i_1} \varphi_{i_2} \rangle \langle \varphi_{i_3} \varphi_{i_4} \rangle \cdots \langle \varphi_{i_{2n-1}} \varphi_{i_{2n}} \rangle$$
$$+ \text{ permutations} \qquad (3.9)$$

It may be useful to quote an example:

$$\langle \varphi_1 \varphi_2 \varphi_3 \varphi_4 \rangle = \langle \varphi_1 \varphi_2 \rangle \langle \varphi_3 \varphi_4 \rangle + \langle \varphi_1 \varphi_3 \rangle \langle \varphi_2 \varphi_4 \rangle$$
$$+ \langle \varphi_1 \varphi_4 \rangle \langle \varphi_2 \varphi_3 \rangle \qquad (3.10)$$

It is important to understand that the total number of terms does not change if some indices are identical:

$$\langle \varphi_1 \varphi_2 \varphi_2 \varphi_4 \rangle = 2 \langle \varphi_1 \varphi_2 \rangle \langle \varphi_2 \varphi_4 \rangle$$
$$+ \langle \varphi_1 \varphi_4 \rangle \langle \varphi_2 \varphi_2 \rangle \qquad (3.11)$$

As an extreme case all indices may be identical:

$$\langle \varphi^{2n} \rangle = (2n - 1)!! \langle \varphi^2 \rangle^n \qquad (3.12)$$

We now come back to our original generating functional (3.4). In the free-field case ($\mathcal{V} = 0$), the integration over $\varphi(x)$ is Gaussian, and generalizing (2.29) we obtain for the free generating functional, after integration by parts:

$$Z_F(\beta; j) = \int \mathcal{D}\varphi \exp\left(-\int_0^\beta d\tau \int d^3x \right.$$

$$\left. \times \left(\varphi(x) \frac{1}{2} \left[-\frac{\partial^2}{\partial \tau^2} - \nabla^2 + m^2 \right] \varphi(x) - j(x)\varphi(x) \right) \right) \qquad (3.13)$$

The integration over $\varphi(x)$ is performed as in (2.29):

$$Z_F(\beta;j) = Z_F(\beta)\exp\left(\frac{1}{2}\int_0^\beta d^4x d^4y \; j(x)\Delta_F(x-y)j(y)\right) \tag{3.14}$$

where the two-point Green function $\Delta_F(x-y)$ is the solution of the partial differential equation:

$$\left(-\frac{\partial^2}{\partial\tau^2} - \nabla^2 + m^2\right)\Delta_F(x-y) = \delta(\tau_x - \tau_y)\delta(\mathbf{x}-\mathbf{y}) \tag{3.15}$$

which is periodic in τ with period β: adding x-dependence does not change the results obtained in the preceding chapter on τ-dependence. Equation (3.15) is solved in Fourier space:

$$(\omega_n^2 + \omega_k^2)\Delta_F(i\omega_n, k) = 1 \tag{3.16}$$

where

$$\omega_n = \frac{2\pi n}{\beta} \qquad \omega_k = (k^2 + m^2)^{1/2} \tag{3.17}$$

From (3.16) we obtain the Matsubara propagator, or the propagator in imaginary-time

$$\Delta_F(i\omega_n, k) = \frac{1}{\omega_n^2 + k^2 + m^2} = \frac{1}{\omega_n^2 + \omega_k^2} \tag{3.18}$$

A very useful quantity is the mixed representation $\Delta_F(\tau, k)$ of Δ_F, which can be read from (2.32):

$$\Delta_F(\tau, k) = T\sum_n e^{-i\omega_n\tau}\Delta_F(i\omega_n, k)$$

$$= \frac{1}{2\omega_k}\left((1 + n(\omega_k))e^{-\omega_k\tau} + n(\omega_k)e^{\omega_k\tau}\right) \tag{3.19}$$

The propagator in x-space $\Delta_F(x) = \Delta_F(\tau, \mathbf{x})$ is obtained by inverse Fourier transformation on the spatial components:

$$\Delta_F(x) = \int \frac{d^3k}{(2\pi)^3}e^{i\mathbf{k}\cdot\mathbf{x}}\Delta_F(\tau, k) \tag{3.20}$$

As a final comment, we note that the partition function of the free scalar field is obtained from (2.76) by integration over all values of $\mathbf{k}$, with

$$\sum_{\mathbf{k}} \rightarrow V\int \frac{d^3k}{(2\pi)^3} \tag{3.21}$$

where V is the volume. Thus the free energy $\Omega = -\beta^{-1}\ln Z(\beta)$ reads, from (2.76):

$$\Omega = V\int \frac{d^3k}{(2\pi)^3}\left[\frac{1}{2}\omega_k + \frac{1}{\beta}\ln\left(1 - e^{-\beta\omega_k}\right)\right] \tag{3.22}$$

The first term in the square bracket of (3.22) is β-independent and leads to a divergent integral; now the divergence comes from the infinite number of modes, after we have controlled the summation over the Matsubara frequencies by dropping the constant term in (2.76). This infinite result is of course nothing other than the zero-point energy of the vacuum, which can be subtracted off, since it is an unobservable constant, although differences in the zero-point energies can be observed (Casimir effect). Ignoring the zero-point energy and setting $m = 0$ in order to perform the integrations analytically, we find

$$\Omega = \frac{V}{2\pi^2 \beta} \int_0^\infty k^2 dk \ln\left(1 - e^{-\beta k}\right) = -\frac{\pi^2 V}{90\beta^4} \tag{3.23}$$

in agreement with (1.7). From this result we compute all thermodynamical quantities, for example the pressure:

$$P = -\frac{\partial \Omega}{\partial V} = \frac{\pi^2 T^4}{90} \tag{3.24}$$

This result for the pressure is that of an ultrarelativistic ideal gas of spinless particles; it is half that of the blackbody radiation, because scalar particles have no spin degrees of freedom, while photons have two transverse polarizations.

3.1.2 First-order corrections: propagator and partition function

Before discussing the general form of the Feynman rules, we shall give elementary examples of perturbative calculations. For the interaction term in (3.2) we take

$$\mathscr{V}(\varphi) = \frac{\lambda}{4!}\varphi^4 \tag{3.25}$$

The full propagator $\Delta(x - y)$ is given by (see (2.18))

$$\Delta(x - y) = \frac{1}{Z(\beta)} \int \mathscr{D}\varphi \, \varphi(x)\varphi(y)e^{-S_E(\beta)} \tag{3.26}$$

with the boundary conditions $\varphi(0, \mathbf{x}) = \varphi(\beta; \mathbf{x})$. Expanding the exponential to first order in λ, we find

$$e^{-S_E(\beta)} \simeq e^{-S_E^{(F)}(\beta)}\left(1 - \frac{\lambda}{4!} \int_0^\beta d^4z \, \varphi^4(z)\right) \tag{3.27}$$

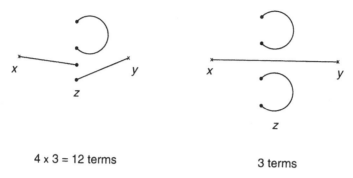

4 x 3 = 12 terms　　　　　　　　3 terms

Fig. 3.1　Wick's theorem.

where $S_E^{(F)}(\beta)$ is the free Euclidean action. Let us evaluate the quantity

$$I(x,y) = \int \mathscr{D}\varphi\, \varphi(x)\varphi(y)\mathrm{e}^{-S_E^{(F)}(\beta)}\left(1 - \frac{\lambda}{4!}\int_0^\beta \mathrm{d}^4 z\, \varphi^4(z)\right) \qquad (3.28)$$

using Wick's theorem. The number of contractions coming from the second term of (3.28) equals the number of contractions of $\varphi(x)$, $\varphi(y)$ and $\varphi^4(z)$. It proves convenient to represent these contractions on a diagram, by drawing two 'external' points x and y, where 'external' means that they refer to the arguments of the propagator, marked by crosses, and an 'internal' or 'vertex' z which stems from expansion into powers of λ, and over which we shall integrate. Because $\varphi(z)$ enters through its fourth power, the point z is drawn as four separate points at first. Every contraction is represented by a line joining the arguments of φ. Two types of term are possible (fig. 3.1) and one checks that the number of terms is correct: $12 + 3 = (6 - 1)!!$. In order to simplify the diagram, the four points z are merged into a single point, with the result shown in fig. 3.2.

The diagrams of fig. 3.2 are called Feynman diagrams (or graphs); one such diagram corresponds to every term, or rather to every group of terms, of the perturbative expansion. The integral I reads

$$I(x,y) = Z_F(\beta)\Big(\Delta_F(x-y) - \frac{1}{2}\lambda \int_0^\beta \mathrm{d}^4 z \Delta_F(x-z)\Delta_F(z=0)$$

$$\times\, \Delta_F(z-y) - \frac{1}{8}\lambda \int_0^\beta \mathrm{d}^4 z\, \Delta_F(x-y)\Delta_F^2(z=0)\Big) \qquad (3.29)$$

where $Z_F(\beta)$ is the free partition function. The factors $1/2$ and $1/8$ in (3.29) stem from the combinatorics of Wick's theorem and are called 'symmetry factors'. In order to obtain the propagator, we still have to

Fig. 3.2 Connected and disconnected graphs.

divide by the full partition function, computed to the same order in λ:

$$Z(\beta) \simeq \int \mathscr{D}\varphi \, e^{-S_E^{(F)}(\beta)} \left(1 - \frac{\lambda}{4!} \int_0^\beta d^4 z \varphi^4(z)\right)$$

$$= Z_F(\beta)\left[1 - \frac{1}{8}\lambda(\Delta_F(x = 0))^2 \int_0^\beta d^4 z\right] \qquad (3.30)$$

The second term in the square bracket of (3.30) is represented by fig. 3.3. To obtain the propagator to first order in λ, we divide (3.29) by (3.30) and the normalization constant $Z_F(\beta)$ cancels from the final result:

$$\Delta(x - y) = \Delta_F(x - y) - \frac{1}{2}\lambda \int_0^\beta d^4 z \Delta_F(x - z)$$

$$\times \Delta_F(z = 0)\Delta_F(z - y) + O(\lambda^2) \qquad (3.31)$$

Equation (3.31) can be represented graphically as in fig. 3.4: the heavy line represents the full, or dressed, propagator to order λ. Note that a factor of $-\lambda$ is associated with the vertex. The disconnected diagram of fig. 3.2 disappears upon division by $Z(\beta)$; this cancellation of disconnected diagrams is a general result, which follows from simple combinatorics; as this is discussed in many textbooks, we shall skip the proof here.

Since a convolution transforms into a product in a Fourier transformation, (3.31) reads in Fourier space

$$\Delta(i\omega_n, k) = \Delta_F(i\omega_n, k) - \frac{\lambda}{2}\Delta_F(i\omega_n, k)$$

$$\times \left(T \sum_m \int \frac{d^3 k'}{(2\pi)^3}\Delta_F(i\omega_m, k')\right)\Delta_F(i\omega_n, k) \qquad (3.32)$$

In general, one defines the self-energy Π through

$$\Delta^{-1}(i\omega_n, k) = \Delta_F^{-1}(i\omega_n, k) + \Pi(i\omega_n, k) \qquad (3.33)$$

To first order in λ, one finds from (3.32) that Π is independent of the external momentum k and is given by

$$\Pi = \frac{\lambda}{2}\Delta_F(x = 0) = \frac{\lambda T}{2} \sum_m \int \frac{d^3 k'}{(2\pi)^3}\Delta_F(i\omega_m, k') \qquad (3.34)$$

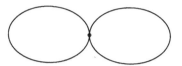

Fig. 3.3   First-order correction to the partition function.

The summation over m has been evaluated in (2.75):

$$\Pi = \frac{\lambda}{2} \int \frac{d^3k}{(2\pi)^3} \frac{1}{2\omega_k}(1 + 2n(\omega_k)) \tag{3.35}$$

To this order in λ, Π can be interpreted as a simple correction to the mass squared: $m^2 \rightarrow m^2 + \Pi$. However, the $T = 0$ part of the integral in (3.35) is ultraviolet divergent. The remedy is well known: one has to renormalize the theory by introducing a mass counter-term in the Euclidean action:

$$\delta S_E = \frac{1}{2}\delta m^2 \int_0^\beta d^4x \, \varphi^2(x) \tag{3.36}$$

with $\delta m^2 = -\Pi|_{T=0}$; we remind the reader that this counter-term is treated as an interaction term in the Euclidean action. Then m is the $(T = 0)$ physical mass of the particle; if the counter-term is taken into account, Π is given by a convergent integral, which is easily evaluated analytically if $m = 0$:

$$\Pi = \delta m_\beta^2 = \frac{\lambda}{2} \int \frac{d^3k}{(2\pi)^3} \frac{n(k)}{k} = \frac{\lambda T^2}{24} \tag{3.37}$$

We have just seen a particular case of a general result (see section 3.5): renormalization of the theory at $T = 0$ is sufficient to make the theory finite at non-zero T.

Let us now look at the free energy; from (3.30) we have

$$\Omega(\beta) = -\frac{1}{\beta} \ln Z_F(\beta) + \frac{1}{8}\lambda V(\Delta_F(x = 0))^2 \tag{3.38}$$

After subtraction of the zero-point energy, $\beta^{-1} \ln Z_F(\beta)$ is given by (3.23); the second term in (3.38) is represented graphically by fig. 3.3. In the computation of $\Delta_F(x = 0)$ we again encounter an ultraviolet divergence linked to the $T = 0$ term (see (3.35)). However, we must remember

Fig. 3.4   First-order correction to the propagator.

to take into account the mass counter-term (3.36), which is treated as an interaction term in the Euclidean action. In order to rewrite the renormalized $\Omega(\beta)$ conveniently, we separate Δ_F into a $T = 0$ and a $T \neq 0$ part:

$$\Delta_F = \Delta_F^{T=0} + \Delta_F^T \qquad (3.39)$$

so that

$$\delta m^2 = -\frac{\lambda}{2}\Delta_F^{T=0}(x = 0) \qquad (3.40)$$

The contribution of the counter-term to $\Omega(\beta)$ is

$$\frac{V}{2}\delta m^2 \Delta_F(x = 0) = -\frac{\lambda V}{4}\Delta_F^{T=0}(x = 0)\Delta_F(x = 0) \qquad (3.41)$$

and this leads to the renormalized $\Omega(\beta)$:

$$\Omega(\beta) = -\frac{1}{\beta}\ln Z_F(\beta) - \frac{\lambda V}{8}(\Delta_F^{T=0}(x = 0))^2 + \frac{\lambda V}{8}(\Delta_F^T(x = 0))^2 \qquad (3.42)$$

The potentially dangerous term proportional to $\Delta_F^{T=0}(x = 0) \times \Delta_F^T(x = 0)$ has been cancelled thanks to mass renormalization. However the result (3.42) is still divergent because of the second term on its RHS; fortunately this term is T-independent and thus harmless: it can be subtracted off. One elegant way to avoid this term is to define the renormalized free energy $\Omega_R(\beta)$ through

$$\Omega_R(T, m^2, \lambda) = \Omega(T, m^2, \lambda) - \Omega(T = 0, m^2, \lambda) \qquad (3.43)$$

with the same renormalized mass in the three terms. From this renormalized partition function, one obtains the correction of order λ to the pressure of the ideal gas; in the $m = 0$ case

$$P = -\frac{1}{\beta}\int\frac{d^3k}{(2\pi)^3}\ln\left(1 - e^{-\beta k}\right) - \frac{\lambda}{8}\left(\int\frac{d^3k}{(2\pi)^3}\frac{n(k)}{k}\right)^2$$

$$= \frac{\pi^2 T^4}{90}\left(1 - \frac{5\lambda}{64\pi^2}\right) \qquad (3.44)$$

As we shall see in the following chapter, the next correction is of order $\lambda^{3/2}$, and not of order λ^2, as could have been expected: this arises from the infrared singular behaviour of the propagator.

3.1.3 Feynman rules

The generating functional $Z(\beta; j)$ can be written as

$$Z(\beta; j) = \mathcal{N}\exp\left(-\int d^4x\,\mathcal{V}\left(\frac{\delta}{\delta j(x)}\right)\right)$$

$$\times \exp\left(\frac{1}{2}\int d^4x\,d^4y\;j(x)\Delta_F(x - y)j(y)\right) \qquad (3.45)$$

This equation is identical to that used to derive Feynman rules at $T = 0$ in Euclidean space; the only difference is that the $T = 0$ propagator is replaced by the $T \neq 0$ Matsubara propagator $\Delta_F(x - y)$. Thus the Feynman rules in x-space are identical to those of the $T = 0$ Euclidean theory, provided one uses the Matsubara propagator and restricts the τ-integration to the range $[0, \beta]$. This feature leads to a modification of the Feynman rules in k-space; since the Matsubara frequencies are discrete, energy conservation will be given by a Kronecker δ-function $\beta\delta(\sum \omega_i)$ at each vertex. We are now in a position to quote the Feynman rules for an interaction $\mathscr{V}(\varphi) = \lambda\varphi^4/4!$

(1) Draw all topologically inequivalent diagrams to a given order of perturbation theory.
(2) Assign a factor $\Delta_F(i\omega_n, k)$ to every line of the diagram.
(3) Assign a factor (the first δ is a Kronecker symbol)

$$-\lambda \beta \, \delta\left(\sum_{i=1}^{4} \omega_i\right)(2\pi)^3 \delta^{(3)}\left(\sum_{i=1}^{4} k_i\right)$$

to every vertex.
(4) Integrate over every internal line with the measure

$$T \sum_n \int \frac{\mathrm{d}^3 k}{(2\pi)^3}$$

(5) Multiply by the symmetry factor.
(6) There will be an overall factor $\beta(2\pi)^3\delta^{(3)}(0) = \beta V$.

3.2 The charged scalar field

The case of the charged scalar field will allow us to introduce, in a simple way, the chemical potential associated with a conserved charge. The Lagrangian density in Minkowski space depends on a complex field $\varphi(x)$:

$$\mathscr{L} = (\partial_\mu\varphi)(\partial^\mu\varphi^*) - m^2\varphi\varphi^* - \frac{\lambda}{4}(\varphi\varphi^*)^2 \tag{3.46}$$

where we have chosen a specific interaction term. This Lagrangian density is invariant under a global gauge (or phase) transformation:

$$\varphi \rightarrow \varphi' = \mathrm{e}^{-i\alpha}\varphi \tag{3.47}$$

which leads to a conserved current:

$$j_\mu = i(\varphi^*\partial_\mu\varphi - \varphi\partial_\mu\varphi^*) \tag{3.48}$$

and to a conserved charge:

$$Q = \int \mathrm{d}^3x \, j_0(x) \tag{3.49}$$

It is convenient to decompose φ into real and imaginary parts: $\varphi = (\varphi_1 + i\varphi_2)/\sqrt{2}$, where φ_1 and φ_2 are real fields. The Hamiltonian density $\mathscr{H}$ and the conserved charge are

$$\mathscr{H} = \frac{1}{2}\Big[\pi_1^2 + \pi_2^2 + (\nabla\varphi_1)^2 + (\nabla\varphi_2)^2$$

$$+ m^2(\varphi_1^2 + \varphi_2^2) + \frac{\lambda}{8}(\varphi_1^2 + \varphi_2^2)^2\Big] \tag{3.50}$$

$$Q = \int d^3x(\varphi_2\pi_1 - \varphi_1\pi_2) \tag{3.51}$$

where π_1 and π_2 are the conjugate momenta of φ_1 and φ_2. Since we have a conserved charge Q, there exists a chemical potential μ conjugate to Q, so that the grand partition function is

$$Z(\beta,\mu) = \text{Tr } e^{-\beta(\hat{H} - \mu\hat{Q})} \tag{3.52}$$

where $\hat{Q}$ is the quantized version of (3.51). Since Q depends on the time derivative of φ, we have to go through the Hamiltonian form of the path integral (exercise 2.1) in order to write a path integral for Z:

$$Z(\beta,\mu) = \int \mathscr{D}(\pi_1,\pi_2) \int \mathscr{D}(\varphi_1,\varphi_2)\exp\Big[\int_0^\beta d^4x\Big(i\pi_1\partial_\tau\varphi_1$$

$$+ i\pi_2\partial_\tau\varphi_2 - \mathscr{H}(\pi_i,\varphi_i) + \mu(\varphi_2\pi_1 - \varphi_1\pi_2)\Big)\Big] \tag{3.53}$$

where $\partial_\tau\varphi_i = \partial\varphi_i/\partial\tau$, and the fields φ_i are periodic in τ: $\varphi_i(0,\mathbf{x}) = \varphi_i(\beta,\mathbf{x})$. Integrating over (π_1,π_2) leads to the following combination in the exponent:

$$\frac{1}{2}(\partial_\tau\varphi_1 - i\mu\varphi_2)^2 + \frac{1}{2}(\partial_\tau\varphi_2 + i\mu\varphi_1)^2 + \frac{1}{2}m^2(\varphi_1^2 + \varphi_2^2)$$

$$= \partial_\tau\varphi\partial_\tau\varphi^* - \mu(\varphi\partial_\tau\varphi^* - \varphi^*\partial_\tau\varphi) - \mu^2\varphi^*\varphi + m^2\varphi^*\varphi$$

$$= -\varphi^*\Big(\frac{\partial^2}{\partial\tau^2} - 2\mu\frac{\partial}{\partial\tau} + \mu^2 - m^2\Big)\varphi \tag{3.54}$$

where for simplicity we have taken $\lambda = 0$ and performed an integration by parts in the last line. Going to frequency space, $\partial_\tau \to -i\omega_n$, we see that the introduction of the chemical potential corresponds to the substitution

$$i\omega_n \to i\omega_n + \mu \tag{3.55}$$

in the Matsubara propagator, which now reads

$$\Delta_F(i\omega_n, k) = \frac{1}{-(i\omega_n + \mu)^2 + k^2 + m^2} \tag{3.56}$$

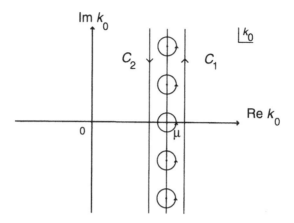

Fig. 3.5   The integration contour in the complex k_0 plane.

As a simple exercise, we compute the grand partition function in the free field case. Following the same steps as in section 2.6, we see that we must evaluate

$$\frac{1}{2\omega}\frac{d}{d\omega}\ln Z(\beta,\mu) = -T\sum_n \frac{1}{(\omega_n - i\mu)^2 + \omega^2} \qquad (3.57)$$

There is a factor of 2 with respect to (2.73) because we integrate over a complex field φ. Instead of using a mixed representation of Δ_F, we take the opportunity of introducing another useful technique in order to perform the summation over n. The trick is to use a contour integral in the complex energy plane. Let $g(k_0)$ be a meromorphic function of k_0, regular on the vertical line $\mathrm{Re}\, k_0 = \mu$, which decreases faster than k_0^{-1} for $|k_0| \to \infty$. We have

$$S = T\sum_{n=-\infty}^{n=+\infty} g(k_0 = i\omega_n + \mu)$$

$$= \int_{C_1 \cup C_2} \frac{dk_0}{2i\pi} g(k_0)\frac{1}{2}\coth\left[\frac{\beta(k_0 - \mu)}{2}\right] \qquad (3.58)$$

where C_1 is a vertical straight line from $\mu + \eta - i\infty$ to $\mu + \eta + i\infty$, while C_2 goes from $\mu - \eta + i\infty$ to $\mu - \eta - i\infty$, $\eta \to 0^+$. In order to prove (3.58), one notes that the function $\frac{\beta}{2}\coth[\beta(k_0 - \mu)/2]$ has poles with unit residue at $k_0 = i\omega_n + \mu$ (fig. 3.5). Furthermore, this function is bounded in all directions of the complex k_0-plane, except on the vertical axis $\mathrm{Re}\, k_0 = \mu$. Closing C_1 and C_2 with large half-circles allows us to pick up the poles of $g(k_0)$:

$$S = -\frac{1}{2}\sum \mathrm{Res}\, g(k_0)\coth\left[\frac{\beta(k_0 - \mu)}{2}\right] \qquad (3.59)$$

where the sum runs over all the values of k_0 such that $g^{-1}(k_0) = 0$ and we have denoted the corresponding residues by Res $g(k_0)$. From our assumption on the asymptotic behaviour of $g(k_0)$ we deduce $\sum$ Res $g(k_0) = 0$, so that (3.59) may be cast in the alternative form

$$S = -\sum \frac{\text{Res } g(k_0)}{e^{\beta(k_0-\mu)} - 1} = -\sum \text{Res } g(k_0) f(k_0 - \mu) \qquad (3.60)$$

Note the $T = 0$ limit of (3.60):

$$\lim_{T=0} S = \sum_{\text{Re } k_0 < \mu} \text{Res } g(k_0) \qquad (3.61)$$

We may now return to (3.57), with

$$g(k_0) = \frac{1}{\omega^2 - k_0^2} \qquad (3.62)$$

Integrating over ω as in section 2.6 and taking the k-integration into account gives for the grand potential $\Omega(\beta, \mu) = -PV = -\beta^{-1} \ln Z(\beta, \mu)$:

$$\Omega(\beta, \mu) = V \int \frac{d^3k}{(2\pi)^3} \left[\omega_k + \frac{1}{\beta} \ln\left(1 - e^{-\beta(\omega_k - \mu)}\right) \right.$$
$$\left. + \frac{1}{\beta} \ln\left(1 - e^{-\beta(\omega_k + \mu)}\right) \right] \qquad (3.63)$$

This is of course the standard result, which can be found in any textbook on statistical mechanics. It is valid only if $|\mu| < m$; for $\mu = m$, one has to take Bose–Einstein condensation into account.

3.3 Real-time formalism

3.3.1 *Path integrals*

For simplicity we work with the neutral scalar field of section 3.1, with zero chemical potential. The field operator in the Heisenberg picture is

$$\hat{\varphi}(x) = e^{it\hat{H}} \hat{\varphi}(0) e^{-it\hat{H}} \qquad (3.64)$$

where the time coordinate $t = x^0$ is allowed to be complex. We are interested in computing thermal Green functions:

$$G_C(x_1, \ldots, x_N) = \langle T_C(\hat{\varphi}(x_1) \ldots \hat{\varphi}(x_N)) \rangle_\beta \qquad (3.65)$$

where the time-ordering T_C is taken along a complex time path C, which will be defined precisely later on; the subscript C refers to the time path. One may give a parametric definition $t = z(v)$ of the path, with v real and monotonically increasing: the ordering along the path will correspond to the ordering in v.

It is convenient to introduce path θ- and δ-functions:

$$\theta_C(t-t') = \theta(v-v') \qquad \delta_C(t-t') = \left(\frac{\partial z}{\partial v}\right)^{-1} \delta(v-v') \qquad (3.66)$$

With these definitions we may write, for example,

$$\begin{aligned}
T_C(\hat{\varphi}(x)\hat{\varphi}(x')) &= \theta_C(t-t')\hat{\varphi}(x)\hat{\varphi}(x') \\
&\quad + \theta_C(t'-t)\hat{\varphi}(x')\hat{\varphi}(x) \qquad (3.67\text{a}) \\
\partial_t T_C(\hat{\varphi}(x)\hat{\varphi}(x')) &= \delta_C(t-t')[\hat{\varphi}(x), \hat{\varphi}(x')] \\
&\quad + T_C(\partial_t\hat{\varphi}(x)\hat{\varphi}(x')) \qquad (3.67\text{b})
\end{aligned}$$

One also extends functional differentiation:

$$\frac{\delta j(x)}{\delta j(x')} = \delta_C(t-t')\delta^{(3)}(\mathbf{x}-\mathbf{x}') \qquad (3.68)$$

for c-number functions $j(x)$ living on the path C.

We wish to construct a generating functional $Z_C(\beta;j)$ which allows us to obtain Green's functions from functional differentiation with respect to sources $j(x)$:

$$G_C(x_1,\ldots,x_N) = \frac{1}{Z(\beta)} \frac{\delta^N Z_C(\beta;j)}{i\delta j(x_1)\ldots i\delta j(x_N)}\bigg|_{j=0} \qquad (3.69)$$

An obvious solution for $Z(\beta;j)$ is

$$Z_C(\beta;j) = \mathrm{Tr}\left[e^{-\beta\hat{H}} T_C \exp\left(i\int_C \mathrm{d}^4 x\, j(x)\hat{\varphi}(x)\right)\right] \qquad (3.70)$$

where the path C must go through all the arguments of the Green function we are interested in; $Z_C(\beta;j=0) = Z(\beta) = \mathrm{Tr}\,\exp(-\beta\hat{H})$ is, as usual, the partition function.

Let us generalize (2.40) by defining the propagator $D_C(x,x')$:

$$D_C(x,x') = \theta_C(t-t')D_C^{>}(x,x') + \theta_C(t'-t)D_C^{<}(x,x') \qquad (3.71)$$

where

$$\begin{aligned}
D_C^{>}(x,x') &= \langle\hat{\varphi}(x)\hat{\varphi}(x')\rangle_\beta \\
D_C^{<}(x,x') &= D_C^{>}(x',x) = \langle\hat{\varphi}(x')\hat{\varphi}(x)\rangle_\beta
\end{aligned} \qquad (3.72)$$

We saw in section 2.2 that $D_C^{>}(x,x')$ and $D_C^{<}(x,x')$ are likely to be analytic functions in the strips $-\beta < \mathrm{Im}(t-t') < 0$ and $0 < \mathrm{Im}(t-t') < \beta$ respectively. Since the limit of an analytic function on the boundary of its analyticity domain is a distribution (generalized function), we may conclude that $D_C^{>}(x,x')$ and $D_C^{<}(x,x')$ are properly defined in the strips $-\beta \leq \mathrm{Im}(t-t') \leq 0$ and $0 \leq \mathrm{Im}(t-t') \leq \beta$ respectively. If we want the propagator in (3.71) to be well defined, we must take paths C such that the imaginary part of t is non-increasing when the parameter v increases.

We may now generalize the results of chapter 2 by first adding space components (trivial) and by using complex values of t in (2.40), rather than real ones. Taking translation invariance into account, we rewrite (3.71) as

$$D_C(x - x') = \theta_C(t - t')\left[D_C^>(x - x') - D_C^<(x - x')\right] + D_C^<(x - x') \quad (3.73)$$

and use (2.44) and (2.45) suitably generalized to four dimensions; instead of a spectral function $\rho(k_0)$, we now have a spectral function $\rho(k) = \rho(k_0, \mathbf{k})$, whose properties will be examined later on (section 3.4). From (3.73) we obtain a very useful representation of the propagator:

$$D_C(x - x') = \int \frac{d^4k}{(2\pi)^4} e^{-ik\cdot(x-x')}\left[\theta_C(t - t') + f(k_0)\right]\rho(k) \quad (3.74)$$

where we recall the definition (2.46) of $f(k_0)$:

$$f(k_0) = \frac{1}{e^{\beta k_0} - 1} \quad (3.75a)$$

and the often used relation

$$1 + f(k_0) + f(-k_0) = 0 \quad (3.75b)$$

The symmetry property

$$D_C(x - x') = D_C(x' - x) \quad (3.76)$$

is easily derived, either from the definition (3.71) or from the representation (3.74). The free propagator D_C^F follows from (3.74) by inserting the free spectral function:

$$\rho_F(k) = 2\pi\varepsilon(k_0)\delta(k^2 - m^2) \quad (3.77)$$

We now turn to the derivation of a path integral representation for the generating functional $Z_C(\beta; j)$ in (3.70). Let $\hat{\varphi}(x) = \hat{\varphi}(t, \mathbf{x})$ be the field operator in the Heisenberg picture and let $|\varphi(\mathbf{x}); t >$ be the state vector which at time t is an eigenstate of $\hat{\varphi}(x)$ with eigenvalue $\varphi(\mathbf{x})$:

$$\hat{\varphi}(x)|\varphi(\mathbf{x}); t >= \varphi(\mathbf{x})|\varphi(\mathbf{x}); t > \quad (3.78)$$

We recall that

$$|\varphi(\mathbf{x}); t >= e^{it\hat{H}}|\varphi(\mathbf{x}); t = 0 > \quad (3.79)$$

and that these states form a complete set for any value of t. We use this property in order to write the thermal average of an operator $\hat{A}$ as

$$
\begin{aligned}
\langle \hat{A} \rangle_\beta &= \frac{1}{Z(\beta)} \mathrm{Tr} \left(\mathrm{e}^{-\beta \hat{H}} \hat{A} \right) \\
&= \frac{1}{Z(\beta)} \int [\mathrm{d}\varphi] < \varphi(\mathbf{x}); t | \mathrm{e}^{-\beta \hat{H}} \hat{A} | \varphi(\mathbf{x}); t > \\
&= \frac{1}{Z(\beta)} \int [\mathrm{d}\varphi] < \varphi(\mathbf{x}); t - i\beta | \hat{A} | \varphi(\mathbf{x}); t >
\end{aligned}
\tag{3.80}
$$

where $[\mathrm{d}\varphi]$ indicates a sum over all field configurations $\varphi(\mathbf{x})$. We may then write $Z(\beta; j)$ in the form

$$
\begin{aligned}
Z_C(\beta; j) = \int [\mathrm{d}\varphi'] < \varphi'(\mathbf{x}); t - i\beta | \\
\times T_C \exp\left(i \int_C \mathrm{d}^4 x\, j(x) \hat{\varphi}(x) \right) | \varphi'(\mathbf{x}); t >
\end{aligned}
\tag{3.81}
$$

where we have chosen for time t the initial time t_i of the path C; then the final time is $t_f = t_i - i\beta$. Following the arguments of the preceding chapter (see (2.14) and (2.15)), we cast $Z_C(\beta; j)$ into the form of a path integral:

$$
Z_C(\beta; j) = \int \mathcal{D}\varphi \exp\left[i \int_C \mathrm{d}^4 x (\mathcal{L}(x) + j(x)\varphi(x)) \right]
\tag{3.82}
$$

with the boundary condition $\varphi(x^0; \mathbf{x}) = \varphi(x^0 - i\beta; \mathbf{x})$. We have assumed for simplicity that the Lagrangian density $\mathcal{L}(x)$ does not contain derivative interactions. The Green functions (3.65) are obtained from functional differentiation of (3.82).

3.3.2 Integration path and propagators

Up to now we have found only mild restrictions on the time path C; it starts from an initial time t_i, ends at a final time $t_i - i\beta$, and between these times the imaginary part of t must be a non-increasing function of the path parameter v. Furthermore, C must contain the real axis, since we are ultimately interested in Green's functions whose time arguments take real values. These restrictions leave open many possibilities for the path C. We shall describe the standard choice, which is illustrated in fig. 3.6.

(i) C starts from a real value t_i, large and negative ($t_i \to -\infty$).
(ii) C follows the real axis up to a large positive value $-t_i$ of t: this part of C is denoted by C_1.
(iii) Then the path goes from $-t_i$ to $-t_i - i\sigma$, with $0 \le \sigma \le \beta$, along a vertical straight line denoted by C_3.

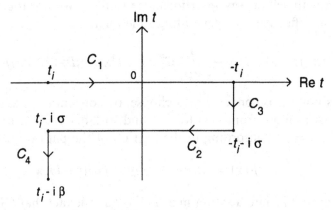

Fig. 3.6   The time path.

(iv) There is a second horizontal straight line C_2 going from $-t_i - i\sigma$ to
$t_i - i\sigma$.

(v) Finally, the path follows a vertical straight line C_4 from $t_i - i\sigma$ to
$t_i - i\beta$.

The perturbative expansion is derived from

$$Z_C(\beta;j) = \exp\left(-i\int_C d^4x \, \mathscr{V}\left(\frac{\delta}{i\delta j(x)}\right)\right) Z_C^F(\beta;j) \qquad (3.83)$$

where the free generating functional $Z_C^F(\beta;j)$ is computed by a Gaussian
integration:

$$Z_C^F(\beta;j) = \mathscr{N} \exp\left[-\frac{1}{2}\int_C d^4x \int_C d^4x' \, j(x)D_C^F(x-x')j(x')\right] \qquad (3.84)$$

In equation (3.84) we have $C = \cup_{a=1}^4 C_a$ and

$$D_C^F(x-x') = \theta_C(t-t')D_C^{>F}(x-x') + \theta_C(t'-t)D_C^{<F}(x-x') \qquad (3.85)$$

It has been argued that the free generating functional $Z_C^F(\beta;j)$ in (3.84)
can be factorized when $t_i \to -\infty$ as

$$Z_C^F(\beta;j) = \mathscr{N}_1 Z_{C_{12}}^F(\beta;j)Z_{C_{34}}^F(\beta;j) \qquad (3.86)$$

with $C_{ab} = C_a \cup C_b$. Then the Green functions we are interested in can be
deduced from $Z_{C_{12}}^F$ only, $Z_{C_{34}}^F$ playing the role of a multiplicative constant.
Although (3.86) should not be taken literally, it does give correct results
for graphs with at least one external line. Without external lines, one needs
an additional prescription, or, better, one computes in the imaginary-time
formalism. Furthermore, some care (use of a regulator) is needed when

dealing with self-energy insertions: see (3.103). We are thus left with the following 'effective' free generating functional:

$$Z_C^F(\beta;j) = \mathcal{N}_2 \exp\left[-\frac{1}{2}\int_{C_{12}} d^4x \int_{C_{12}} d^4x'\, j(x)D_C^F(x-x')j(x')\right] \quad (3.87)$$

At this point it is convenient to choose, by convention, t and t' to be real variables running from $-\infty$ to $+\infty$ and to label the source $j(x)$ with an index a, $a = 1,2$, according to the part C_a of the path on which it lives

$$j_1(x) = j(t,\mathbf{x}) \qquad j_2(x) = j(t-i\sigma,\mathbf{x}) \quad (3.88a)$$

In order to take into account in a simple way the fact that C_2 was oriented originally from $+\infty$ to $-\infty$, it is convenient to abandon our convention (3.68) on functional differentiation for the benefit of

$$\frac{\delta j_a(x)}{\delta j_b(x')} = \delta_{ab}\delta^{(4)}(x-x') \quad (3.88b)$$

With these conventions we rewrite (3.87) as

$$Z_C^F(\beta;j) = \mathcal{N}_2 \exp\left(-\frac{1}{2}\int_{-\infty}^{\infty} d^4x \int_{-\infty}^{\infty} d^4x'\, j_a(x)D_{ab}^F(x-x')j_b(x')\right) \quad (3.89)$$

Equation (3.89) involves the four components of the real-time propagator

$$D_{11}^F(t-t') = D_F(t-t') \quad (3.90a)$$
$$D_{22}^F(t-t') = D_F^*(t-t') \quad (3.90b)$$
$$D_{12}^F(t-t') = D_C^F(t-(t'-i\sigma)) = D^<(t-t'+i\sigma) \quad (3.90c)$$
$$D_{21}^F(t-t') = D_C^F((t-i\sigma)-t') = D^>(t-t'-i\sigma) \quad (3.90d)$$

For notational simplicity, we have indicated only the time arguments in (3.90). The second equation stems from $\theta_C(t) = \theta(-t)$ on C_2, while the last two equations follow by noting that 'times' on C_2 are always 'later' than 'times' on C_1. Taking the change of sign on C_2 due to our convention (3.89) into account, we arrive at the final form of the effective generating functional:

$$Z_C(\beta;j) = \mathcal{N}_3 \exp\left(-i\int_{-\infty}^{\infty} d^4x\left[\mathcal{V}\left(\frac{\delta}{i\delta j_1(x)}\right)\right.\right.$$
$$\left.-\mathcal{V}\left(\frac{\delta}{i\delta j_2(x)}\right)\right]\right)\exp\left(-\frac{1}{2}\int_{-\infty}^{\infty} d^4x\right.$$
$$\left.\times\int_{-\infty}^{\infty} d^4x'\, j_a(x)D_{ab}^F(x-x')j_b(x')\right) \quad (3.91)$$

This equation is equivalent to the path integral representation:

$$
Z_C(\beta; j) = \int \mathscr{D}\varphi_1 \mathscr{D}\varphi_2 \exp\left(-\frac{1}{2}\int_{-\infty}^{\infty} d^4x\, d^4x'\, \varphi_a(x)\right.
$$
$$
\times (D_F^{-1})_{ab}(x - x')\varphi_b(x') - i\int_{-\infty}^{\infty} d^4x\Big(\mathscr{V}(\varphi_1)
$$
$$
\left. - \mathscr{V}(\varphi_2)\Big) + i\int_{-\infty}^{\infty} d^4x\, j_a(x)\varphi_a(x)\right) \tag{3.92}
$$

One notes that (3.92) may be interpreted by identifying φ_2 as a 'ghost' field living on C_2; we thus arrive at a doubling of the field degrees of freedom. This complication, which does not arise in the Matsubara formalism, is unavoidable if one wishes to use a formalism involving real times from the outset (see Appendix B). Of course only the 'physical' field φ_1 occurs on the external lines of Green's functions, which are obtained from functional differentiation with respect to $j_1(x)$. However, the ghost field induces a modification of the naïve Feynman rules, since the propagators in (3.90) have off-diagonal elements.

The explicit expression of the free propagator (3.90) in k-space is easily derived from (3.74):

$$
D_{11}^F(k) = \frac{i}{k^2 - m^2 + i\eta} + n(k_0)2\pi\delta(k^2 - m^2)
$$
$$
= (D_{22}^F(k))^* \tag{3.93a}
$$
$$
D_{12}^F(k) = e^{\sigma k_0} f(k_0)\rho_F(k)
$$
$$
= e^{\sigma k_0}[n(k_0) + \theta(-k_0)]2\pi\delta(k^2 - m^2) \tag{3.93b}
$$
$$
D_{21}^F(k) = e^{-\sigma k_0}(1 + f(k_0))\rho_F(k)
$$
$$
= e^{-\sigma k_0}[n(k_0) + \theta(k_0)]2\pi\delta(k^2 - m^2) \tag{3.93c}
$$

$D_{11}^F(k)$ is nothing other than the generalization of (2.71). Off-diagonal elements depend on σ, and useful choices for σ are as follows:

(i) the symmetrical choice $\sigma = \beta/2$, leading to

$$
D_{12}^F(k) = D_{21}^F(k) = e^{\beta|k_0|/2}n(k_0)2\pi\delta(k^2 - m^2) \tag{3.94}
$$

(ii) the choice $\sigma = 0$, where

$$
D_{12}^F(k) = D_F^{<}(k_0) = \left(\theta(-k_0) + n(k_0)\right)2\pi\delta(k^2 - m^2) \tag{3.95a}
$$
$$
D_{21}^F(k) = D_F^{>}(k_0) = \left(\theta(k_0) + n(k_0)\right)2\pi\delta(k^2 - m^2) \tag{3.95b}
$$

We remind the reader that, contrary to $f(k_0)$, $n(k_0)$ is defined with an absolute value of k_0:

$$n(k_0) = \frac{1}{e^{\beta|k_0|} - 1} \tag{3.96}$$

It can easily be checked (exercise 3.3) that the physical results are independent of the choice of σ.

3.3.3 *Feynman rules with the symmetric propagator*

Assume that we want to compute real-time Green's functions $G(p_1, \dots, p_N)$ to a given order of perturbation theory. We have on the one hand vertices linked to external momenta, which are of type 1, and on the other hand internal vertices. Remember from (3.92) that vertices do not mix the fields φ_1 and φ_2; the mixing occurs via off-diagonal elements of the propagator. Any internal vertex can be either of type 1 or type 2. Given a configuration of internal vertices, we have to join them by the corresponding propagators: D_{11} links two vertices of type 1, D_{12} a vertex of type 1 with a vertex of type 2, etc., and we must sum over all possibilities. With $\mathscr{V}(\varphi) = \lambda \varphi^4/4!$, to each vertex of type 1 is associated a factor $-i\lambda(2\pi)^4\delta(\sum_{i=1}^4 k_i)$, and to each vertex of type 2 is associated a factor $+i\lambda(2\pi)^4\delta(\sum_{i=1}^4 k_i)$. Finally, as in the $T = 0$ Feynman rules, we must integrate over all internal lines with the measure $\int d^4k/(2\pi)^4$ and extract an overall factor $(2\pi)^4\delta(\sum_{i=1}^N p_i)$ corresponding to energy-momentum conservation.

For the sake of definiteness, we make the choice $\sigma = \beta/2$ and we write the symmetric propagator in matrix form, the indices 1 and 2 corresponding to the 'physical' (φ_1) and 'ghost' (φ_2) fields respectively. In many calculations it is useful to 'diagonalize' the matrix D_{ab}:

$$D_{ab}^F(k) = U_{ac}(k) \begin{pmatrix} D_0^F(k) & 0 \\ 0 & D_0^{F*}(k) \end{pmatrix}_{cd} U_{db}(k) \tag{3.97}$$

where $D_0^F(k)$ is the $T = 0$ Feynman propagator

$$D_0^F(k) = \frac{i}{k^2 - m^2 + i\eta} = i\mathbf{P}\frac{1}{k^2 - m^2} + \pi\delta(k^2 - m^2) \tag{3.98}$$

In (3.98) $\mathbf{P}$ denotes a principal value and the matrix U is given by

$$U(k) = \begin{pmatrix} \sqrt{1 + n(k_0)} & \sqrt{n(k_0)} \\ \sqrt{n(k_0)} & \sqrt{1 + n(k_0)} \end{pmatrix} \tag{3.99}$$

In order to show that U^{-1} 'diagonalizes' D^F, it is convenient to decompose D^F into real and imaginary parts:

$$D^F = \begin{pmatrix} i\mathbf{P}\frac{1}{k^2-m^2} & 0 \\ 0 & -i\mathbf{P}\frac{1}{k^2-m^2} \end{pmatrix}$$

$$+ 2\pi\delta(k^2 - m^2)\begin{pmatrix} \frac{1}{2} + n(k_0) & e^{\beta|k_0|/2}n(k_0) \\ e^{\beta|k_0|/2}n(k_0) & \frac{1}{2} + n(k_0) \end{pmatrix} \tag{3.100}$$

and to remark that the first matrix is not affected by the transformation.

Let us illustrate the necessity of the doubling of degrees of freedom on a simple example. Assume that we have a free Lagrangian describing particles of mass $m^2 + \mu^2$, but that we treat μ^2 as a perturbation:

$$\mathscr{V}(\varphi) = \frac{1}{2}\mu^2\varphi_1^2 - \frac{1}{2}\mu^2\varphi_2^2 \tag{3.101}$$

where φ_1 is the physical field and φ_2 the ghost field. To first order in perturbation theory we have

$$D_{11} = D_{11}^F(m^2) - i\mu^2[(D_{11}^F(m^2))^2 - (D_{12}^F(m^2))^2] + \dots \tag{3.102}$$

where D_{12}^F is given by (3.94). In order to interpret (3.102), we need a regularization of δ-functions; because we always start from the representation (2.63) of the θ-function, a correct regularization is

$$\delta(x) = \lim_{\varepsilon \to 0} \frac{1}{\pi}\frac{\varepsilon}{x^2 + \varepsilon^2} \tag{3.103}$$

which gives the formal relation:

$$\frac{1}{x + i\varepsilon}\delta(x) = -\frac{1}{2}\delta'(x) - i\pi(\delta(x))^2 \tag{3.104}$$

The regularization (3.103) is also useful at $T = 0$, and is not specific of finite temperature. One discovers that the unwanted δ^2 disappear in (3.102), which may be rewritten as

$$D_{11} = D_{11}^F(m^2) + \mu^2\frac{\partial D_{11}^F}{\partial m^2} + \dots = D_{11}^F(m^2 + \mu^2) \tag{3.105}$$

as it should. This formula is known as the mass-derivative formula and was used as a heuristic trick before the doubling of degrees of freedom was discovered. Note that the derivation of (3.105) holds because we have used $n(k_0)$ and not $n(\omega_k)$: this prescription can be justified by taking the vertical part of the contour into account.

3.4 The self-energy in the real-time formalism

The above example was given in order to illustrate the doubling of the number of degrees of freedom. There is of course a much faster (and better) way of handling the calculation, which makes use of the diagonal form (3.97) of the propagator. We start from Dyson's equation in matrix form and write the full propagator $D_{ab}(k)$ by introducing the self-energy $\Pi_{ab}(k)$ (our notation for the self-energy is Π for bosons and Σ for fermions)

$$D_{ab}(k) = D^F_{ab}(k) + D^F_{ac}(k)(-i\Pi_{cd}(k))D_{db}(k) \qquad (3.106)$$

The same matrix $U(k)$ which diagonalizes the free propagator $D^F_{ab}(k)$ also diagonalizes the full propagator $D_{ab}(k)$. Indeed, from (3.74) one readily obtains

$$\mathrm{Re}D_{11}(k) = \left(\frac{1}{2} + f(k_0)\right)\rho(k) \qquad (3.107a)$$

$$D_{12}(k) = e^{\beta k_0/2} f(k_0)\rho(k) \qquad (3.107b)$$

The ratio $D_{12}/\mathrm{Re}D_{11}$ is obviously the same as in the free case, and the matrix U^{-1} which diagonalizes the second matrix in (3.100) also diagonalizes the corresponding matrix in the case of the full propagator. Then $U(k)$ diagonalizes Π_{ab}:

$$-i\Pi_{ab}(k) = U^{-1}_{ac}(k)\begin{pmatrix} -i\overline{\Pi}(k) & 0 \\ 0 & (-i\overline{\Pi}(k))^* \end{pmatrix}_{cd} U^{-1}_{db}(k) \qquad (3.108)$$

The calculation at the end of the preceding subsection could have been performed at once by noting that $\overline{\Pi} = \mu^2$ and by making the following substitution in the diagonal form of the propagator:

$$\frac{i}{k^2 - m^2 + i\eta} \rightarrow \frac{i}{k^2 - (m^2 + \mu^2) + i\eta} \qquad (3.109)$$

The matrix equation (3.108) leads to relations between various matrix elements:

$$\mathrm{Re}\,\Pi_{11}(k) = \mathrm{Re}\,\overline{\Pi}(k) \qquad (3.110a)$$

$$\mathrm{Im}\,\Pi_{11}(k) = \coth\left(\frac{1}{2}\beta|k_0|\right)\mathrm{Im}\,\overline{\Pi}(k)$$

$$= \varepsilon(k_0)(1 + 2f(k_0))\mathrm{Im}\,\overline{\Pi}(k) \qquad (3.110b)$$

$$\Pi_{12}(k) = \frac{-i}{\sinh(\beta|k_0|/2)}\mathrm{Im}\,\overline{\Pi}(k) \qquad (3.110c)$$

The diagonal form of the propagator shows that products of δ-functions with the same argument, which would of course be catastrophic for the theory, do not occur in self-energy insertions, although a superficial examination of the Feynman rules would seem to imply such products. Indeed, let us insert a string of n self-energies $\overline{\Pi}(k)$ on a line with momentum k (fig. 3.7). The corresponding propagator $^{(n)}\tilde{D}(k)$ reads in matrix form

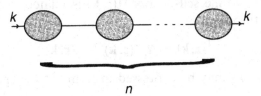

Fig. 3.7   A string of self-energies.

(in the following three equations we denote matrices by tildes in order to avoid any confusion)

$$^{(n)}\tilde{D}(k) = \left(\prod_{i=1}^{n}\left[\tilde{D}^{F}(k)(-i\tilde{\Pi}(k))\right]\right)\tilde{D}^{F}(k) \tag{3.111}$$

Using (3.97) and (3.98), equation (3.111) can be rewritten as

$$^{(n)}\tilde{D}(k) = U(k)\begin{pmatrix}^{(n)}D(k) & 0 \\ 0 & ^{(n)}D(k)^{*}\end{pmatrix}U(k) \tag{3.112}$$

with

$$^{(n)}D(k) = \left(\prod_{i=1}^{n}\left[D_{0}^{F}(k)(-i\overline{\Pi}(k))\right]\right)D_{0}^{F}(k) \tag{3.113}$$

and $\overline{\Pi}(k)$ is the diagonal element in (3.108). These expressions show that $D_0(k)$ and $D_0^{*}(k)$ never mix, and pinching singularities arising, for example, from the product $D_0(k)D_0^{*}(k)$ are always absent. From (3.112) and (3.113) one may easily compute the matrix elements of $^{(n)}\tilde{D}(k)$ by using the identity

$$\frac{1}{(x+i\eta)^{n+1}} = \mathbf{P}\frac{1}{x^{n+1}} - i\pi\frac{(-1)^{n}}{n!}\delta^{(n)}(x) \tag{3.114}$$

We quote only the result for $^{(n)}D_{12}(k)$, leaving the more complicated expression of $^{(n)}D_{11}(k)$ as exercise 3.4:

$$\begin{aligned}^{(n)}D_{12}(k) &= \frac{1}{\sinh(\beta|k_0|/2)}\operatorname{Re}{}^{(n)}D(k) \\ &= \frac{1}{\sinh(\beta|k_0|/2)}\left[\frac{(-1)^{n}}{n!}\pi\delta^{(n)}(k^2-m^2)\operatorname{Re}\left(\overline{\Pi}(k)\right)^{n}\right. \\ &\quad \left. -\mathbf{P}\frac{1}{(k^2-m^2)^{n+1}}\operatorname{Im}\left(\overline{\Pi}(k)\right)^{n}\right]\end{aligned} \tag{3.115}$$

Finally, using (2.68) and (2.69) one derives the relation between the real- and imaginary-time self-energies

$$\overline{\Pi}(k) = \Pi(k_0 + i\eta k_0, \mathbf{k}) \tag{3.116}$$

where the imaginary-time self-energy $\Pi(z, \mathbf{k})$ is related to the imaginary-time propagator through

$$\Delta^{-1}(z, \mathbf{k}) = \Delta_F^{-1}(z, \mathbf{k}) + \Pi(z, \mathbf{k}) \tag{3.117}$$

We recall that $\Delta(z, \mathbf{k})$ may be expressed in terms of the spectral function $\rho(k_0, \mathbf{k})$ (cf.(2.60)):

$$\Delta(z, \mathbf{k}) = \int_{-\infty}^{+\infty} \frac{dk_0}{2\pi} \frac{\rho(k_0, \mathbf{k})}{k_0 - z} \tag{3.118}$$

and that the function $\rho(k_0, \mathbf{k})$ obeys the positivity condition $\varepsilon(k_0)\,\rho(k_0, \mathbf{k}) \geq 0$ and the sum rule (cf.(2.51)):

$$\int_{-\infty}^{+\infty} \frac{dk_0}{2\pi} k_0 \rho(k_0, \mathbf{k}) = 1 \tag{3.119}$$

which follows from the equal-time canonical commutation relations of the field $\hat{\varphi}(t, \mathbf{x})$ and its time derivative:

$$\left[\hat{\varphi}(t, \mathbf{x}), \frac{d}{dt'} \hat{\varphi}(t', \mathbf{x}') \right]\Big|_{t'=t} = i\delta^{(3)}(\mathbf{x} - \mathbf{x}') \tag{3.120}$$

In relativistic field theory, sum rules such as (3.119) are in general useless due to renormalization problems: $\Pi(k)$ has to be subtracted in order to be made finite. In relativistic thermal field theories, such sum rules may hold in the limit of high temperatures, because in many important cases (see chapter 6) the leading term of Π is proportional to T^2 and the $T = 0$ terms are negligible. This leading term is finite in perturbation theory, at least to one-loop order, and the sum rule is then relevant.

Finally, we note that the relation between real- and imaginary-time Green's functions has been investigated recently by several authors and relations generalizing (3.116) have been established for three- and four-point Green's functions. The main advantage of the real-time formalism is, of course, that no analytical continuation is needed. On the other hand, the price to be paid is a more intricate formalism, due to the doubling of the field degrees of freedom.

3.5 Renormalization at non-zero temperature

In computing loop diagrams at finite temperature, one encounters divergences which must be taken care of. Fortunately, it is easy to see that renormalization at zero temperature suffices to make the theory finite at non-zero T. The reason is not difficult to understand: temperature does not modify the theory at distances $\ll 1/T$, and thus the ultraviolet (short-distance) singularities are the same as at $T = 0$. A more formal argument is as follows: let us start from the expression of the $T = 0$ free propagator in Euclidean space:

$$\Delta_F(\tau, \mathbf{x}; T = 0) = \int \frac{\mathrm{d}^4 k}{(2\pi)^4} \frac{\mathrm{e}^{i(k_4\tau + \mathbf{k}\cdot\mathbf{x})}}{k_4^2 + \omega_k^2} \tag{3.121}$$

with $\omega_k^2 = \mathbf{k}^2 + m^2$, and let us compute the sum

$$\sum_{n=-\infty}^{+\infty} \Delta_F(\tau + n\beta, \mathbf{x}; T = 0) = \int \frac{\mathrm{d}^3 k}{(2\pi)^3} \mathrm{e}^{i\mathbf{k}\cdot\mathbf{x}}$$

$$\times \sum_{n=-\infty}^{+\infty} \Delta_F(\tau + n\beta, \mathbf{k}; T = 0) \tag{3.122}$$

A straightforward calculation gives

$$\sum_{n=-\infty}^{+\infty} \Delta_F(\tau + n\beta, \mathbf{k}; T = 0) = \sum_{n=-\infty}^{+\infty} \frac{1}{2\omega_k} \mathrm{e}^{-\omega_k|\tau + n\beta|}$$

$$= \frac{1}{2\omega_k} \left((1 + n(\omega_k))\mathrm{e}^{-\omega_k\tau} + n(\omega_k)\mathrm{e}^{\omega_k\tau} \right) = \Delta_F(\tau, \mathbf{k}; T) \tag{3.123}$$

where we have used (3.19) in order to identify the thermal propagator in the second line of (3.123). We thus discover that

$$\Delta_F(\tau, \mathbf{x}; T) = \sum_{n=-\infty}^{+\infty} \Delta_F(\tau + n\beta, \mathbf{x}; T = 0) \tag{3.124}$$

This expression can also be obtained by appealing to Poisson's summation formula, or by noting that $\Delta_F(\tau, \mathbf{x}; T)$ is the only solution of the equation

$$(-\partial_\tau^2 - \nabla^2 + m^2)\Delta_F(\tau, \mathbf{x}; T) = \delta(\tau)\delta^{(3)}(\mathbf{x}) \tag{3.125}$$

with the correct periodicity property (2.27) in the interval $]-\beta, \beta[$ (exercise 3.5). Since the ultraviolet singularities of the Euclidean propagator in x-space occur for $\tau^2 + \mathbf{x}^2 \to 0$, one sees from (3.124) that the short-distance singularities of the thermal propagator $\Delta_F(\tau + n\beta, \mathbf{x}; T)$ come from $n = 0$ only, and that they are identical to those of the $T = 0$ propagator.

In k-space, one can start from the real-time expression of the free propagator D^F (3.93) with $\sigma = \beta/2$: this propagator can be split into a $T = 0$ part and a T-dependent part, which contains a factor decreasing at least as

$$\mathrm{e}^{-\beta|k_0|/2}\delta(k^2 - m^2) \tag{3.126}$$

Writing $D^F = D^F_{T=0} + D^F_\beta$, we see that divergences can only arise from loop integrations $\int \mathrm{d}^4 k$ which do not contain D^F_β, since the δ-function in (3.126) puts the k-line on mass-shell, and the $\mathrm{d}^3 k$-integration is then damped exponentially. Of course, divergences could come from integrations over other loop momenta which do not contain D_β^F factors, but the corresponding divergences are taken care of by $T = 0$ counter-terms. In conclusion, we see that the $T \neq 0$ theory is renormalized by temperature-independent counter-terms.

References and further reading

Standard references on quantum field theory at zero temperature are Itzykson and Zuber (1980); Ramond (1980); Parisi (1988); or Zinn-Justin (1989); the last two deal almost exclusively with Euclidean field theory. The derivation of Feynman rules in Euclidean space follows Le Bellac (1992), chap. 5, to which the reader is referred for further details. The book by Brown (1992) also contains an introduction to finite T quantization. Quantization at non-zero temperature in a non-relativistic context is treated in many textbooks, for example Kadanoff and Baym (1962) or Fetter and Walecka (1971). More recent references are Blaizot and Ripka (1986) and Negele and Orland (1988). The discussion in section 3.2 is borrowed from Kapusta (1989), chap. 2, the original work being due to Haber and Weldon (1982). The form (3.74) of the propagator was first derived by Mills (1969). The role of the vertical part of the contour is discussed by Niégawa (1989), Evans (1994), Pearson (1994) and Gelis (1994, unpublished). There are several articles on the problem of analytical continuation to real time of Green's functions with $N \geq 3$: Kobes (1990, 1991), Evans (1992), Aurenche and Becherrawy (1992) (who proposed the Retarded/Advanced (R/A) formalism), van Eijck and van Weert (1992), Guérin (1994). The real-time formalism is reviewed in detail by Landsman and van Weert (1987), who give a complete list of references to the earlier literature; see also Rivers (1987), chaps. 14 and 15. Renormalization at non-zero temperature is treated by Collins (1984), chap. 11 and by Landsman and van Weert (1987).

Exercises

3.1 Compute the mixed representation $\Delta_F(\tau, \mathbf{k})$ of the imaginary-time propagator, taking the inverse Fourier transform of $\Delta_F(i\omega_n, \mathbf{k})$ and using contour integration in the complex k_0-plane. Show that one may write for $\tau > 0$

$$\Delta_F(\tau, \mathbf{k}) = - \int \frac{\mathrm{d}k_0}{2i\pi} \frac{\mathrm{e}^{-k_0\tau}}{\mathrm{e}^{-\beta k_0} - 1} \frac{1}{\omega_k^2 - k_0^2}$$

using the contour of section 3.2 with $\mu = 0$. What function should be used instead of $(\mathrm{e}^{-\beta k_0} - 1)^{-1}$ in the case $\tau < 0$? Generalize to non-zero chemical potential.

3.2 Show that the mixed representation $D_C(t, \mathbf{k})$ of the real-time propagator is

$$D_C(t, \mathbf{k}) = \frac{1}{2\omega_k} \Big[\theta_C(t) e^{-i\omega_k t} + \theta_C(-t) e^{i\omega_k t}$$
$$+ n(\omega_k)(e^{-i\omega_k t} + e^{i\omega_k t}) \Big]$$

3.3 Use energy conservation to show that Green's functions are independent of the contour parameter σ in real-time perturbation theory (hint: work out a simple example).

3.4 Compute $^{(n)}D_{11}(k)$ from (3.113) and (3.114).

3.5 Alternative proofs of (3.123). (a) Let $\Delta_0(\tau)$ be the $T = 0$ propagator of the harmonic oscillator in imaginary-time

$$\Delta_0(\tau) = \frac{1}{2\omega} e^{-\omega|\tau|}$$

and define

$$F(\tau) = \sum_{n=-\infty}^{n=\infty} \Delta_0(\tau + n\beta)$$

Show that the Fourier coefficients of $F(\tau)$ are

$$A_n = \int_{-\infty}^{\infty} d\tau \, e^{-i\omega_n \tau} F(\tau)$$

(b) Show that $F(\tau)$ is, in the interval $]-\beta, +\beta[$, the unique solution of

$$(-\partial_\tau^2 + \omega^2) F(\tau) = \delta(\tau)$$

with the periodicity property $F(\tau - \beta) = F(\tau)$.

4

Simple applications of perturbation theory

Having established the Feynman rules at finite temperature, we perform a few simple calculations in order to obtain some familiarity with these rules and to understand their physical meaning. In the first section of this chapter, we shall examine the partition function of the φ^4-model to the next order of perturbation theory; we shall discover that in fact, due to infrared divergences, perturbation theory breaks down and the next order correction to the calculations of the preceding chapter is not of order λ^2, but of order $\lambda^{3/2}$ (this breakdown of perturbation theory will be examined at length in the physically relevant case of gauge theories, in chapters 6–8). Then in section 4.2, we undertake some one-loop calculations, either in imaginary-time or in real-time formalism, and we show how their results are related. In section 4.3 we generalize the Cutkosky, or cutting, rules which are so important at zero temperature. Although the finite T results are more complicated, it is still possible to derive useful formulae. Finally, in the last section, we briefly discuss the physical interpretation of our calculations.

4.1 Ring diagrams in the φ^4-model

We have shown in section 3.1.2 that the self-energy of the φ^4-model, computed to first order in λ, is independent of the external momentum; after mass renormalization, we find in the massless case $m = 0$, or $m \ll T$ (see (3.37)):

$$\Pi_1 = \frac{\lambda T^2}{24} \tag{4.1}$$

where the subscript 1 denotes the first order in λ. This result can be interpreted as the occurrence of a thermal mass $\delta m_\beta^2 = \Pi_1$, of the order of λT^2. Indeed, from (3.33) the inverse propagator reads

$$\Delta^{-1}(i\omega_n, \mathbf{k}) = \Delta_F^{-1}(i\omega_n, \mathbf{k}) + \delta m_\beta^2 = \omega_n^2 + \mathbf{k}^2 + \delta m_\beta^2 \qquad (4.2)$$

This expression may be continued to real values of the energy, for example by using $i\omega_n \to k_0 + i\eta$ if one is interested in the retarded propagator. Assume we now consider the propagator of a particle with a 'soft momentum', which means that its energy k_0 and its momentum $\mathbf{k}$ are $\sim \lambda^{1/2}T$. For such a 'soft particle', the inverse free propagator is of order λT^2, but the first order correction Π_1 to Π (see (4.2)) is also of order λT^2. This means that we cannot limit ourselves to a naïve perturbative expansion, but that we must use some kind of resummation. The solution in this simple case is obvious: we have to use the propagator defined by (4.2). An even better choice for Π is given by the self-consistent equation

$$\Pi = \frac{\lambda}{2} T \sum_n \int \frac{\mathrm{d}^3 p}{(2\pi)^3} \frac{1}{\omega_n^2 + \mathbf{k}^2 + \Pi} \qquad (4.3)$$

Having performed the summation over n as in (2.75) and the $T = 0$ mass renormalization, we are led for the T-dependent part Π_β of Π to the equation

$$1 = \frac{\lambda}{4\pi^2} \int_1^\infty \mathrm{d}x \frac{\sqrt{x^2 - 1}}{\exp\left(\beta\Pi_\beta^{1/2}x\right) - 1} \qquad (4.4)$$

The expansion in powers of λ is obtained from that of the function $F(u)$:

$$F(u) = \int_1^\infty \mathrm{d}x \frac{\sqrt{x^2 - 1}}{e^{ux} - 1} = \frac{2\pi^2}{u^2}\left(\frac{1}{12} - \frac{u}{4\pi} + O(u^2 \ln u)\right) \qquad (4.5)$$

The results are plotted in fig. 4.1, where one can see that the deviation with respect to the first-order calculation is rather small. However an expansion in powers of λ reveals an interesting feature:

$$\Pi_\beta = \frac{\lambda T^2}{24}\left[1 - 3\left(\frac{\lambda}{24\pi^2}\right)^{1/2} + \ldots\right] \qquad (4.6)$$

We see that the next term in Π_β is not in λ^2, as we would expect from naïve perturbation theory, but, rather, it is in $\lambda^{3/2}$. This reflects the breakdown of perturbation theory due to infrared divergences.

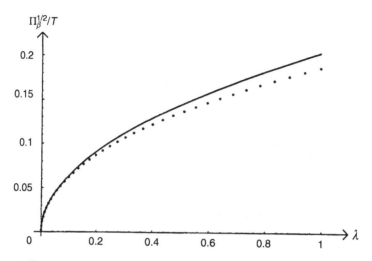

Fig. 4.1. $\Pi_\beta^{1/2}/T$ as a function of λ. Solid line: one-loop result. Dotted line: result from (4.4).

The same feature also appears in the perturbative expansion of the partition function. The resummation method leads to so-called 'ring diagrams' (fig. 4.2), which contribute the following to $\Omega = -\frac{1}{\beta}\ln Z(\beta)$:

$$
\begin{aligned}
\Omega_{\text{ring}} &= -\frac{1}{2}VT\sum_n \int \frac{d^3k}{(2\pi)^3} \\
&\times \sum_{N=2}^{\infty} \frac{1}{N}\left(-\Pi_1(i\omega_n,\mathbf{k})\Delta_F(i\omega_n,\mathbf{k})\right)^N \\
&= \frac{1}{2}VT\sum_n \int \frac{d^3k}{(2\pi)^3}\left[\ln\left(1+\Pi_1(i\omega_n,\mathbf{k})\Delta_F(i\omega_n,\mathbf{k})\right)\right. \\
&\left. - \Pi_1(i\omega_n,\mathbf{k})\Delta_F(i\omega_n,\mathbf{k})\right]
\end{aligned}
\tag{4.7}
$$

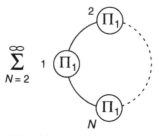

Fig. 4.2   Ring diagrams.

The factor of $1/N$ in (4.7) is a symmetry factor which can be understood in the following way: there is a factor of $3!$ for connecting two lines at each vertex, a factor of 2 for how the remaining two lines at each vertex are connected to the adjacent vertices; a factor of $(N-1)!/2$ for the number of ways of ordering the vertices along the circle, and a factor of $1/N!$ from the expansion of the exponential of the interaction in (3.45). The summation over N begins at $N = 2$, because the graph of fig. 3.3 has already been taken into account at first order (see (3.38)); $\Pi_1 = \lambda T^2/24$ is the first order temperature-dependent self-energy, and the explicit expression of (4.7) is

$$\Omega_{\text{ring}} = \frac{1}{2}VT\sum_n \int \frac{d^3k}{(2\pi)^3}\left(\ln\left(1+\frac{\lambda T^2/24}{\omega_n^2+\mathbf{k}^2}\right) - \frac{\lambda T^2/24}{\omega_n^2+\mathbf{k}^2}\right) \tag{4.8}$$

If one restricts oneself to the $n = 0$ term in (4.8), also called the static mode, the integral is easily evaluated after integration by parts of the logarithm and one finds for (4.8)

$$\Omega_{\text{ring}} \simeq -\frac{VT^4}{12\pi}\left(\frac{\lambda}{24}\right)^{3/2} \tag{4.9}$$

This result again exhibits a breakdown of perturbation theory due to infrared divergences. It can easily be seen that the non-static modes $n \neq 0$ lead to terms that are proportional to λ^2: infrared divergences come from the static modes only, as $2\pi nT$ acts as a mass term in the propagator for $n \neq 0$. From the explicit evaluation (4.9) to order $\lambda^{3/2}$, one obtains the expression for the pressure:

$$P = \frac{\pi^2 T^4}{90}\left(1 - \frac{15}{8}\left(\frac{\lambda}{24\pi^2}\right) + \frac{15}{2}\left(\frac{\lambda}{24\pi^2}\right)^{3/2} + \cdots\right) \tag{4.10}$$

The first term corresponds to the free boson gas, the second one to the graph of fig. 3.3, and the last one to the ring diagrams of fig. 4.2.

4.2 Some one-loop diagrams

In order to gain some practice with finite-temperature Feynman rules, we shall compute some one-loop diagrams; these calculations will find important applications in the last five chapters of this book. We begin with a computation in the imaginary-time formalism, and then compare this with the same computation in the real-time formalism.

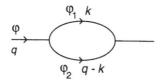

Fig. 4.3   One-loop diagram for the self-energy.

4.2.1 Self-energy diagram in the imaginary-time formalism

Our goal is to compute the diagram of fig. 4.3, which occurs typically in one-loop approximations of the self-energy. We take three scalar fields φ, φ_1 and φ_2 with an interaction described by the Lagrangian density:

$$\mathscr{L}(x) = -g\, \varphi(x)\varphi_1(x)\varphi_2(x) \tag{4.11}$$

Figure 4.3 represents the one-loop approximation to the self-energy of the field $\varphi(x)$. It is convenient to write the mixed representation (3.19) of the imaginary-time free propagator $\Delta(\tau, E_i)$ of the field $\varphi(x)$ as

$$\Delta(\tau, E_i) = \sum_{s=\pm 1} \frac{s}{2E_i}(1 + f(sE_i))e^{-sE_i\tau} = \sum_{s=\pm 1} \Delta_s(\tau, E_i) \tag{4.12}$$

where, for notational simplicity, we have dropped the subscript F, and we have labelled the momentum dependence with the energy E_i:

$$E_i = \left(\mathbf{k}_i^2 + m_i^2\right)^{1/2} \tag{4.13}$$

rather than with the three-momentum $\mathbf{k}_i$; the function $f(x)$ is defined in (3.75a). It is easily checked that (4.12) is equivalent to (3.19). We may also write (4.12) in frequency space:

$$\Delta(i\omega_n, E_i) = \sum_{s=\pm 1} \Delta_s(i\omega_n, E_i) = -\sum_{s=\pm 1} \frac{s}{2E_i}\frac{1}{i\omega_n - sE_i} \tag{4.14}$$

In the computation of the diagram in fig. 4.3, we need to perform a frequency sum by following the method explained in section 2.6 (see (2.77)–(2.79))

$$T\sum_n \Delta_{s_1}(i\omega_n, E_1)\Delta_{s_2}(i(\omega - \omega_n), E_2)$$

$$= -\frac{s_1 s_2}{4E_1 E_2}\frac{1 + f(s_1 E_1) + f(s_2 E_2)}{i\omega - s_1 E_1 - s_2 E_2} \tag{4.15}$$

where ω is the Matsubara frequency of the external line with momentum q. Note that in Euclidean space, ω is nothing other than minus the fourth component of the four-vector q: $\omega = -q_4$; furthermore, $\mathbf{k}_1 = \mathbf{k}$ and

$k_2 = q - k$. Summing over s_1 and s_2 and writing the k integral explicitly leads to the following expression for the graph of fig. 4.3:

$$\Pi(i\omega, \mathbf{q}) = g^2 \int \frac{d^3k}{(2\pi)^3} \frac{1}{4E_1 E_2} \left[(1 + n_1 + n_2) \right.$$
$$\times \left(\frac{1}{i\omega - E_1 - E_2} - \frac{1}{i\omega + E_1 + E_2} \right) - (n_1 - n_2)$$
$$\left. \times \left(\frac{1}{i\omega - E_1 + E_2} - \frac{1}{i\omega + E_1 - E_2} \right) \right] \tag{4.16}$$

where

$$n_i = n(E_i) = \frac{1}{e^{\beta E_i} - 1} \tag{4.17}$$

Equation (4.16) is a rather obvious generalization of (2.82). The overall sign in (4.16) is fixed by our definition (3.33). $\Pi(i\omega, \mathbf{q})$ is defined *a priori* only for values of ω equal to Matsubara frequencies; in order to obtain, for example, the retarded self-energy Π_R, one has to perform the analytic continuation $i\omega \to q_0 + i\eta$, as explained in section 2.4. We emphasize once more that this continuation is possible only after the frequency sum has been performed. Note that the weights $n(E_i)$ involve the on-shell energy E_i of the internal line (i) of the diagram.

The physical interpretation of the various terms in (4.16) is made clearer if one looks at the imaginary part of Π_R:

$$\text{Im}\,\Pi_R(q_0, \mathbf{q}) = \frac{1}{2i} \text{Disc}\,\Pi(q_0 + i\eta, \mathbf{q})$$
$$= \frac{1}{2i} \left[\Pi(q_0 + i\eta, \mathbf{q}) - \Pi(q_0 - i\eta, \mathbf{q}) \right] \tag{4.18}$$

whose explicit expression, which generalizes (2.85), is from (4.16),

$$\text{Im}\,\Pi_R(q_0, \mathbf{q}) = -\pi g^2 \int \frac{d^3k}{(2\pi)^3} \frac{1}{4E_1 E_2} \left[(1 + n_1 + n_2) \right.$$
$$\times (\delta(q_0 - E_1 - E_2) - \delta(q_0 + E_1 + E_2))$$
$$\left. - (n_1 - n_2)(\delta(q_0 - E_1 + E_2) - \delta(q_0 + E_1 - E_2)) \right] \tag{4.19}$$

Let us look at the Bose–Einstein factor $(1 + n_1 + n_2)$ in the first term of (4.19) for $q_0 > 0$. The discontinuity of Π_R is proportional to the *difference*

$$\Gamma = \Gamma^> - \Gamma^< \tag{4.20}$$

between the decay and creation rates of particle φ; the decay process, for $q_0 = E_1 + E_2$, is $\varphi \to \varphi_1 + \varphi_2$, and the creation process is the inverse process $\varphi_1 + \varphi_2 \to \varphi$. The Bose–Einstein factor is thus

$$(1 + n_1)(1 + n_2) - n_1 n_2 = 1 + n_1 + n_2 \tag{4.21}$$

At zero temperature, one obtains $\Gamma^< = 0$, of course, and $\Gamma^>$ is proportional to the factor of one in (4.21). Similarly, the second term in (4.19), with $q_0 = E_1 - E_2$, corresponds to the reactions $\varphi_1 \to \varphi + \bar{\varphi}_2$ (creation) and $\bar{\varphi}_2 + \varphi \to \varphi_1$ (decay), the Bose–Einstein factor being

$$n_2(1 + n_1) - n_1(1 + n_2) = n_2 - n_1 \tag{4.22}$$

In all cases we have, of course, the detailed balance condition

$$\Gamma^> = e^{\beta q_0} \Gamma^< \tag{4.23}$$

In the case of a small deviation from thermal equilibrium, Γ must be interpreted as the inverse of the relaxation time which governs the approach to equilibrium (see section 4.4).

The terms with $q_0 = E_1 - E_2$ or $q_0 = E_2 - E_1$ correspond to the so-called 'Landau damping' mechanism: particles disappear or are created through scattering in the bath, and not via the processes which are avaible at zero temperature.

In addition to the frequency sum (4.15), we shall also need the following in later chapters:

$$J(i\omega, \mathbf{q}) = T \sum_n \omega_n \Delta(i\omega_n, E_1) \Delta(i(\omega - \omega_n), E_2) \tag{4.24}$$

In order to perform the summation in (4.24), we integrate the Fourier representation of $\Delta(\tau)$ by parts and use the periodicity in τ to obtain

$$\omega_n \Delta(i\omega_n) = \int_0^\beta d\tau \left(i \frac{\partial \Delta}{\partial \tau} \right) e^{i\omega_n \tau} \tag{4.25a}$$

with

$$i \frac{\partial \Delta}{\partial \tau} = \sum_{s=\pm 1} (-isE_i) \Delta_s(\tau, E_i) \tag{4.25b}$$

Thus in the summation of (4.24), ω_n is replaced by $(-is_1 E_1)$ and

$$J(i\omega, \mathbf{q}) = T \sum_{s_1, s_2} \sum_n (-is_1 E_1) \Delta_{s_1}(i\omega_n, E_1) \Delta_{s_2}(i(\omega - \omega_n), E_2)$$

$$= \sum_{s_1, s_2} \frac{is_2}{4E_2} \frac{1 + f(s_1 E_1) + f(s_2 E_2)}{i\omega - s_1 E_1 - s_2 E_2} \tag{4.26}$$

The final result is

$$T \sum_n \omega_n \Delta(i\omega_n, E_1) \Delta(i(\omega - \omega_n), E_2)$$

$$= \frac{i}{4E_2} \left[(1 + n_1 + n_2) \left(\frac{1}{i\omega - E_1 - E_2} + \frac{1}{i\omega + E_1 + E_2} \right) \right.$$

$$\left. - (n_1 - n_2) \left(\frac{1}{i\omega - E_1 + E_2} + \frac{1}{i\omega + E_1 - E_2} \right) \right] \tag{4.27}$$

It can easily be checked that the results (4.16) and (4.27) are real, as they should be.

4.2.2 Self-energy diagram in the real-time formalism

As a further exercise, we compute the diagram of fig. 4.3 in the real-time formalism. This calculation will provide us with a check of equation (3.116), which allows us to compare real- and imaginary-time formalisms. We begin with the element Π_{11} of the self-energy, which is given by

$$\Pi_{11}(q) = -ig^2 \int \frac{d^4k}{(2\pi)^4} D_{11}^F(k, m_1^2) D_{11}^F(q-k, m_2^2) \qquad (4.28)$$

The overall sign is determined from (3.106). It is convenient to compute the real and imaginary parts of Π_{11} separately by writing (cf (3.100))

$$D_{11}^F(k, m_1^2) = i\mathbf{P}\frac{1}{k^2 - m_1^2} + \left(\frac{1}{2} + n(k_0)\right) 2\pi\delta(k^2 - m_1^2) \qquad (4.29)$$

Then we obtain for $\operatorname{Re}\Pi_{11}(q)$ ($= \operatorname{Re}\overline{\Pi}$ from (3.110a))

$$\operatorname{Re}\Pi_{11}(q) = g^2 \int \frac{d^4k}{(2\pi)^3} \left[\left(\frac{1}{2} + n(k_0)\right)\delta(k^2 - m_1^2)\, \mathbf{P}\frac{1}{(q-k)^2 - m_2^2} \right.$$
$$\left. + \left(\frac{1}{2} + n(q_0 - k_0)\right)\delta((q-k)^2 - m_1^2)\,\mathbf{P}\frac{1}{k^2 - m_2^2} \right] \qquad (4.30)$$

Let us evaluate the first term in the square bracket of (4.30) by using partial fractioning:

$$\frac{1}{(q-k)^2 - m_2^2} = \frac{1}{2E_2}\left(\frac{1}{q_0 - k_0 - E_2} - \frac{1}{q_0 - k_0 + E_2}\right) \qquad (4.31)$$

The integration over k_0 is performed using the δ-function, and we obtain

$$g^2 \int \frac{d^3k}{(2\pi)^3} \frac{1}{4E_1E_2}\left(\frac{1}{2} + n_1\right)\left[\frac{1}{q_0 - E_1 - E_2} - \frac{1}{q_0 - E_1 + E_2}\right.$$
$$\left. + \frac{1}{q_0 + E_1 - E_2} - \frac{1}{q_0 + E_1 + E_2}\right] \qquad (4.32)$$

The second term in the square bracket of (4.30) is obtained by interchanging indices 1 and 2; adding this second term leads immediately to the real part of (4.16) once the analytic continuation $i\omega \to q_0 + i\eta$ has been perfomed:

$$\operatorname{Re}\Pi_{11}(q) = \operatorname{Re}\overline{\Pi}(q) = \operatorname{Re}\Pi_R(q) = \operatorname{Re}\Pi(q_0 + i\eta, \mathbf{q}) \qquad (4.33)$$

In order to compute the imaginary part, it is simplest to use Π_{12} and the following form of the symmetric propagator (see (3.93b))

$$D_{12}^F(k) = e^{\beta k_0/2} f(k_0)\varepsilon(k_0) 2\pi\delta(k^2 - m_1^2) \qquad (4.34)$$

Using the identity

$$f(k_0)f(q_0 - k_0) = f(q_0)(1 + f(k_0) + f(q_0 - k_0)) \tag{4.35}$$

yields

$$
\begin{aligned}
\Pi_{12} = ig^2 e^{\beta q_0/2} \int \frac{d^4k}{(2\pi)^2} f(q_0)(1 + f(k_0) + f(q_0 - k_0)) \\
\times \frac{1}{4E_1 E_2} [\delta(k_0 - E_1) - \delta(k_0 + E_1)] \\
\times [\delta(q_0 - k_0 - E_2) - \delta(q_0 - k_0 + E_2)]
\end{aligned}
\tag{4.36}
$$

From (3.110c) we find $\mathrm{Im}\,\overline{\Pi}$:

$$
\begin{aligned}
\mathrm{Im}\,\overline{\Pi}(q_0, \mathbf{q}) = -\pi g^2 \varepsilon(q_0) \int \frac{d^3k}{(2\pi)^3} \frac{1}{4E_1 E_2} \Big[(1 + n_1 + n_2) \\
\times (\delta(q_0 - E_1 - E_2) - \delta(q_0 + E_1 + E_2)) \\
- (n_1 - n_2)(\delta(q_0 - E_1 + E_2) - \delta(q_0 + E_1 - E_2)) \Big]
\end{aligned}
\tag{4.37}
$$

and comparing with (4.19) we discover that

$$\mathrm{Im}\,\overline{\Pi}(q_0, \mathbf{q}) = \varepsilon(q_0)\mathrm{Im}\,\Pi_R(q_0, \mathbf{q}) = \varepsilon(q_0)\mathrm{Im}\,\Pi(q_0 + i\eta, \mathbf{q}) \tag{4.38}$$

The sign function $\varepsilon(q_0)$ occurs because in Π_R the analytic continuation is $i\omega \to q_0 + i\eta$, while the continuation giving the time-ordered self-energy is $i\omega \to q_0 + i\eta q_0$. Note that in all cases the real part of Π is an even function of q_0, while the imaginary part is an odd (even) function of q_0 in the retarded (time-ordered) case, and that (4.33) and (4.38) provide a check on (3.116). As a final comment, let us note that real-time calculations are much simpler if one writes off-diagonal elements of the propagator in terms of $f(k_0)$ rather than in terms of $n(k_0)$.

4.2.3 *The three-point vertex at one-loop*

We now turn to the one-loop approximation to the three-point vertex. The kinematics is defined in fig. 4.4, and, in the imaginary time formalism, the expression to be computed is

$$\Gamma(p_1, p_2) = -g^3 T \sum_n \int \frac{d^3k}{(2\pi)^3} \Delta(k) \Delta(p_1 - k) \Delta(p_2 - k) \tag{4.39}$$

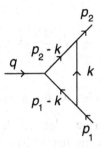

Fig. 4.4 One-loop diagram for the vertex.

Let ω_1 and ω_2 be the Matsubara frequencies corresponding to the external lines with momenta p_1 and p_2 respectively, and let us define

$$E = \left(\mathbf{k}^2 + m^2\right)^{1/2} \quad E_1 = \left((\mathbf{p_1} - \mathbf{k})^2 + m_1^2\right)^{1/2}$$
$$E_2 = \left((\mathbf{p_2} - \mathbf{k})^2 + m_2^2\right)^{1/2} \tag{4.40}$$

where m, m_1, m_2 are the masses associated with the internal lines of momenta k, p_1-k, p_2-k respectively. The frequency sum to be evaluated is

$$S = T \sum_n \Delta(i\omega_n, E)\Delta(i(\omega_1 - \omega_n), E_1)\Delta(i(\omega_2 - \omega_n), E_2)$$

$$= -\frac{T}{8EE_1E_2} \sum_{n,s,s_1,s_2}$$

$$\times \frac{ss_1s_2}{(i\omega_n - sE)(i(\omega_1 - \omega_n) - s_1E_1)(i(\omega_2 - \omega_n) - s_2E_2)} \tag{4.41}$$

We use partial fractioning to write (4.41) in the form

$$S = T \sum_{n,s,s_1,s_2} \frac{1}{i(\omega_1 - \omega_2) - s_1E_1 + s_2E_2}$$

$$\times \left[\frac{s_2}{2E_2} \Delta_s(i\omega_n, E)\Delta_{s_1}(i(\omega_1 - \omega_n), E_1) \right.$$

$$\left. - \frac{s_1}{2E_1} \Delta_s(i\omega_n, E)\Delta_{s_2}(i(\omega_2 - \omega_n), E_2) \right] \tag{4.42}$$

The frequency sum is performed using (4.15):

$$S = -\sum_{s,s_1,s_2} \frac{ss_1s_2}{8EE_1E_2} \frac{1}{i(\omega_1 - \omega_2) - s_1E_1 + s_2E_2}$$

$$\times \left[\frac{1 + f(sE) + f(s_1E_1)}{i\omega_1 - sE - s_1E_1} - \frac{1 + f(sE) + f(s_2E_2)}{i\omega_2 - sE - s_2E_2} \right] \tag{4.43}$$

and Γ is finally given by

$$\Gamma(p_1, p_2) = -g^3 \int \frac{\mathrm{d}^3 k}{(2\pi)^3} S(\mathbf{k}, \mathbf{p}_1, \mathbf{p}_2) \qquad (4.44)$$

4.3 Cutting rules at finite temperature

In this section, our goal is to derive rules which give the imaginary part of Feynman diagrams in the real-time formalism, when all external particles are of type 1; these rules, first derived by Kobes and Semenoff, generalize the zero-temperature cutting rules.

4.3.1 Derivation of the rules

Let us recall briefly how one derives cutting rules (or Cutkosky's rules) at zero temperature. At a given order of perturbation theory we consider a particular Feynman graph $G(x_1, \ldots, x_p)$ contributing to a zero-temperature-connected Green function amputated of its external legs. The analytic expression of the diagram is written in x-space; the p vertices are labelled $x_1, \ldots, x_p$, and are linked by free (Feynman) propagators $D_0^F(x_i - x_j)$:

$$\begin{aligned} D_0^F(x) &= \theta(x_0) D_0^{>F}(x) + \theta(-x_0) D_0^{<F}(x) \\ &= < 0| T(\hat{\varphi}_{\mathrm{in}}(x) \hat{\varphi}_{\mathrm{in}}(0))|0 > \end{aligned} \qquad (4.45)$$

where the subscript 0 refers to zero temperature and $\hat{\varphi}_{\mathrm{in}}$ denotes a free field; for the sake of definiteness, we use a $g\varphi^3$-model. From $G(x_1, \ldots, x_p)$ we build new quantities $F(x_1, \ldots, \underline{x_i}, \ldots, x_p)$ having the same topology as G, but where some of the vertices are 'underlined'; in the graphical representation (Feynman graph) of F, underlined x_i's will be represented by circled vertices. If in the original expression two vertices are linked by the propagator $D_0^F(x_i - x_j)$, the expression for the new quantities is built according to the following rules.

(i) If the vertices x_i and x_j are not underlined, one keeps the original propagator $D_0^F(x_i - x_j)$.
(ii) If x_i is underlined and x_j is not underlined, the two vertices are linked by $D_0^{>F}(x_i - x_j)$.
(iii) If x_j is underlined and x_i is not underlined, the two vertices are linked by $D_0^{<F}(x_i - x_j)$.
(iv) If both vertices are underlined, one uses the complex conjugate $(D_0^F(x_i - x_j))^*$ of the propagator.
(v) A factor of $-ig$ is associated with the ordinary vertices, a factor of $+ig$ with the underlined vertices.

Rules (i)–(v) may be summarized by introducing a 'propagator' $\overline{D}_0^F$ for the quantities F:

$$\overline{D}_0^F(x_i - x_j) = D_0^F(x_i - x_j) \qquad (4.46a)$$

$$\overline{D}_0^F(\underline{x}_i - x_j) = D_0^{>F}(x_i - x_j) \qquad (4.46b)$$

$$\overline{D}_0^F(x_i - \underline{x}_j) = D_0^{<F}(x_i - x_j) \qquad (4.46c)$$

$$\overline{D}_0^F(\underline{x}_i - \underline{x}_j) = (D_0^F(x_i - x_j))^* \qquad (4.46d)$$

The original expression G coincides with that of F when no vertex is circled. Note that the rules are consistent with the symmetry property $D_0^F(x_i - x_j) = D_0^F(x_j - x_i)$:

$$\overline{D}_0^F(x_j - \underline{x}_i) = D_0^{<F}(x_j - x_i) = D_0^{>F}(x_i - x_j)$$
$$= \overline{D}_0^F(\underline{x}_i - x_j) \qquad (4.47)$$

This construction leads to the so-called 'largest time equation': assume that the time coordinate x_i^0 is larger than any x_j^0. Then

$$\overline{D}_0^F(x_i - x_j) = D_0^{>F}(x_i - x_j) = \overline{D}_0^F(\underline{x}_i - x_j) \quad \forall j \qquad (4.48a)$$

$$\overline{D}_0^F(\underline{x}_i - x_j) = D_0^{<F}(x_i - x_j) = \overline{D}_0^F(x_i - \underline{x}_j) \quad \forall j \qquad (4.48b)$$

where we have used $D_F^{>*}(x - x') = D_F^{<}(x - x')$. The largest time equation reads

$$F(x_1, \ldots, \underline{x}_h, \ldots, \underline{x}_i, \ldots, x_p) = -F(x_1, \ldots, \underline{x}_h, \ldots, x_i, \ldots, x_p) \qquad (4.49)$$

where the two sides differ only in that x_i is underlined (not underlined) on the LHS (RHS), the minus sign stemming from the replacement $-ig \to ig$ at the vertex x_i. This entails the following corollary:

$$\sum F(x_1, \ldots, \underline{x}_i, \ldots, x_p) = 0 \qquad (4.50)$$

where the sum is over all possible ways of underlining all vertices. This equation allows us to write down the imaginary part of iG. Indeed, we have $G(x_1, \ldots, x_p) = F(x_1, \ldots, x_p)$ with no vertex underlined, and $G^*(x_1, \ldots, x_p) = F(\underline{x}_1, \ldots, \underline{x}_p)$ with all vertices underlined: if all vertices are underlined, the propagators are the complex conjugates of the original ones, and the same is true for the vertices. Equation (4.50) can be cast in the form

$$\text{Im}\,(iG(x_1, \ldots, x_p)) = -\frac{1}{2} \sum_{(x)} F(x_1, \ldots, \underline{x}_i, \ldots, x_p) \qquad (4.51)$$

where the sum on the RHS runs over all possibilities of underlining vertices, except for the case with no vertex or all vertices underlined.

Let us now examine the situation in Fourier space, where our convention is that $\exp(-i(k \cdot (x_i - x_j)))$ corresponds to a momentum k flowing into vertex i; we have for the 'cut propagators'

$$D_0^{>F}(k) = \theta(k_0)\, 2\pi\delta(k^2 - m^2) \tag{4.52a}$$

$$D_0^{<F}(k) = \theta(-k_0)\, 2\pi\delta(k^2 - m^2) \tag{4.52b}$$

From (4.52) we see that energy always flows from uncircled to circled vertices; at $T = 0$, this condition leads to many Fs vanishing, as energy must be conserved at each vertex, and entails that circled and uncircled vertices must form connected sets. This observation is at the basis of the $T = 0$ cutting rules.

At finite temperature, the situation is more complicated because

$$D_F^>(k) = (\theta(k_0) + n(k_0))\, 2\pi\delta(k^2 - m^2)$$
$$= \varepsilon(k_0)(1 + f(k_0))\, 2\pi\delta(k^2 - m^2) \tag{4.53a}$$
$$D_F^<(k) = (\theta(-k_0) + n(k_0))\, 2\pi\delta(k^2 - m^2)$$
$$= \varepsilon(k_0)f(k_0)\, 2\pi\delta(k^2 - m^2) \tag{4.53b}$$

and energy can flow in both directions. One useful remark is that, from (3.93), $D_F^>$ and $D_F^<$ coincide with the off-diagonal elements D_{21} and D_{12} of the real-time propagator when one makes the choice $\sigma = 0$. With this choice, uncircled vertices may be identified with vertices of type 1 and circled vertices with vertices of type 2.

In our Feynman graph $G(x_1, \ldots, x_p)$, let us denote by y_i, $i = 1, \ldots, n$ the vertices linked to external lines and by z_j, $j = 1, \ldots, m$ the internal vertices: $p = m + n$. Vertices y_i are always of type 1; vertices z_j are either of type 1 or of type 2. The relevant amplitude is

$$\mathcal{G}(y, z) = \sum_{z_j \in \{1,2\}} G(y_1, \ldots, y_n; z_1, \ldots, z_m) \tag{4.54}$$

since one has to sum over all possibilities for internal vertices in order to obtain the full amplitude.

As in the $T = 0$ case, one defines quantities

$$F(y_1, \ldots, \underline{y_i}, \ldots, y_n; z_1, \ldots, \underline{z_j}, \ldots, z_m)$$

built according to the rules (4.48), now using the $T \neq 0$ propagators. Since the construction parallels that of $T = 0$, these quantities obey the largest time equation (4.50). From a preceding remark we may write

$$\mathcal{G}(y, z)|_{\sigma=0} = \sum_{z_j \in \{1,2\}} F(y_1, \ldots, y_i, \ldots, y_n; z_1, \ldots, \underline{z_j}, \ldots, z_m) \tag{4.55}$$

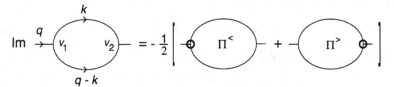

Fig. 4.5 Kobes–Semenoff rules for a one-loop self-energy diagram.

Let us try to obtain $\mathcal{G}^*(y,z)$ from $\mathcal{G}(y,z)$ by interchanging circled and uncircled vertices; this operation leads to the desired result: for example $D_F^>(x_i - x_j)$ is changed to $D_F^<(x_i - x_j) = (D_F^>(x_i - x_j))^*$. This entails

$$\mathcal{G}^*(y,z) = \sum_z F(\underline{y}_1,\ldots,\underline{y}_i,\ldots,\underline{y}_n; z_1,\ldots,\underline{z}_j,\ldots, z_m) \qquad (4.56)$$

Now using the largest time equation (4.50), we obtain the finite T generalization of (4.51)

$$\mathrm{Im}\,(i\mathcal{G}(y,z)) = -\frac{1}{2}\sum_{(y),z} F(y_1,\ldots,\underline{y}_i,\ldots, y_n; z_1,\ldots,\underline{z}_j,\ldots, z_m) \qquad (4.57)$$

In (4.57), the sum is over all possible circlings of vertices y and z, except for the cases with no y-vertex or all y-vertices circled. Equation (4.57) is the basic Kobes–Semenoff relation. In momentum space, one takes the Fourier transform of the largest time equation in order to obtain the analogue of (4.57).

4.3.2 *Application to the self-energy*

The most important application of (4.57) has up to now been the computation of imaginary parts of self-energies. Let us denote by v_1 and v_2 the two external vertices of Π, and by w the internal vertices. Equation (4.57) becomes (note that, according to our previous conventions, a factor of i has been included in the definition of Π_{ab}; see fig. 4.5 for the graphical representation of (4.58)):

$$\mathrm{Im}\,\Pi_{11}(q) = -\frac{1}{2}\sum_w \left(\Pi(\underline{v}_1, v_2; w) + \Pi(v_1, \underline{v}_2; w)\right)$$

$$= -\left(\frac{1}{2}\Pi^<(q) + \frac{1}{2}\Pi^>(q)\right) \qquad (4.58)$$

For example, in the one-loop graph of fig. 4.5, the expression of $\Pi^>(q)$ is

$$\Pi^>(q) = g^2 \int \frac{d^4k}{(2\pi)^2}(1 + f(k_0))(1 + f(q_0 - k_0))\varepsilon(k_0)$$

$$\times \varepsilon(q_0 - k_0)\delta(k^2 - m^2)\delta((q - k)^2 - m^2) \qquad (4.59)$$

In general, we have, from (2.43),

$$\Pi^>(q) = e^{\beta q_0} \Pi^<(q) \tag{4.60}$$

which entails, from (3.110b),

$$\text{Im } \Pi(q) = -\frac{1}{2}\varepsilon(q_0)(1 - e^{-\beta q_0})\Pi^>(q) \tag{4.61}$$

We may thus compute $\text{Im }\Pi(q)$ from $\Pi^>(q)$, where only the vertex v_2 is circled; an example is given in fig. 4.6. The main difference with the $T = 0$ case is that one loses the notion of cutting a graph, as the vertices do not necessarily form connected sets (fig. 4.6c).

In the case of self-energy insertions, it may be more convenient to start from (3.115) rather than from the cutting rules. Assume, for example, that we want to compute the graph given in fig. 4.6. Let us call $^{(1)}D(k)$ the propagator of the upper line, with a self-energy insertion $\Pi(k)$; we may write

$$\Pi^>(q) = \int \frac{d^4k}{(2\pi)^4} {}^{(1)}D^>(k)D_F^{\gtrless}(q-k)$$

$$= -2(1 + f(q_0)) \int \frac{d^4k}{(2\pi)^4}(1 + f(k_0) + f(q_0 - k_0))$$

$$\times \varepsilon(k_0)\varepsilon(q_0 - k_0)\left[\pi\delta'(k^2 - m^2)\text{Re }\Pi(k)\right.$$

$$\left. + \mathbf{P}\frac{1}{(k^2 - m^2)^2}\text{Im }\Pi(k)\right]2\pi\delta((q-k)^2 - m^2) \tag{4.62}$$

It is a good exercise to recover this formula by using the rules of fig. 4.6 and the regularization (3.103) of δ-functions. Equation (4.62) can be generalized with fully dressed cut propagators $D^>(k)$ and $D^>(q-k)$; from (2.80) we obtain

$$\Pi^>(q) = 2\pi \int \frac{d^3k}{(2\pi)^3}\frac{dk_0}{2\pi}\frac{dk_0'}{2\pi}(1 + f(k_0))(1 + f(k_0'))$$

$$\times \rho(k_0)\rho'(k_0')\delta(q_0 - k_0 - k_0') \tag{4.63}$$

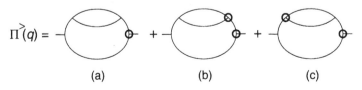

$$\Pi^>(q) =$$
(a)   (b)   (c)

Fig. 4.6   Kobes–Semenoff rules for a two-loop self-energy diagram.

which, from (2.45), is equivalent to

$$\Pi^>(q) = \int \frac{\mathrm{d}^4 k}{(2\pi)^4} D^>(k) D'^>(q-k) \tag{4.64}$$

It is instructive to derive (4.62) as an approximation to (4.63).

4.4 Physical interpretation of discontinuities

In the preceding section we learned how to compute discontinuities, or imaginary parts, of self-energies. At zero temperature, it is well-known that these imaginary parts are directly related to decay rates. We would like to ascertain whether a similar physical interpretation can be given at non-zero temperature. Before moving on the derivation of a general formula, we give two examples which provide an intuitive interpretation of the result.

4.4.1 Production of a weakly interacting massive particle

Our first example deals with the production of a massive particle, which interacts only weakly with the particles of the thermal bath: thus it escapes the bath once it has been produced, and it is never thermalized. We denote by Φ the corresponding (scalar) particle, and by φ_1 and φ_2 the particles of the thermal bath which are assumed, for simplicity, to be massless. The production of the massive particle occurs through the reaction

$$\varphi_1 + \varphi_2 \to \Phi \tag{4.65}$$

which is governed by a small coupling constant λ; φ_1 and φ_2 interact strongly, the strong coupling constant being denoted by g, but we need not give the explicit form of this interaction. We are clearly looking for a scalar model for the production of a heavy virtual photon γ^* in a quark–gluon plasma, with the following correspondence: $\Phi \to \gamma^*$, $(\varphi_1, \varphi_2) \to (q, \bar{q})$, $\lambda \to e$, $g \to$ QCD coupling constant.

To zeroth order in g, the production rate of Φ per unit of time and volume is given by

$$\frac{\mathrm{d}\Gamma}{\mathrm{d}^3 q} = \frac{\lambda^2}{2q_0} \int \frac{\mathrm{d}^3 k_1}{(2\pi)^3 2E_1} \int \frac{\mathrm{d}^3 k_2}{(2\pi)^3 2E_2}$$
$$\times n(E_1) n(E_2) (2\pi)^4 \delta^{(4)}(q - k_1 - k_2) \tag{4.66}$$

where q ($q_0 > 0$) denotes the four-momentum of Φ and $E_1 = \mathbf{k}_1, E_2 = \mathbf{k}_2$.

Let us compare this expression with that giving the imaginary part of the Φ-particle self-energy, $\operatorname{Im}\Pi(q)$, to the same order of perturbation theory. Since $k_0 > 0$ and $(q_0 - k_0) > 0$, we have from (4.37), with the substitution $g^2 \to \lambda^2$,

$$
\begin{aligned}
\text{Im}\,\Pi(q_0, \mathbf{q}) = &-\pi\lambda^2 \int \frac{d^3 k}{(2\pi)^3} \frac{1}{4E_1 E_2} \\
&\times (1 + n(E_1) + n(E_2))\delta(q_0 - E_1 - E_2)
\end{aligned}
\tag{4.67}
$$

and, using (4.35), we establish the relation

$$
q_0 \frac{d\Gamma}{d^3 q} = -\frac{\text{Im}\,\Pi(q_0, \mathbf{q})}{(2\pi)^3 (e^{\beta q_0} - 1)} = \frac{\Pi^<(q_0, \mathbf{q})}{2(2\pi)^3}
\tag{4.68}
$$

$\text{Im}\,\Pi$ contains both annihilation and creation rates (see (4.21)); $f(q_0)$ corrects for that. Alternatively, one may use $\Pi^<$.

4.4.2 *Production of a weakly interacting massless particle*

Our next example is slightly more involved, but also more instructive; it is, for example, a scalar model for axion production in a QED plasma. In the plasma we have particles of mass m ('electrons') and particles of zero mass ('photons') which interact strongly, with a coupling constant g; a massless particle ('axion') interacts weakly with the 'photons', with a small coupling constant λ. In this case the 'axion' cannot be produced directly by the fusion of two photons because of the lack of phase space: this can be seen by assuming an infinitesimal mass to the photon. The simplest mechanisms are described by the scattering and annihilation graphs of fig. 4.7, in which the kinematics is defined.

Let us look first at the scattering graph of fig. 4.7(a); we may write the production rate as

$$
\begin{aligned}
q_0 \frac{d\Gamma_a}{d^3 q} = &\frac{\lambda^2 g^2}{2} \int \frac{d^3 k_1}{(2\pi)^3 2k_1} n(k_1) \\
&\times \int \frac{d^3 p_1}{(2\pi)^3 2E_1} n(E_1) \int \frac{d^3 p_2}{(2\pi)^3 2E_2} (1 + n(E_2)) \\
&\times \frac{1}{(k^2)^2} 2\pi \delta^{(4)}(q + p_2 - k_1 - p_1)
\end{aligned}
\tag{4.69}
$$

with, as usual,

$$
E_i = \left(\mathbf{p}_i^2 + m^2 \right)^{1/2}
\tag{4.70}
$$

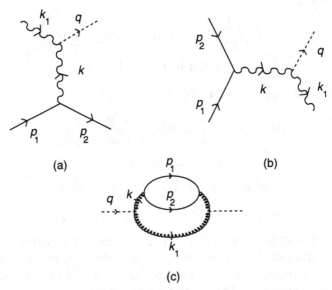

Fig. 4.7.   Scalar model for 'axion' production. Solid lines: 'electrons'. Wavy lines: 'photons'. Dashed lines: 'axions'. (a) Scattering graph. (b) Annihilation graph. (c) Axion self-energy.

It is convenient to use $\mathbf{p}_1$ and $\mathbf{k} = \mathbf{p}_1 - \mathbf{p}_2 = \mathbf{q} - \mathbf{k}_1$ as integration variables, which leads to

$$
q_0 \frac{\mathrm{d}\Gamma_a}{\mathrm{d}^3 q} = \frac{\lambda^2 g^2}{2(2\pi)^3} \int \frac{\mathrm{d}^3 k}{(2\pi)^3 2|\mathbf{q} - \mathbf{k}|} n(|\mathbf{q} - \mathbf{k}|) \frac{1}{(k^2)^2}
$$
$$
\times \int \frac{\mathrm{d}^3 p_1}{(2\pi)^3} \frac{1}{4E_1 E_2} n(E_1)(1 + n(E_2)) 2\pi\delta(k_0 - E_1 + E_2)
$$

(4.71a)

In an analogous way, the annihilation graph of fig. 4.7(b) contributes

$$
q_0 \frac{\mathrm{d}\Gamma_b}{\mathrm{d}^3 q} = \frac{\lambda^2 g^2}{4(2\pi)^3} \int \frac{\mathrm{d}^3 k}{(2\pi)^3 2|\mathbf{q} - \mathbf{k}|} (1 + n(|\mathbf{q} - \mathbf{k}|)) \frac{1}{(k^2)^2}
$$
$$
\times \int \frac{\mathrm{d}^3 p_1}{(2\pi)^3} \frac{1}{4E_1 E_2} n(E_1) n(E_2) 2\pi\delta(k_0 - E_1 - E_2)
$$

(4.71b)

where now $\mathbf{k} = \mathbf{p}_1 + \mathbf{p}_2 = \mathbf{k}_1 + \mathbf{q}$ and the additional factor of $1/2$ arises from Bose statistics.

Let us denote by $\Pi(q)$ the self-energy of the 'axion', and by $\tilde{\Pi}(k)$ that of the 'photon'. Using (4.62), we have, to the same order of perturbation theory (fig. 4.7c),

$$\Pi^>(q) = -2\lambda^2 \int \frac{d^4k}{(2\pi)^4} (1 + f(k_0))(1 + f(q_0 - k_0)) \frac{1}{(k^2)^2} \tag{4.72a}$$
$$\times \, \varepsilon(k_0)\varepsilon(q_0 - k_0) 2\pi\delta((q-k)^2) \text{Im}\,\tilde{\Pi}(k)$$

with $\text{Im}\,\tilde{\Pi}(k)$ being given by (4.37):

$$\text{Im}\,\tilde{\Pi}(k_0, \mathbf{k}) = -\frac{1}{2}\pi g^2 \varepsilon(k_0) \int \frac{d^3 p_1}{(2\pi)^3} \frac{1}{4E_1 E_2}$$
$$\times \left[(1 + n(E_1) + n(E_2))(\delta(k_0 - E_1 - E_2) \right.$$
$$- \delta(k_0 + E_1 + E_2)) - (n(E_1) - n(E_2))$$
$$\left. \times (\delta(k_0 - E_1 + E_2) - \delta(k_0 + E_1 - E_2)) \right] \tag{4.72b}$$

The term proportional to $\text{Re}\,\Pi(k)$ in (4.62) does not contribute, because phase space vanishes; the factor of $1/2$ again stems from Bose statistics. A glance at kinematics allows some simplifications in (4.72). Indeed, we have $q = k + k_1$ with $q^2 = k_1^2 = 0$ and $q_0 > 0$. Thus,

$$k^2 = -2qk_{10}(1 - \varepsilon(k_{10})\hat{\mathbf{q}} \cdot \hat{\mathbf{k}}_1) \tag{4.73}$$

which implies that the sign of k^2 is opposite to that of $k_{10} = q_0 - k_0$. Then if $(q_0 - k_0) < 0$, $k^2 > 0$ and $k_0 > 0$; it is easily checked that in this situation $k_0 = E_1 + E_2$. On the contrary, if $k^2 < 0$, we must have either $k_0 = E_1 - E_2$ or $k_0 = E_2 - E_1$; the term proportional to $\delta(k_0 + E_1 + E_2)$ never contributes. It is clear that $k^2 > 0$ corresponds to the annihilation graph and $k^2 < 0$ to the scattering graph in fig. 4.7(c): the imaginary part of the self-energy takes care of both mechanisms.

In order to compare (4.71) and (4.72), one has to perform simple manipulations on the statistical factors:

$$(1 + f(k_0))(n(E_1) - n(E_2))\delta(k_0 - E_1 + E_2)$$
$$= -e^{\beta k_0} n(E_1)(1 + n(E_2))\delta(k_0 - E_1 + E_2) \tag{4.74a}$$

$$(1 + f(k_0))(1 + n(E_1) + n(E_2))\delta(k_0 - E_1 - E_2)$$
$$= e^{\beta k_0} n(E_1) n(E_2)\delta(k_0 - E_1 - E_2) \tag{4.74b}$$

From these results it is easy to check that

$$q_0 \frac{d\Gamma_a}{d^3 q} + q_0 \frac{d\Gamma_b}{d^3 q} = \frac{1}{2(2\pi)^3} e^{-\beta q_0} \Pi^>(q_0, \mathbf{q})$$
$$= \frac{1}{2(2\pi)^3} \Pi^<(q_0, \mathbf{q}) \tag{4.75}$$

Thus, the production rate is again given by the same expression as in (4.68). It is interesting to remark that the factor of $1/2$ due to Bose statistics disappears in the calculation of $d\Gamma_a/d^3q$ because the two δ-functions which multiply $(n(E_1) - n(E_2))$ in (4.72b) give identical contributions; however this factor survives, as it should, in $d\Gamma_b/d^3q$. Note also that the rate for particle creation is directly related to $\Pi^<(q)$, as will become clear shortly.

4.4.3 *General proof*

Let us finally give a general proof of (4.68), or, equivalently, (4.75). We denote by $\Phi(x)$ the field corresponding to the weakly interacting particle and by $\varphi(x)$ the field corresponding to the particles in the heat bath; as in the preceding examples λ (g) is the weak (strong) coupling constant. We assume a $\lambda\Phi(x)\varphi^2(x)$ interaction between the weakly interacting particle and the particles in the heat bath and a $g\varphi^4(x)$ interaction between the particles in the heat bath; q ($q_0 > 0$) is the momentum of the produced particles. To lowest order in λ, the S-matrix element for the transition $(i) \rightarrow (q, f)$, where i and f are physical states of particles in the heat bath, reads

$$S_{fi}(q) = i\lambda \int d^4x \, e^{iq\cdot x} < f|\hat{\varphi}^2(x)|i > \tag{4.76}$$

Using translation invariance, summing over final states and over initial states with weight $[Z^{-1}(\beta)\exp(-\beta E_i)]$ yields

$$\frac{1}{Z(\beta)} \sum_{i,f} e^{-\beta E_i}|S_{fi}(q)|^2 = \lambda^2\Omega \int d^4x \, e^{iq\cdot x} \langle\hat{\varphi}^2(0)\hat{\varphi}^2(x)\rangle_\beta \tag{4.77}$$

Ω is the space-time volume where the interaction takes place, while the Fourier transform in (4.77) is the φ^2-correlation $\tilde{\Pi}^<(q)$:

$$\tilde{\Pi}^<(q) = \int d^4x \, e^{iq\cdot x} \langle\hat{\varphi}^2(0)\hat{\varphi}^2(x)\rangle_\beta \tag{4.78}$$

The total production rate per unit of time and volume Γ_{tot} is

$$\Gamma_{\text{tot}} = \frac{1}{\Omega}\frac{1}{Z(\beta)} \int \sum_{i,f} e^{-\beta E_i}|S_{fi}(q)|^2 \frac{d^3q}{2q_0(2\pi)^3}$$

$$= \frac{1}{Z(\beta)} \int \sum_{i,f} (2\pi)^4\delta^{(4)}(p_f + q - p_i)e^{-\beta E_i}$$

$$\times |< f|\hat{\varphi}^2(0)|i >|^2 \frac{d^3q}{2q_0(2\pi)^3} \tag{4.79}$$

Now, to lowest order in λ, but to all orders in g with our choice of a φ^4-interaction, $\tilde{\Pi}(q)$ is nothing other than one-particle irreducible self-energy $\Pi(q)$ of the Φ-particle, up to a factor of λ^2. The situation would

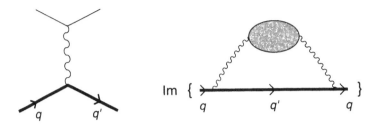

Fig. 4.8   Scalar model for the energy loss of a heavy fermion.

be more involved with a φ^3-interaction, but we may limit ourselves to the simplest case because the above identification will also hold in the physically relevant cases of section 5.3. We finally write the differential rate per unit of time and volume in terms of the Φ self-energy:

$$q_0 \frac{d\Gamma}{d^3 q} = \frac{1}{2} \frac{\Pi^<(q_0, \mathbf{q})}{(2\pi)^3} = -\frac{f(q_0) \text{Im} \,\Pi(q_0, \mathbf{q})}{(2\pi)^3} \tag{4.80}$$

We emphasize that (4.80) holds to all orders in g, but is only valid to lowest order in the weak coupling constant λ.

We conclude this section by studying a case which will be of interest in section 8.1; we examine a scalar model for the energy loss of a heavy fermion propagating in a QED plasma. Here our heavy fermion is a massive scalar particle Φ of mass m, which interacts weakly with massless scalar 'photons', while the thermal bath is composed of these 'photons' and scalar 'electrons'; once more our weak and strong coupling constants are denoted respectively by λ and g. We want to write down the interaction rate $\Gamma(q_0)$ of the massive particle, namely the number of its interactions per unit of time, when it propagates through the bath with an energy q_0. To lowest order in perturbation theory, the interaction occurs via the scattering graph of fig. 4.8. It would be easy to relate $\Gamma(q_0)$ and the imaginary part of the massive particle self-energy, $\text{Im} \,\Pi$, to this order of perturbation theory, along the lines of the calculation performed in section 4.4.2 . Since this calculation may be found in section 8.1, in the 'realistic' case, we leave it as exercise 4.4 for the reader.

In general, the interaction rate can be related to the imaginary part $\text{Im} \,\Pi(q)$ of the self-energy of the Φ-particle to lowest order in λ, but to all orders in g. Assume a $\lambda A(x) \Phi^2(x)$ interaction between the 'photon' and the heavy 'fermion'. The S-matrix element for the transition $(q, i) \to (q', f)$ reads, to lowest order in λ,

$$S_{fi}(q', q) = i\lambda \int d^4 x \, e^{i(q'-q)\cdot x} < f|\hat{A}(x)|i> \tag{4.81}$$

Using translation invariance, summing over final states and over initial states with weight $[Z^{-1}(\beta)\exp(-\beta E_i)]$ yields

$$\frac{1}{\Omega}\frac{1}{Z(\beta)}\sum_{i,f}e^{-\beta E_i}|<q',f|S|q,i>|^2$$

$$=\lambda^2\int d^4x\,e^{i(q-q')\cdot x}\langle\hat{A}(x)\hat{A}(0)\rangle_\beta$$

$$=\lambda^2\tilde{D}^>(q-q') \tag{4.82}$$

where $\tilde{D}(q)$ is the (fully dressed) propagator of the A-particle. This leads to the following expression for the interaction rate $\Gamma(q_0)$:

$$\Gamma(q_0)=\frac{1}{\Omega}\frac{1}{Z(\beta)}\int\sum_{i,f}e^{-\beta E_i}|S_{fi}(q',q)|^2\frac{d^3q}{2q_0(2\pi)^3}$$

$$=\frac{\lambda^2}{2q_0}\int\frac{d^4q'}{(2\pi)^4}\,2\pi\delta_+(q'^2-m^2)\tilde{D}^>(q-q')$$

$$=\frac{\lambda^2}{2q_0}\int\frac{d^4q'}{(2\pi)^4}\,D_F^{\gtrless}|_{T=0}(q')\tilde{D}^>(q-q') \tag{4.83}$$

where $D_F^{\gtrless}|_{T=0}(q')$ is the free $T=0$ cut propagator of the Φ-particle. The crucial point in the calculation is that, since the particle Φ is not thermalized, the cut propagator $D^>(q')$ to be used in (4.83) is the $T=0$ cut propagator (4.52a), and not the thermal propagator as in (4.64). To lowest order in λ, but to all orders in g, the RHS of (4.83) is nothing other than the self-energy $\Pi^>(q_0,\mathbf{q})$ of the Φ-particle and we have for the interaction rate

$$\Gamma(q_0)=\frac{1}{2q_0}\Pi^>(q_0,\mathbf{q})=-\frac{1}{q_0}(1+f(q_0))\mathrm{Im}\,\Pi(q_0,\mathbf{q}) \tag{4.84}$$

That this interaction rate is directly related to $\Pi^>(q)$ is easy to understand: $\Pi^>(q)$ describes a decay rate, the rate at which particles leave the initial phase space. The relations

$$\Pi^>(q)=e^{\beta q_0}\Pi^<(q) \tag{4.85a}$$

or

$$D^>(q)=e^{\beta q_0}D^<(q) \tag{4.85b}$$

are nothing other than detailed balance (exercise 4.5).

4.4.4 Small deviations from equilibrium

When λ is not small, particle Φ is in thermal equilibrium and the interpretation of $\mathrm{Im}\,\Pi$ is different. Let us examine first the case where Φ is an unstable particle in the vacuum. As explained earlier in this chapter (see equations (4.20)–(4.22)), the statistical factors $(1 + n_1 + n_2)$ and $(n_1 - n_2)$ which appear in the expression (4.19) of $\operatorname{Im} \Pi$ can be interpreted as the difference between decay rates and creation rates of particle Φ. Let us assume that at time $t = 0$ the number of Φs with energy q_0 is given by an arbitrary non-equilibrium distribution $f(q_0, t = 0)$; at a later time, this distribution will be $f(q_0, t)$. The function $f(q_0, t)$ obeys the differential equation

$$\frac{\mathrm{d}f}{\mathrm{d}t} = -f\Gamma^> + (1 + f)\Gamma^< \qquad (4.86)$$

where $\Gamma^>$ and $\Gamma^<$ are the decay and creation rates: (4.86) is clearly reminiscent of Boltzmann's equation. To lowest order in λ, these rates are directly related to $\Pi^>$ and $\Pi^<$, as we have just seen. In higher orders in λ, Φ itself can appear in the graphs which occur in the perturbative expansion of its self-energy, so that $\Gamma^>$ and $\Gamma^<$ depend on f, and (4.86) is a non-linear equation. However, for small departures from equilibrium, one can use the exact equilibrium distributions in the calculation of Π, and the solution of (4.86) reads

$$f(q_0, t) = \frac{\Gamma^<}{\Gamma^> - \Gamma^<} + c(q_0)\mathrm{e}^{-(\Gamma^> - \Gamma^<)t} \qquad (4.87)$$

where $c(q_0)$ is an arbitrary function depending on the initial conditions. Because of detailed balance

$$\Gamma^> = \mathrm{e}^{\beta q_0}\Gamma^< \qquad (4.88)$$

and $f(q_0, t)$ tends to the limiting equilibrium distribution $f(q_0)$, independently of the initial conditions. The rate of approach to equilibrium, $\Gamma = (\Gamma^> - \Gamma^<)$, is the inverse of the relaxation time τ and is directly related to $\operatorname{Im} \Pi$

$$\Gamma(q_0) = \frac{1}{\tau} = -\frac{1}{q_0}\operatorname{Im} \Pi(q_0) \qquad (4.89)$$

The mean free path l of a Φ-particle with velocity v will be given by $l = v/\Gamma^>$, since the first interaction is necessarily an absorption.

If Φ is a stable particle associated to a conserved quantum number as in the last example of section 4.4.3, then the situation is different. As in the case of the unstable particle, the mean free path is given by $l = v/\Gamma^>$. However, if instead of considering an unthermalized Φ-particle propagating in the plasma, we assume that it is close to thermal equilibrium, then the relaxation time has a more complicated expression than (4.89), because the RHS of (4.86) depends on n: see section 8.1 and exercise 8.4.

References and further reading

The self-consistent equation (4.4) is written and solved in Altherr (1990); the evaluation of ring diagrams follows Kapusta (1989), chap. 3. The $T = 0$ cutting rules are derived in Le Bellac (1992), chap. 10, following t'Hooft and Veltman (1974). The $T \neq 0$ cutting rules were first derived by Kobes and Semenoff (1986); see also Ashida, Nakkagawa, Niégawa and Yokota (1992a,b). Proofs of (4.68) were first given by Weldon (1990) and Gale and Kapusta (1991), while the interpretation of $1/\Gamma$ as a relaxation time is discussed by Weldon (1983); see Ma (1985) for clear explanations on detailed balance.

Exercises

4.1 Compute directly $\Pi(i\omega, \mathbf{q})$ in (4.16) at $T = 0$, by using an integration (rather than a discrete summation) over ω_n, which is a continuous variable at $T = 0$. Compare with the $T = 0$ limit of (4.16).

4.2 Show that for T much larger than q_0 and q, the T-dependent part of (4.16) behaves as $\ln T$. What is the dependence on T of the first-order correction to the vertex in $\lambda\varphi^4$-theory for large T?

4.3 Use the Kobes–Semenoff cutting rules to derive (4.62).

4.4 We wish to compute to lowest order in perturbation theory the interaction rate $\Gamma(q_0)$ of a heavy scalar particle in a plasma of 'photons' and 'electrons' (see the second half of section 4.4.3). Compute $\Gamma(q_0)$ to lowest order of perturbation theory by taking into account the 'electron'–heavy particle collisions mediated by 'photon' exchange. Then compute the heavy particle self-energy $\Pi^>(q_0, \mathbf{q})$ to lowest order in $(\lambda g)^2$ and check (4.84) to this order of perturbation theory (see also figs. 8.1 and 8.2).

4.5 The probability for a scalar particle Φ to propagate through a heat bath of scalar particles φ with energy $q_0 > 0$ decreases at the rate

$$\Gamma^>(q_0) = \frac{1}{2q_0} \int d\Omega \, |M(\Phi, 1, \ldots, a \to 1', \ldots, b')|^2$$
$$\times n_1 \ldots n_a (1 + n_1') \ldots (1 + n_b')$$

and increases at the rate

$$\Gamma^<(q_0) = \frac{1}{2q_0} \int d\Omega \, |M(1', \ldots, b' \to \Phi, 1, \ldots, a)|^2$$
$$\times n_1' \ldots n_b' (1 + n_1) \ldots (1 + n_a)$$

where $\int d\Omega$ denotes phase space integration. If CP invariance holds, then the two matrix elements squared are equal. Show that $\Gamma^> = e^{\beta q_0} \Gamma^<$: this is the property of detailed balance.

5
Dirac and gauge fields
at finite temperature

Up to now we have been living in a world of scalar particles. Of course this is not sufficient when dealing with the real world, where spin one-half and spin one particles play a prominent role, and we have to learn how to quantize Dirac and gauge fields at finite temperature. The quantization of Dirac fields is a rather straightforward generalization of what we have already learned. As in the zero-temperature case, the quantization of gauge fields is more subtle, and we shall study in some detail the gauge field propagator and the role of the unphysical degrees of freedom. In the final section of this chapter we shall generalize the results of section 4.4, by deriving the rate for photon and lepton pair production from a quark–gluon plasma.

5.1 The Dirac field at finite temperature

5.1.1 *Coherent fermion states and path integrals*

As a preliminary step, we generalize to the case of fermions the path integral formalism which was set up in chapters 2 and 3. We first consider an elementary case, which is the transposition to fermions of the simple harmonic oscillator of chapter 2. Since we want to describe particles which obey Fermi statistics, we introduce two Hermitian conjugate creation and annihilation operators a and $a^\dagger$, which obey the anticommutation relations

$$\left\{a, a^\dagger\right\} = a\, a^\dagger + a^\dagger\, a = 1 \qquad (5.1)$$

together with $a^2 = a^{\dagger 2} = 0$. The corresponding Hilbert space is two-dimensional, and a convenient basis is given by the vacuum state $|0>$ and the one-particle state $|1>$:

$$a|0> = 0 \qquad a^\dagger|0> = |1> \qquad (5.2)$$

86

On this basis, a and $a^\dagger$ have the following representation:

$$a = \begin{pmatrix} 0 & 0 \\ 1 & 0 \end{pmatrix} \qquad a^\dagger = \begin{pmatrix} 0 & 1 \\ 0 & 0 \end{pmatrix} \qquad (5.3)$$

The occupation number operator is $N = a^\dagger a$.

We now recall the concept of a Grassmann algebra $\mathscr{A}$: such an algebra is generated by n 'anticommuting c-numbers' $\eta_1, ..., \eta_n$ with the following properties:

(i) $\{\eta_i, \eta_j\} = 0$.
(ii) If η_i and $\eta_j \in \mathscr{A}$, then $\lambda \eta_i + \mu \eta_j \in \mathscr{A}$, where λ and μ are arbitrary complex numbers.
(iii) If η_i and $\eta_j \in \mathscr{A}$, then $\eta_i \eta_j \in \mathscr{A}$.

Let η and η^* be the generators of a two-dimensional Grassmann algebra; by assumption they anticommute with a and $a^\dagger$. We define 'coherent fermionic states' $|\eta>$ by

$$|\eta> = e^{-\eta a^\dagger}|0> = |0> -\eta|1> = (1 - \eta a^\dagger)|0> \qquad (5.4)$$

which obey

$$a|\eta> = \eta|\eta> = \eta|0> \qquad (5.5)$$

In the bosonic case, analogous coherent states, first introduced as 'quasi-classical states' in the case of the harmonic oscillator, are also familiar and prove to be useful in many problems. The bra $<\eta|$ conjugate to the ket $|\eta>$ is, by convention,

$$<\eta| = <0|e^{-a\eta^*} = <0|- <1|\eta^* \qquad (5.6)$$

which is equivalent to choosing

$$(\eta\, a)^\dagger = a^\dagger \eta^* \qquad (5.7)$$

We also note the following identities:

$$<\eta|0> = <0|\eta> = 1 \qquad <1|\eta> = <\eta|1>^* = -\eta \qquad (5.8)$$

and

$$<\eta'|\eta> = \exp(\eta'^* \eta) \qquad (5.9)$$

The rule for the integration of a polynomial in η, η^*, with complex coefficients a_0, a_1, a_1', a_{12}, is, by convention,

$$\int d\eta^* d\eta \, (a_0 + a_1\eta + a_1'\eta^* + a_{12}\eta^*\eta) = -a_{12} \qquad (5.10)$$

from which we deduce

$$\int d\eta^* d\eta \, \exp(-\lambda\eta^*\eta) = \lambda \qquad (5.11)$$

As is well known, equation (5.11) generalizes to

$$\int \prod_{i=1}^{n} d\eta_i^* d\eta_i \exp\left(-\sum_{i,j=1}^{n} \eta_i^* M_{ij} \eta_j\right) = \det M \tag{5.12}$$

The following identities are proved by inspection, using (5.8) and (5.9):

$$\int d\eta^* d\eta \exp(-\eta^* \eta) |\eta ><\eta|$$
$$= |0 >< 0| + |1 >< 1| = 1 \tag{5.13}$$

$$\int d\eta^* d\eta \exp(-\eta^* \eta) < -\eta|A|\eta >$$
$$= < 0|A|0 > + < 1|A|1 >= \operatorname{Tr} A \tag{5.14}$$

where A is any operator in our two-dimensional space (exercise 5.1).

After these preliminaries, we are ready to write a path integral for the partition function, the Hamiltonian being given by

$$H(a^\dagger, a) = \omega \, a^\dagger \, a \tag{5.15}$$

We start from (5.14) with $A = \exp(-\beta H)$, and, as in chapter 2, we split the interval $[0, \beta]$ into $(n+1)$ intervals of equal length $\varepsilon = \beta/(n+1)$ and insert at point i, $i = 1, \ldots, n$ a complete set of states in the form

$$\int d\eta_i^* d\eta_i \exp(-\eta_i^* \eta_i) |\eta_i >< \eta_i| \tag{5.16}$$

From (5.5) and (5.9) this procedure yields matrix elements

$$< \eta_i | e^{-\varepsilon H(a^\dagger, a)} | \eta_{i-1} > \,\simeq\, < \eta_i | (1 - \varepsilon H(a^\dagger, a)) | \eta_{i-1} >$$
$$= e^{\eta_i^* \eta_{i-1}} (1 - \varepsilon H(\eta_i^*, \eta_{i-1})) \simeq e^{\left(\eta_i^* \eta_{i-1} - \varepsilon H(\eta_i^*, \eta_{i-1})\right)} \tag{5.17}$$

The partition function now reads

$$Z = \int d\eta^* d\eta \prod_{i=1}^{n} d\eta_i^* d\eta_i \, e^{-S(\eta_i^*, \eta_i)} \tag{5.18}$$

where the argument of the exponential is given by

$$S(\eta_i^*, \eta_i) = \eta^* (\eta_n + \eta) + \eta_n^* (\eta_n - \eta_{n-1}) + \ldots$$
$$+ \eta_1^* (\eta_1 - \eta) + \varepsilon H(-\eta^*, \eta_n)$$
$$+ \varepsilon H(\eta_n^*, \eta_{n-1}) + \ldots + \varepsilon H(\eta_1^*, \eta) \tag{5.19}$$

This expression is rendered more transparent if we use the antiperiodic boundary condition $\eta_0 = \eta$, $\eta_{n+1} = -\eta$ (see (5.14)); then S may be rewritten as

$$S = \sum_{i=1}^{n+1} \eta_i^* (\eta_i - \eta_{i-1}) + \varepsilon \sum_{i=1}^{n+1} H(\eta_i^*, \eta_{i-1}) \tag{5.20}$$

From (5.20) one sees that S is a discrete approximation of

$$S = \int_{0;\eta(\beta)=-\eta(0)}^{\beta} \left(\eta^*(\tau)\partial_\tau\eta(\tau) + H(\eta^*(\tau),\eta(\tau))\right)d\tau \qquad (5.21)$$

and we obtain the path integral representation of Z:

$$Z = \int_{\eta(\beta)=-\eta(0)} \mathscr{D}(\eta^*,\eta)$$
$$\times \exp\left(-\int_0^\beta \left[\eta^*(\tau)\partial_\tau\eta(\tau) + H(\eta^*(\tau),\eta(\tau))\right]d\tau\right) \qquad (5.22)$$

The formal expression (5.22) is defined as the $\varepsilon \to 0$ limit of (5.18). The origin of the antiperiodicity condition $\eta(\beta) = -\eta(0)$ is of course the bra $< -\eta|$ in expression (5.14) of the trace. Using (5.12) we may check the expression of Z:

$$Z = \lim_{\varepsilon\to 0} \det S_{ij} \qquad (5.23)$$

where, from (5.20), S_{ij} is the matrix (written here for $n = 3$)

$$S_{ij} = \begin{pmatrix} 1 & 0 & 0 & 1-\varepsilon\omega \\ -1+\varepsilon\omega & 1 & 0 & 0 \\ 0 & -1+\varepsilon\omega & 1 & 0 \\ 0 & 0 & -1+\varepsilon\omega & 1 \end{pmatrix} \qquad (5.24)$$

One recovers the result of elementary statistical mechanics:

$$Z = \lim_{n\to\infty} \left(1 + \left(1 - \frac{\beta\omega}{n+1}\right)^{n+1}\right) = 1 + e^{-\beta\omega} \qquad (5.25)$$

Expression (5.22) is easily generalized to a set of N fermionic oscillators:

$$\{a_i, a_j\} = \{a_i^\dagger, a_j^\dagger\} = 0 \quad \{a_i, a_j^\dagger\} = \delta_{ij} \qquad (5.26)$$

One finds

$$Z = \int_{\eta_i(\beta)=-\eta_i(0)} \mathscr{D}(\eta_i^*,\eta_i) \exp\left(-\int_0^\beta\right.$$
$$\left. \times \sum_{i,j} \left[\eta_i^*(\tau)\partial_\tau\eta_i(\tau) + H(\eta_i^*(\tau),\eta_j(\tau))\right]d\tau\right) \qquad (5.27)$$

provided the Hamiltonian $H(a_i^\dagger, a_j)$ is written in normal form, with all creation operators positioned to the left of all annihilation operators.

5.1.2 Quantization of the Dirac field

We shall come back to the functional formalism later, but we revert for the moment to the operator approach. In order to define our notations precisely, we recall that the Dirac equation reads

$$(i\gamma^\mu \partial_\mu - m)\psi(x) = (i\slashed{\partial} - m)\psi(x) = 0 \qquad (5.28)$$

where the Dirac matrices γ^μ obey the algebra

$$\{\gamma^\mu, \gamma^\nu\} = 2g^{\mu\nu} \qquad (5.29)$$

A standard representation is

$$\gamma^0 = \begin{pmatrix} 1 & 0 \\ 0 & -1 \end{pmatrix} \qquad \gamma^i = \begin{pmatrix} 0 & -\sigma_i \\ \sigma_i & 0 \end{pmatrix} \qquad (5.30)$$

where 1 stands for the unit 2×2 matrix and the σ_is are Pauli matrices. The free Dirac field can be Fourier analyzed (in order to avoid cumbersome notations, we omit circumflexes on field operators ψ and $\overline{\psi}$)

$$\psi(x) = \sum_{r=1}^{2} \int \frac{d^3 p}{(2\pi)^3 2E_p}$$
$$\times \left[b_r(\mathbf{p}) u_r(\mathbf{p}) e^{-ip\cdot x} + d_r^\dagger(\mathbf{p}) v_r(\mathbf{p}) e^{ip\cdot x} \right] \qquad (5.31a)$$

$$\overline{\psi}(x) = \psi^\dagger(x)\gamma^0 = \sum_{r=1}^{2} \int \frac{d^3 p}{(2\pi)^3 2E_p}$$
$$\times \left[b_r^\dagger(\mathbf{p}) \overline{u}_r(\mathbf{p}) e^{ip\cdot x} + d_r(\mathbf{p}) \overline{v}_r(\mathbf{p}) e^{-ip\cdot x} \right] \qquad (5.31b)$$

where $u_r(\mathbf{p})$ $(v_r(\mathbf{p}))$ is a positive (negative) energy Dirac spinor, r is a spin index, $E_p = (\mathbf{p}^2 + m^2)^{1/2}$, $p \cdot x = E_p x^0 - \mathbf{p} \cdot \mathbf{x}$, while the operators $b_r(\mathbf{p})$ and $d_r(\mathbf{p})$ obey the anticommutation relations

$$\{b_r(\mathbf{p}), b_s(\mathbf{p}')\} = \{d_r(\mathbf{p}), d_s(\mathbf{p}')\} = 0 \qquad (5.32a)$$
$$\{b_r(\mathbf{p}), b_s^\dagger(\mathbf{p}')\} = \{d_r(\mathbf{p}), d_s^\dagger(\mathbf{p}')\}$$
$$= (2\pi)^3 \, 2E_p \, \delta_{rs}\delta^{(3)}(\mathbf{p} - \mathbf{p}') \qquad (5.32b)$$

In the general (interacting) case, the Dirac field obeys the equal-time canonical anticommutation relations

$$\left\{\psi_\alpha(x), \psi_\beta^\dagger(x')\right\}\Big|_{t=t'} = \delta_{\alpha\beta}\delta^{(3)}(\mathbf{x} - \mathbf{x}') \qquad (5.33)$$

As in the spin zero case, we define two-point functions $S^>$ and $S^<$:

$$S_{\alpha\beta}^>(x, x') = \langle \psi_\alpha(x)\overline{\psi}_\beta(x') \rangle_\beta \qquad (5.34a)$$
$$S_{\alpha\beta}^<(x, x') = -\langle \overline{\psi}_\beta(x')\psi_\alpha(x) \rangle_\beta \qquad (5.34b)$$

In (5.34) $\langle A \rangle_\beta$ represents as usual a thermal average, and we first restrict ourselves for simplicity to zero chemical potential; note the minus sign in (5.34b) which reflects the fermionic nature of the field. As in the bosonic case, we use the fact that $\exp(-\beta H)$ is an evolution operator in imaginary time together with the cyclicity of the trace to derive the KMS relation:

$$S_{\alpha\beta}^>(t, \mathbf{x}; t', \mathbf{x}') = -S_{\alpha\beta}^<(t + i\beta, \mathbf{x}; t', \mathbf{x}') \tag{5.35}$$

Defining the Fourier transform:

$$S_{\alpha\beta}^>(p) = \int d^4x \, e^{ip \cdot x} S_{\alpha\beta}^>(x) \tag{5.36}$$

and using (5.35) yields

$$S_{\alpha\beta}^<(p_0) = -e^{-\beta p_0} S_{\alpha\beta}^>(p_0) \tag{5.37}$$

As in the scalar case, these relations imply that one may write $S^>$ and $S^<$ in terms of a spectral function $\rho(p)$:

$$S_{\alpha\beta}^>(p) = (1 - \tilde{f}(p_0))\rho_{\alpha\beta}(p) \tag{5.38a}$$

$$S_{\alpha\beta}^<(p) = -\tilde{f}(p_0)\rho_{\alpha\beta}(p) \tag{5.38b}$$

$$\rho_{\alpha\beta}(p) = S_{\alpha\beta}^>(p) - S_{\alpha\beta}^<(p) \tag{5.38c}$$

with

$$\tilde{f}(p_0) = \frac{1}{e^{\beta p_0} + 1} \tag{5.39}$$

It is important to note that equations (5.38) can be deduced from the bosonic case (2.45) with the substitution $f(p_0) \rightarrow -\tilde{f}(p_0)$. From rotational invariance, we write the following decomposition of $\rho(p)$:

$$\rho(p) = \gamma_0 \rho_0(p) - \gamma \cdot \mathbf{p} \, \rho_p(p) + \rho_m(p) \tag{5.40}$$

As in the bosonic case, we deduce a sum rule from the anticommutation relations (5.33):

$$\int_{-\infty}^{\infty} \frac{dp_0}{2\pi} \rho_0(p) = \frac{1}{4} \int_{-\infty}^{\infty} \frac{dp_0}{2\pi} \text{Tr} \, (\gamma_0 \rho(p)) = \frac{1}{4} \int d^4x \, \delta(t) \, e^{-ip \cdot x}$$

$$\times \langle \psi_\alpha(x)\psi_\alpha^\dagger(0) + \psi_\alpha^\dagger(0)\psi_\alpha(x) \rangle_\beta = 1 \tag{5.41}$$

It is straightforward to derive the free spectral function $\rho_F(p)$ from the Fourier decomposition (5.31) of the free Dirac field and the anticommutation relation (5.32). One obtains (exercise 5.2)

$$\rho_F(x) = S_F^>(x) - S_F^<(x)$$

$$= \int \frac{d^3p}{(2\pi)^3 2E_p} \left[\Lambda_+(\mathbf{p})e^{-ip \cdot x} - \Lambda_-(\mathbf{p})e^{ip \cdot x} \right] \tag{5.42}$$

where $\Lambda_+(\mathbf{p})$ and $\Lambda_-(\mathbf{p})$ are the projectors on the positive and negative energy states:

$$\Lambda_+(\mathbf{p}) = \sum_{r=1}^{2} u_r \bar{u}_r = \gamma_0 E_p - \gamma \cdot \mathbf{p} + m \tag{5.43a}$$

$$\Lambda_-(\mathbf{p}) = -\sum_{r=1}^{2} v_r \bar{v}_r = -\gamma_0 E_p + \gamma \cdot \mathbf{p} + m \tag{5.43b}$$

Equation (5.42) is equivalent to

$$\rho_F(p) = 2\pi \varepsilon(p_0)(\not{p} + m)\delta(p^2 - m^2) \tag{5.44}$$

which yields, from (5.38), the cut propagators

$$S_F^>(p) = 2\pi(1 - \tilde{f}(p_0))\varepsilon(p_0)(\not{p} + m)\delta(p^2 - m^2) \tag{5.45a}$$
$$S_F^<(p) = -2\pi \tilde{f}(p_0)\varepsilon(p_0)(\not{p} + m)\delta(p^2 - m^2) \tag{5.45b}$$

The Matsubara propagator (see (2.22)):

$$\begin{aligned}
S_{\alpha\beta}(\tau, \mathbf{x}; \tau', \mathbf{x}') &= \langle T(\psi_\alpha(-i\tau, \mathbf{x})\,\overline{\psi}_\beta(-i\tau', \mathbf{x}')) \rangle_\beta \\
&= \theta(\tau - \tau') S_{\alpha\beta}^>(\tau, \mathbf{x}; \tau', \mathbf{x}') \\
&\quad + \theta(\tau' - \tau) S_{\alpha\beta}^<(\tau, \mathbf{x}; \tau', \mathbf{x}')
\end{aligned} \tag{5.46}$$

obeys the antiperiodicity condition

$$S_{\alpha\beta}(\tau - \beta, \mathbf{x}; \tau', \mathbf{x}') = -S_{\alpha\beta}(\tau, \mathbf{x}; \tau', \mathbf{x}') \tag{5.47}$$

for $0 \le \tau \le \beta$. Because of (5.47), the Matsubara frequencies are given by $\omega_n = (2n + 1)\pi T$, with n integer; as in (2.58) one derives the following representation of the free Matsubara propagator in Fourier space:

$$S_F(i\omega_n, \mathbf{p}) = -\int_{-\infty}^{\infty} \frac{dp_0}{2\pi} \frac{\rho_F(p_0, \mathbf{p})}{i\omega_n - p_0} = \frac{m - \not{p}}{\omega_n^2 + E_p^2} \tag{5.48}$$

In writing (5.48), we have shifted from Minkowski to Euclidean space, and at this point it may be useful to give some details on the analytical continuation from Minkowski space to Euclidean space. The time component of a four-vector is transformed as in $t \to -i\tau = -ix_4$: $p_0 \to -ip_4$; then the Matsubara frequencies ω or ω_n are to be identified with $-p_4$ (this minus sign is somewhat unfortunate, but we have chosen to stick to the usual convention for the Matsubara frequencies, so that $i\omega \to q_0$). Our convention for Dirac matrices is that γ-matrices in Euclidean space obey anticommutation relations ($\mu, \nu = 1, \dots, 4$)

$$\{\gamma_\mu, \gamma_\nu\} = -2\delta_{\mu\nu} \tag{5.49}$$

with $\gamma_4 = i\gamma_0$; thus in (5.48),

$$\not{p} = \gamma_4 p_4 + \gamma \cdot \mathbf{p} = -\gamma_4 \omega_n + \gamma \cdot \mathbf{p} \tag{5.50}$$

Note the following systematic rule for shifting from Minkowski space to Euclidean space: spatial components ($\mu = i = 1, 2, 3$) remain unchanged:

$$g_{\mu\nu} \to -\delta_{\mu\nu} \qquad A_\mu B^\mu \to -A_\mu B_\mu$$
$$A\!\!\!/ \to -A\!\!\!/ \qquad \partial^\mu A_\mu \to \partial_\mu A_\mu \tag{5.51}$$

The position of the indices will often suffice to distinguish between the two cases.

5.1.3 Frequency sums

In computing Feynman diagrams with internal fermion lines, we shall encounter frequency sums, and we have to learn how to evaluate them. Let us denote the quantity $(\omega_n^2 + E_p^2)^{-1}$, which appears in the free propagator (5.48), by $\tilde{\Delta}(i\omega_n, E_p)$, and its mixed representation by $\tilde{\Delta}(\tau, E_p)$, suppressing the subscript F for notational simplicity:

$$\tilde{\Delta}(\tau, E_p) = T \sum_n e^{-i\omega_n \tau} \tilde{\Delta}(i\omega_n, E_p) \tag{5.52}$$

It can easily be checked (exercise 5.3) that

$$\tilde{\Delta}(\tau, E_p) = \frac{1}{2E_p}\left[(1 - \tilde{n}(E_p))e^{-E_p\tau} - \tilde{n}(E_p)e^{E_p\tau}\right]$$
$$= \sum_{s=\pm 1} \frac{s}{2E_p}(1 - \tilde{f}(sE_p))e^{-sE_p\tau} \tag{5.53}$$

where the Fermi–Dirac distribution $\tilde{n}(p_0)$ is

$$\tilde{n}(p_0) = \frac{1}{e^{\beta|p_0|} + 1} \tag{5.54}$$

One should note the absolute value of p_0 in (5.54), in contrast with the definition (5.39) of $\tilde{f}(p_0)$. Note also that

$$\tilde{f}(E) = \tilde{n}(E) \qquad \tilde{f}(-E) = 1 - \tilde{n}(E)$$
$$1 - \tilde{f}(E) - \tilde{f}(-E) = 0 \tag{5.55}$$

In frequency space, the formula corresponding to (5.53) is

$$\tilde{\Delta}(i\omega_n, E_p) = \frac{1}{\omega_n^2 + E_p^2} = \sum_{s=\pm 1} \tilde{\Delta}_s(i\omega_n, E_p)$$
$$= \sum_{s=\pm 1} -\frac{s}{2E_p}\frac{1}{i\omega_n - sE_p} \tag{5.56}$$

The frequency sums are performed by following the methods explained in sections 2.6 and 4.2 in the bosonic case, remembering, however, that $\exp(i\omega_n\beta) = -1$. Let us give the generalizations of (4.15) and (4.26).

(i) Fermion–boson case:

$$T\sum_n \Delta_{s_1}(i\omega_n, E_1)\tilde{\Delta}_{s_2}(i(\omega - \omega_n), E_2)$$

$$= -\frac{s_1 s_2}{4E_1 E_2}\frac{1 + f(s_1 E_1) - \tilde{f}(s_2 E_2)}{i\omega - s_1 E_1 - s_2 E_2} \tag{5.57}$$

and

$$T\sum_n \omega_n\Delta_{s_1}(i\omega_n, E_1)\tilde{\Delta}_{s_2}(i(\omega - \omega_n), E_2)$$

$$= \frac{i s_2}{4E_2}\frac{1 + f(s_1 E_1) - \tilde{f}(s_2 E_2)}{i\omega - s_1 E_1 - s_2 E_2} \tag{5.58}$$

(ii) Fermion–antifermion case:

$$T\sum_n \tilde{\Delta}_{s_1}(i\omega_n, E_1)\tilde{\Delta}_{s_2}(i(\omega - \omega_n), E_2)$$

$$= -\frac{s_1 s_2}{4E_1 E_2}\frac{1 - \tilde{f}(s_1 E_1) - \tilde{f}(s_2 E_2)}{i\omega - s_1 E_1 - s_2 E_2} \tag{5.59}$$

and

$$T\sum_n \omega_n\tilde{\Delta}_{s_1}(i\omega_n, E_1)\tilde{\Delta}_{s_2}(i(\omega - \omega_n), E_2)$$

$$= \frac{i s_2}{4E_2}\frac{1 - \tilde{f}(s_1 E_1) - \tilde{f}(s_2 E_2)}{i\omega - s_1 E_1 - s_2 E_2} \tag{5.60}$$

One should note that formulae (5.57)–(5.60) may be obtained from (4.15) and (4.26) by a simple substitution: replace $f(sE)$ for a bosonic line by $-\tilde{f}(sE)$ for a fermionic line. As has already been pointed out, the substitution $f(p_0) \rightarrow -\tilde{f}(p_0)$ gives a systematic rule for substituting a fermionic line for a bosonic line in an arbitrary Feynman graph (compare, for example, (4.12) and (5.53)); in terms of $n(E)$ the substitution is $n(E) \rightarrow -\tilde{n}(E)$.

It is also interesting to quote the generalization of equation (4.64), for instance in the fermion–antifermion case with bare propagators (exercise 5.4):

$$\text{Im } T\sum_n \int \frac{d^3 p}{(2\pi)^3}\text{Tr}\left[S_F(i\omega_n, E_p)S_F(i(\omega_n - \omega), E_{|\mathbf{q}-\mathbf{p}|})\right]$$

$$= \frac{1}{2}\left(1 - e^{-\beta q_0}\right)\int \frac{d^4 p}{(2\pi)^4}\text{Tr}\left[S_F^>(p)S_F^<(p - q)\right]$$

$$= -\frac{1}{2}\left(1 - e^{-\beta q_0}\right)\Pi^>(q) \tag{5.61}$$

where the minus sign in the last line is associated with the fermion loop; this equation is in agreement with (4.38) and (4.61). Note that the momentum of the fermionic lines is oriented in the same direction as the charge flow, hence the choice $S_F^<(p-q)$ in (5.61).

5.1.4 Non-zero chemical potential

We now introduce a non-zero chemical potential μ. In order to derive the expression of the propagator, we shall compare the path integral for the partition function in the cases $\mu = 0$ and $\mu \neq 0$. The free Dirac Hamiltonian is

$$\hat{H} = \int d^3x\, \psi^\dagger(x)\gamma^0(-i\gamma \cdot \nabla + m)\psi(x) \qquad (5.62)$$

and from (5.22) we immediately write a path integral representation for Z in the $\mu = 0$ case:

$$Z = \int_{\psi_\alpha(\beta)=-\psi_\alpha(0)} \mathscr{D}(\psi_\alpha^*, \psi_\alpha)$$
$$\times \exp\left(-\int_0^\beta \left[\psi_\alpha^* \partial_\tau \psi_\alpha + H(\psi_\alpha^*, \psi_\alpha)\right] d\tau\right) \qquad (5.63)$$

Remember that in (5.63) ψ_α^* and ψ_α are Grassmann variables, and not operators as in (5.62)! Furthermore, the Hamiltonian has to be written in normal form in order to obtain (5.63) by substituting Grassmann variables to operators. From (5.62) and (5.63) we deduce that the two-point Green function $S_F(\tau, \mathbf{x}; \tau', \mathbf{x}')$:

$$S_F(\tau, \mathbf{x}; \tau', \mathbf{x}') = \langle T(\psi(-i\tau, \mathbf{x})\overline{\psi}(-i\tau', \mathbf{x}'))\rangle_\beta \qquad (5.64)$$

obeys the partial differential equation

$$(\gamma^0 \partial_\tau - i\gamma \cdot \nabla + m)S_F(\tau, \mathbf{x}; \tau', \mathbf{x}') = \delta^{(4)}(x - x') \qquad (5.65)$$

When μ is non-zero, the free Dirac Hamiltonian must be replaced by

$$\hat{H}' = \hat{H} - \mu\hat{Q} = \hat{H} - \mu\int d^3x\, \psi^\dagger\gamma^0\psi \qquad (5.66)$$

The charge $\hat{Q}$ is conserved and obeys commutation relations which remain valid in the interacting case:

$$[\hat{Q}, \psi(x)] = -\psi(x) \qquad [\hat{Q}, \overline{\psi}(x)] = \overline{\psi}(x) \qquad (5.67)$$

The functional integral (5.63) becomes

$$Z = \int_{\psi_\alpha(\beta)=-\psi_\alpha(0)} \mathcal{D}(\psi_\alpha^*, \psi_\alpha)$$

$$\times \exp\left(-\int_0^\beta \left[\psi_\alpha^*(\partial_\tau - \mu)\psi_\alpha + H(\psi_\alpha^*, \psi_\alpha)\right] d\tau\right) \qquad (5.68)$$

so that we find the same result as in the bosonic case (section 3.2):

$$\partial_\tau \psi \to (\partial_\tau - \mu)\psi \qquad \partial_\tau \psi^* \to (\partial_\tau + \mu)\psi^* \qquad (5.69)$$

or

$$i\omega_n \to i\omega_n + \mu = i\frac{\pi(2n+1)}{\beta} + \mu \qquad (5.70)$$

in the fermion propagator: see exercise 5.5 for further comments.

Finally, we have to generalize our frequency sums. We follow the method of section 3.2, using an integration contour in the complex p_0-plane. Let $g(p_0)$ be a meromorphic function of p_0, regular on the vertical line $\mathrm{Re}\, p_0 = \mu$, which decreases faster than p_0^{-1} for $|p_0| \to \infty$. The function $\frac{\beta}{2}\tanh[\beta(p_0-\mu)/2]$ has poles with unit residue at $p_0 = i\omega_n + \mu$ and is bounded in all directions of the complex p_0-plane, except on the vertical axis $\mathrm{Re}\, p_0 = \mu$. This allows us to write

$$S = T \sum_{n=-\infty}^{n=+\infty} g(p_0 = i\omega_n + \mu)$$

$$= \int_{C_1 \cup C_2} \frac{dp_0}{2i\pi} g(p_0)\frac{1}{2}\tanh\left[\frac{\beta(p_0-\mu)}{2}\right] \qquad (5.71)$$

where C_1 is a vertical straight line from $\mu + \eta - i\infty$ to $\mu + \eta + i\infty$, while C_2 goes from $\mu - \eta + i\infty$ to $\mu - \eta - i\infty$, $\eta \to 0^+$ (see fig. 3.5). Closing C_1 and C_2 with large half-circles allows us to pick up the poles of $g(p_0)$:

$$S = -\frac{1}{2}\sum \mathrm{Res}\, g(p_0)\tanh\left[\frac{\beta(p_0-\mu)}{2}\right] \qquad (5.72)$$

where the sum runs over all the values of p_0 such that $g^{-1}(p_0) = 0$ and we have denoted by $\mathrm{Res}\, g(p_0)$ the corresponding residues. From our assumption on the asymptotic behaviour of $g(p_0)$ we deduce $\sum \mathrm{Res}\, g(p_0) = 0$, so that (5.72) may be cast in the alternative form

$$S = \sum \frac{\mathrm{Res}\, g(p_0)}{e^{\beta(p_0-\mu)}+1} = \sum \mathrm{Res}\, g(p_0)\tilde{f}(p_0 - \mu) \qquad (5.73)$$

where we note once more the rule $f(p_0) \to -\tilde{f}(p_0)$ for going from the bosonic case (3.60) to the fermionic one (5.73). The $T = 0$ limit of (5.73) is

$$\lim_{T=0} S = \sum_{\mathrm{Re}\, p_0 < \mu} \mathrm{Res}\, g(p_0) \qquad (5.74)$$

We limit ourselves to the $\mu \neq 0$ generalization of (5.59), leaving the other cases as exercises. Defining the fermion and antifermion Fermi–Dirac distributions $\tilde{f}_+$ and $\tilde{f}_-$:

$$\tilde{f}_\pm(sE) = \frac{1}{e^{\beta(sE \mp \mu)} + 1} \tag{5.75a}$$

which obey

$$1 - \tilde{f}_+(E) - \tilde{f}_-(-E) = 0 \tag{5.75b}$$

we obtain

$$T \sum_n \tilde{\Delta}_{s_1}(i\omega_n + \mu, E_1)\tilde{\Delta}_{s_2}(i(\omega - \omega_n) - \mu, E_2)$$

$$= -\frac{s_1 s_2}{4E_1 E_2} \frac{1 - \tilde{f}_+(s_1 E_1) - \tilde{f}_-(s_2 E_2)}{i\omega - s_1 E_1 - s_2 E_2} \tag{5.76}$$

This equation leads to

$$T \sum_n \tilde{\Delta}(i\omega_n + \mu, E_1)\tilde{\Delta}(i(\omega - \omega_n) - \mu, E_2)$$

$$= -\frac{1}{4E_1 E_2}\left[\frac{1 - \tilde{n}_+(E_1) - \tilde{n}_-(E_2)}{i\omega - E_1 - E_2} - \frac{\tilde{n}_-(E_1) - \tilde{n}_-(E_2)}{i\omega + E_1 - E_2} \right.$$
$$\left. + \frac{\tilde{n}_+(E_1) - \tilde{n}_+(E_2)}{i\omega - E_1 + E_2} - \frac{1 - \tilde{n}_-(E_1) - \tilde{n}_+(E_2)}{i\omega + E_1 + E_2} \right] \tag{5.77}$$

with $(E > 0!)$

$$\tilde{n}_\pm(E) = \frac{1}{e^{\beta(E \mp \mu)} + 1} \tag{5.78}$$

5.2 Quantization of gauge fields at non-zero temperature

As in the $T = 0$ case, we have to face the problem of gauge invariance. In the computation of Green's functions, there are only minor modifications to the Feynman rules of the $T = 0$ theory: it is enough to transpose to the case of gauge fields what has already been learned from the study of scalar fields. The result of a perturbative calculation, for example that of the self-energy, will nevertheless differ qualitatively from that of the $T = 0$ case because Lorentz invariance is lost. This will be discussed in detail in the following chapter.

The new features which appear in the calculation of the partition function are more unexpected, as one discovers that the Faddeev–Popov (F–P) ghosts play a surprisingly important role. We shall begin with a brief overview of the partition function, referring the reader to Kapusta's book for further details.

5.2.1 The partition function

If one attempts to compute the partition function from the standard expression Tr [exp($-\beta H$)], it is clear that the result cannot be gauge-independent (by 'Hamiltonian' we mean the quantity which can be derived from the effective Lagrangian after fixing the gauge for quantization). In a physical gauge, there are two transverse degrees of freedom, and in the free-field case, one recovers the correct formula for blackbody radiation at once. In a covariant gauge, in addition to transverse photons, one finds longitudinal and time-like photons, and a naïve calculation gives twice the free energy of the blackbody radiation, which is obviously incorrect. The origin of the error lies in the fact that F–P ghosts contribute to the partition function a term which exactly cancels the contribution of longitudinal and time-like photons. One is familiar with F–P ghosts in non-Abelian gauge theories, but it is surprising to find that they are also necessary in the QED case, where they are usually neglected. However, one must remember that F–P ghosts do occur, even in QED, in non-linear gauges, and in linear gauges they may be neglected for the calculation of Green's functions only because their contribution yields a multiplicative constant in the generating functional. This constant may not be neglected in the calculation of the partition function, and it cancels the unwanted contribution of the unphysical degrees of freedom precisely.

For the sake of simplicity, we first examine the QED (Abelian) case. Let us call $A_\mu(x)$ the gauge field and

$$F_{\mu\nu} = \partial_\mu A_\nu - \partial_\nu A_\mu \tag{5.79}$$

the field strength tensor. For a free field, the action reads

$$S = -\frac{1}{4} \int d^4x \, F_{\mu\nu} F^{\mu\nu} \tag{5.80}$$

This action is invariant under gauge transformations:

$$A_\mu(x) \rightarrow A'_\mu(x) = A_\mu(x) + \partial_\mu \omega(x) \tag{5.81}$$

where $\omega(x)$ is a scalar function which parametrizes the gauge transformations. The momenta conjugate to the space components of $A^i(x)$ are, up to a sign, the components $E_i(x) = \mathbf{E}_i(x)$ of the electric field

$$\pi_i = -E_i = -F_{0i} \tag{5.82}$$

while the magnetic field $\mathbf{B}(x)$ is

$$B_i = \varepsilon_{ijk} \partial_j A^k \tag{5.83}$$

We work in an axial gauge, $A^3 = 0$ to be specific. The momenta π_1 and π_2 are independent variables; E_3 is not an independent variable, but it is a function of π_1 and π_2, which may be computed from Gauss's law $\nabla \cdot \mathbf{E} = 0$. There are thus two dynamical variables A^1 and A^2 with conjugate momenta π_1 and π_2. We define $\pi_3 = -E_3(\pi_1, \pi_2)$, but π_3 is not to be interpreted as a conjugate momentum. The partition function is written as a Hamiltonian path integral (exercise 2.1):

$$Z = \int \mathscr{D}(\pi_1, \pi_2) \int_{A^i(0) = A^i(\beta)} \mathscr{D}(A^1, A^2) \exp \left[\int_0^\beta \mathrm{d}^4 x \right.$$
$$\left. \times \left(i\pi_1 \partial_\tau A^1 + i\pi_2 \partial_\tau A^2 - \mathscr{H} \right) \right] \tag{5.84}$$

where we have used our notational convention

$$\int_0^\beta \mathrm{d}^4 x = \int_0^\beta \mathrm{d}\tau \int \mathrm{d}^3 x \tag{5.85}$$

while the Hamiltonian density $\mathscr{H}$ is

$$\mathscr{H} = \frac{1}{2}(\mathbf{E}^2 + \mathbf{B}^2) = \frac{1}{2}\left(\pi_1^2 + \pi_2^2 + E_3^2(\pi_1, \pi_2) + \mathbf{B}^2 \right) \tag{5.86}$$

Equation (5.84) is then transformed by using

$$1 = \int \mathscr{D}\pi_3 \, \delta(\pi_3 + E_3(\pi_1, \pi_2)) \tag{5.87}$$

and

$$\delta(\pi_3 + E_3(\pi_1, \pi_2)) = \delta(\nabla \cdot \pi) \det \left(\frac{\delta(\nabla \cdot \pi)}{\delta \pi_3} \right)$$
$$= \det \left[\partial_3 \delta^{(3)}(\mathbf{x} - \mathbf{y}) \right] \delta(\nabla \cdot \pi) \tag{5.88}$$

In the following step, one inserts an integral representation of $\delta(\nabla \cdot \pi)$:

$$\delta(\nabla \cdot \pi) = \int \mathscr{D}A_4 \exp \left[i \int_0^\beta \mathrm{d}^4 x \, A_4 \left(\nabla \cdot \pi \right) \right] \tag{5.89}$$

where $A_4 = iA_0$: from now on we work in Euclidean space: $x_\mu = (\mathbf{x}, x_4) = (\mathbf{x}, \tau)$, $A_\mu = (\mathbf{A}, A_4)$. Performing the π-integration, we are left with

$$Z = \int \mathscr{D}(A_1, A_2, A_4) \det \left[\partial_3 \delta^{(3)}(\mathbf{x} - \mathbf{y}) \right]$$
$$\times \exp \left[\int_0^\beta \mathrm{d}^4 x \left(\frac{1}{2}(i\partial_\tau \mathbf{A} - i\nabla A_4)^2 - \frac{1}{2}\mathbf{B}^2 \right) \right] \tag{5.90}$$

where $\mathbf{A} = (A_1, A_2, 0)$. The A-integration is rendered more aesthetic by inserting

$$1 = \int \mathscr{D}A_3 \, \delta(A_3) \tag{5.91}$$

which yields

$$Z = \int_{A_\mu(0)=A_\mu(\beta)} \mathcal{D}A_\mu \delta(A_3) \det\left[\partial_3 \delta^{(3)}(\mathbf{x}-\mathbf{y})\right]$$

$$\times \exp\left[-\int_0^\beta d^4x \mathcal{L}_E(x)\right] \tag{5.92}$$

where $\mathcal{L}_E$ is the (positive definite) Euclidean Lagrangian density

$$\mathcal{L}_E = \frac{1}{4} F_{\mu\nu} F_{\mu\nu} = \frac{1}{2}(\mathbf{E}^2 + \mathbf{B}^2) \tag{5.93}$$

The final result (5.92) can also be obtained from the Faddeev–Popov construction which we now recall in the general, non-Abelian case. We use Hermitian generators for the Lie algebra of the gauge group (the group index a can appear either as a subscript or as a superscript, as dictated by convenience of writing). These generators are denoted by t_a in the fundamental representation and by T_a in the adjoint representation; they obey the commutation relations

$$[t_a, t_b] = if_{abc}t_c \qquad [T_a, T_b] = if_{abc}T_c \tag{5.94}$$

where the f_{abc}s are the structure constants of the group and $(T_a)_{bc} = -if_{abc}$. The infinitesimal gauge transformations read

$$\delta A_\mu^a(x) = D_\mu^{ab}(x)\omega^b(x) = (\partial_\mu^x \delta^{ab} + gf^{abc}A_\mu^c(x))\omega^b(x) \tag{5.95}$$

where D_{ab}^μ is the covariant derivative, g is the coupling constant and the functions $\omega_c(x)$ are infinitesimal scalar functions which parametrize the gauge transformations. Starting from the manifestly gauge-invariant, but ill-defined, expression for Z:

$$Z = \int \mathcal{D}A_\mu^a \exp\left[-\int_0^\beta d^4x \mathcal{L}_E(x)\right] \tag{5.96}$$

which is a straightforward generalization of the scalar field result, one inserts, for a gauge condition $f^a(A_\mu^b(x)) = 0$

$$1 = \Delta_f(A)\Delta_f^{-1}(A)$$

$$\Delta_f^{-1}(A) = \int \prod_x dh(x) \prod_{x,a} \delta\left[f^a({}^hA(x))\right] \tag{5.97}$$

Here $h(x)$ is the x-dependent group element and ${}^hA(x)$ is the gauge transform of $A(x)$. It is then possible to integrate over $h(x)$, namely, over gauge orbits, with the result

$$Z = \int \mathcal{D}A_\mu^a \Delta^f(A) \prod_{x,a} \delta(f_a(A(x))) \exp\left[-\int_0^\beta d^4x \mathcal{L}_E(x)\right] \tag{5.98}$$

$\Delta_f(A)$ is the F–P determinant

$$\Delta_f(A) = \det M_f^{ab}(x, y) = \det \left[\frac{\delta f^a({}^{\omega}A(x))}{\delta \omega^b(y)} \right]$$

$$= \det \left[\frac{\partial f^a}{\partial A_{\mu}^c(x)} D_{\mu}^{cb}(x) \delta^{(4)}(x - y) \right] \qquad (5.99)$$

Expression (5.98) for Z is gauge-invariant by construction; in the case of the axial gauge $A_3 = 0$ it reduces to (5.92) as

$$\frac{\delta f^a(x)}{\delta A_{\mu}^c(z)} = \delta_{\mu 3} \delta^{ac} \delta^{(4)}(x - z) \qquad (5.100)$$

In the general case the determinant in (5.99) depends on A_{μ}^a; it can be expressed as a functional integral over Grassmann fields $\eta(x)$ and $\bar{\eta}(x)$ by (5.12)

$$\det M_f^{ab}(x, y) = \int_{\text{periodic}} \mathcal{D}(\eta, \bar{\eta}) \exp \left[- \int_0^{\beta} d^4x\, d^4y \right.$$

$$\left. \times \bar{\eta}_a(x) M_f^{ab}(x, y) \eta_b(y) \right] \qquad (5.101)$$

The F–P determinant is defined on the space of periodic funtions; thus, in spite of their fermionic character, the Grassmann fields $\eta(x)$ and $\bar{\eta}(x)$ obey periodic boundary conditions $\eta(0) = \eta(\beta)$ and $\bar{\eta}(0) = \bar{\eta}(\beta)$.

Let us illustrate the F–P construction in the case of covariant (or Lorentz) gauges:

$$f^a(A(x)) = \partial_{\mu} A_{\mu}^a(x) - \lambda_a(x) = 0 \qquad (5.102)$$

where the λs are scalar functions; note that Δ_f is independent of $\lambda_a(x)$. One inserts in Z a constant in the form of

$$\int \mathcal{D}\lambda(x) \exp \left[-\frac{1}{2\xi} \int_0^{\beta} d^4x\, \lambda_a^2(x) \right] \qquad (5.103)$$

where ξ is the gauge-fixing parameter; this leads to the following form of Z:

$$Z = \int_{\text{periodic}} \mathcal{D}(A_{\mu}^a, \eta_b, \bar{\eta}_c) \exp \left[- \int_0^{\beta} d^4x\, \mathcal{L}_{\text{eff}}(x) \right] \qquad (5.104)$$

where the effective Lagrangian density is

$$\mathcal{L}_{\text{eff}}(x) = \frac{1}{4} F_{\mu\nu}^a(x) F_{\mu\nu}^a(x) + \frac{1}{2\xi} (\partial_{\mu} A_{\mu}^a(x))^2$$

$$+ \bar{\eta}_a(x) \left[\partial^2 \delta_{ab} + g f_{abc} A_{\mu}^c(x) \partial_{\mu} \right] \eta_b(x) \qquad (5.105)$$

and $\partial^2 = \sum_{\mu=1}^{4} \partial_{\mu}^2$.

Let us now revert to the QED case (or the non-Abelian case with $g = 0$); for the sake of simplicity, we work in Feynman gauge $\xi = 1$:

$$\mathscr{L}_{\text{eff}}(x) = -\frac{1}{2} A_\mu(x) \partial^2 A_\mu(x) + \bar{\eta}(x) \partial^2 \eta(x) \qquad (5.106)$$

The partition function is proportional to

$$(\det \partial^2)^{-2} \det \partial^2 = (\det \partial^2)^{-1} \qquad (5.107)$$

where the first factor on the LHS of (5.107) comes from the four degrees of freedom of the photon and the second factor from the F–P ghosts. Equation (5.107) shows explicitly how the contribution from the F–P ghosts cancels the unwanted contribution from the unphysical degrees of freedom of the photon. In order to recover the formula for blackbody radiation, one should evaluate the proportionality constant which is hidden in (5.107); this constant can be obtained through a careful evaluation of the functional integral (exercise 2.2).

5.2.2 *The gauge boson propagator in linear gauges*

As was already mentioned, the Feynman rules for the calculation of Green's functions follow immediately from the $T = 0$ rules, provided one transposes to gauge fields what has been learned in chapter 3 for scalar fields. More precisely, in the imaginary-time formalism, one uses the Feynman rules of the Euclidean theory, the only modification with respect to the $T = 0$ case being the replacements $-p_4 \to \omega_n = 2\pi n/\beta$ and $\int \mathrm{d}p_4 \to T \sum_n$. In real time, one uses the rules which have been derived from the time paths of chapter 3 and the matrix form of the propagator; however there are some interesting features of the unphysical degrees of freedom which will be examined later on. At zero temperature, covariant gauges have a definite advantage over non-covariant gauges such as the Coulomb or axial gauges. Calculations are simplified considerably due to Lorentz invariance, and, furthermore, the renormalization program can be implemented in practice only in covariant gauges. At non-zero temperature, Lorentz invariance is broken because the heat bath defines a privileged reference frame, and renormalization problems are of secondary importance, so that non-covariant gauges may present useful alternatives to covariant gauges. Since non-covariant gauges are not as familiar as covariant gauges, we devote this section to a short review of their properties, and in particular those of the propagator, in linear gauges of the form $f^\mu A_\mu^a(x) = 0$, where f^μ is a vector built from a fixed vector and derivatives (more generally we could take f^μ as a matrix in group space). These gauges include the Lorentz gauge $\partial_\mu A_a^\mu = 0$, the Coulomb gauge $\partial_i A_a^i = 0$ and axial gauges $n_\mu A_a^\mu = 0$. We first consider the $T = 0$ case and work in Minkowski space; for simplicity, group indices are suppressed, except when explicitly needed.

From now on, in order to avoid any confusion between three- and four-momenta, we shall denote four-momenta by upper case letters while energies and three-momenta are denoted by the corresponding lower case letters:

$$Q_\mu = (q_0, \mathbf{q}) \tag{5.108}$$

We decompose f_μ into a vector parallel to Q_μ, where Q_μ is the gauge boson momentum, and a vector n_μ orthogonal to Q_μ

$$f_\mu = n_\mu + \frac{Q \cdot f}{Q^2} Q_\mu \tag{5.109}$$

with

$$n_\mu = P_{\mu\nu} f^\nu = \left(g_{\mu\nu} - \frac{Q_\mu Q_\nu}{Q^2} \right) f^\nu \tag{5.110}$$

where $P_{\mu\nu}$ is the projector on a plane orthogonal to Q_μ. Some care is necessary in obtaining the covariant gauge as a limiting case,[*] as the first term on the RHS of (5.109) is absent and $f_\mu = Q_\mu$. Let us introduce the tensors $A_{\mu\nu}$, $B_{\mu\nu}$, $C_{\mu\nu}$, $E_{\mu\nu}$:

$$
\begin{aligned}
B_{\mu\nu} &= \frac{n_\mu n_\nu}{n^2} & C_{\mu\nu} &= n_\mu Q_\nu + n_\nu Q_\mu \\
E_{\mu\nu} &= \frac{Q_\mu Q_\nu}{Q^2} & A_{\mu\nu} &= P_{\mu\nu} - B_{\mu\nu}
\end{aligned}
\tag{5.111}
$$

We may then decompose $f_\mu f_\nu$ as follows:

$$f_\mu f_\nu = n^2 B_{\mu\nu} + \frac{Q \cdot f}{Q^2} C_{\mu\nu} + \frac{(Q \cdot f)^2}{Q^2} E_{\mu\nu} \tag{5.112}$$

If we use a gauge-fixing term of the form

$$\mathscr{L}_{\text{gf}} = -\frac{1}{2\xi} \int d^4 x (f^\mu A_\mu)^2 \tag{5.113}$$

the bare inverse propagator $D_{F,\mu\nu}^{-1}$ reads

$$-i D_{F,\mu\nu}^{-1}(Q) = (Q^2 g_{\mu\nu} - Q_\mu Q_\nu) + f_b B_{\mu\nu} + f_c C_{\mu\nu} + f_e E_{\mu\nu} \tag{5.114}$$

where we have defined the gauge parameters f_b, f_c, f_e:

$$f_b = \frac{n^2}{\xi} \qquad f_c = \frac{(Q \cdot f)}{\xi Q^2} \qquad f_e = \frac{(Q \cdot f)^2}{\xi Q^2} \tag{5.115}$$

---

[*] In order to obtain the covariant gauge as a limiting case without difficulty, a better procedure would be to introduce a fixed vector $\tilde{n}_\mu$ and to write f_μ as a linear combination of $\tilde{n}_\mu$ and Q_μ.

Note that for a covariant gauge $f_b = f_c = 0$, $f_e = Q^2/\xi$. The gauge parameters obey the relation

$$f_b f_e = n^2 Q^2 f_c^2 \tag{5.116}$$

The physical interpretation of (5.116) is that the gauge-fixing term does not modify the transverse part of the propagator. In order to find the bare propagator $D^F_{\mu\nu}$ we have to compute the inverse of

$$X_{\mu\nu} = \alpha A_{\mu\nu} + \beta B_{\mu\nu} + \gamma C_{\mu\nu} + \eta E_{\mu\nu} \tag{5.117}$$

A straightforward, but cumbersome, calculation gives

$$X^{-1}_{\mu\nu} = \alpha' A_{\mu\nu} + \beta' B_{\mu\nu} + \gamma' C_{\mu\nu} + \eta' E_{\mu\nu} \tag{5.118}$$

with

$$\alpha' = \frac{1}{\alpha} \quad \beta' = \frac{\eta}{\delta} \quad \gamma' = -\frac{\gamma}{\delta} \quad \eta' = \frac{\beta}{\delta} \quad \delta = \beta\eta - Q^2 n^2 \gamma^2 \tag{5.119}$$

We then obtain

$$iD^F_{\mu\nu} = \frac{1}{Q^2}(A_{\mu\nu} + B_{\mu\nu}) - \frac{f_c}{f_e Q^2} C_{\mu\nu} + \frac{f_b + Q^2}{f_e Q^2} E_{\mu\nu} \tag{5.120}$$

The three important cases are

(i) The Lorentz gauge $\partial^\mu A_\mu = 0$:

$$iD^F_{\mu\nu} = \frac{1}{Q^2}(P_{\mu\nu} + \xi E_{\mu\nu}) \tag{5.121}$$

(ii) The Coulomb gauge $\partial^i A_i = 0$:

$$iD^F_{\mu\nu} = -\frac{g_{0\mu} g_{0\nu}}{q^2} + \frac{1}{Q^2} A_{\mu\nu} + \frac{\xi Q^2}{(q^2)^2} E_{\mu\nu} \tag{5.122}$$

(iii) The temporal axial gauge $A_0 = 0$

$$iD^F_{\mu\nu} = \frac{1}{Q^2} A_{\mu\nu} - \frac{\hat{q}_i \hat{q}_j}{q_0^2} + \frac{\xi Q^2}{q_0^2} E_{\mu\nu} \tag{5.123}$$

In the last two cases, the vector n_μ has been defined by

$$n_\mu = P_{\mu\nu} \tilde{n}^\nu \quad \tilde{n}_\mu = (-q_0, \mathbf{0}) : \text{Coulomb} \quad \tilde{n}_\mu = (1, \mathbf{0}) : \text{Axial} \tag{5.124}$$

so that in both cases

$$A_{00} = A_{0i} = 0 \quad A_{ij} = g_{ij} + \hat{q}_i \hat{q}_j \tag{5.125}$$

Then $A_{\mu\nu}$ is, up to a sign, the projector $P^T_{\mu\nu}$ on the plane orthogonal to Q_μ and $\mathbf{q}$:

$$A_{\mu\nu} = -P^T_{\mu\nu} \quad P^T_{00} = P^T_{0i} = 0 \quad P^T_{ij} = \delta_{ij} - \hat{\mathbf{q}}_i \hat{\mathbf{q}}_j \tag{5.126}$$

Let us now turn to the interacting case; we define self-energies Π_a, Π_b, Π_c, Π_e by

$$i\Pi_{\mu\nu} = (D^{-1})_{\mu\nu} - (D_F^{-1})_{\mu\nu}$$
$$= i\left(\Pi_a A_{\mu\nu} + \Pi_b B_{\mu\nu} + \Pi_c C_{\mu\nu} + \Pi_e E_{\mu\nu}\right) \qquad (5.127)$$

From (5.119) the dressed propagator takes the form

$$iD_{\mu\nu} = \frac{1}{Q^2 + \Pi_a}A_{\mu\nu} + \frac{1}{b(Q)}B_{\mu\nu} - \frac{1}{b(Q)}\frac{f_c + \Pi_c}{f_e + \Pi_e}C_{\mu\nu}$$
$$+ \frac{1}{b(Q)}\frac{Q^2 + \Pi_b + f_b}{f_e + \Pi_e}E_{\mu\nu} \qquad (5.128)$$

with

$$b(Q) = Q^2 + \Pi_b + f_b - n^2 Q^2 \frac{(f_c + \Pi_c)^2}{f_e + \Pi_e} \qquad (5.129)$$

The four functions $\Pi_a, ..., \Pi_e$ are not all independent; in the general, non-Abelian, case they are constrained by BRS invariance. Let us recall the definition of the BRS transformation:

$$\delta_{\text{BRS}}A_a^\mu(x) = D_{ab}^\mu(x)\eta_b(x)\zeta \qquad (5.130a)$$

$$\delta_{\text{BRS}}\overline{\eta}_a(x) = \frac{1}{\xi}f_a(A)\zeta \qquad (5.130b)$$

$$\delta_{\text{BRS}}\eta_a(x) = -\frac{g}{2}f_{abc}\eta_b(x)\eta_c(x)\zeta \qquad (5.130c)$$

where ζ is the BRS parameter, which is an x-independent Grassmann variable anticommuting with η, $\overline{\eta}$ and commuting with A. The transformation (5.130) leaves invariant the effective action

$$S_{\text{eff}} = -\int d^4x\left[\frac{1}{4}F_{\mu\nu}^a(x)F_a^{\mu\nu}(x) + \frac{1}{2\xi}(f_\mu A_a^\mu(x))^2\right.$$
$$\left. + \overline{\eta}_a(x)f_\mu(\partial_x^\mu\delta_{ab} + gf_{abc}A_c^\mu(x))\eta_b(x)\right] \qquad (5.131)$$

and the integration measure $\mathscr{D}(A_a^\mu, \eta_a, \overline{\eta}_a)$. From the obvious relation $\langle T(A_a^\mu(x)\overline{\eta}_b(y))\rangle = 0$ we obtain for our linear gauge condition

$$0 = \frac{\delta_{\text{BRS}}}{\delta\zeta}\langle T(f_\mu A_a^\mu(x)\overline{\eta}_b(y))\rangle = \frac{1}{\xi}\langle T(f_\mu A_a^\mu(x)f_\nu A_b^\nu(y))\rangle$$
$$- \langle T(f_\mu D_{ac}^\mu(x)\eta_c(x)\overline{\eta}_b(y))\rangle \qquad (5.132)$$

where $\langle T(...)\rangle$ is a formal way of writing an average of products of fields taken with the weight $\exp(iS_{\text{eff}})$. The functional derivative of the effective action (5.131) with respect to $\overline{\eta}_a(x)$ gives

$$\frac{\delta S_{\text{eff}}}{\delta\overline{\eta}_a(x)} = -f_\mu D_{ab}^\mu(x)\eta_b(x) \qquad (5.133)$$

An equation of motion follows from (5.133):

$$\left\langle T\left(\frac{\delta S_{\text{eff}}}{\delta\bar{\eta}_a(x)}\bar{\eta}_b(y)\right)\right\rangle = -i\delta_{ab}\delta^{(4)}(x-y) \tag{5.134}$$

We use (5.134) to transform the RHS of (5.132) and obtain in momentum space

$$f^\mu f^\nu D_{\mu\nu} = f^\mu f^\nu D^F_{\mu\nu} = -i\xi \tag{5.135}$$

because the second term in (5.132) is, from (5.134), the same as in the non-interacting case. At zero temperature, in covariant gauges, (5.135) together with Lorentz invariance allows us to derive the stronger relation

$$Q^\mu D_{\mu\nu} = Q^\mu D^F_{\mu\nu} = \frac{-i\xi\, Q_\nu}{Q^2} \tag{5.136}$$

but this relation does not generalize at non-zero T. From (5.135) one derives a non-linear relation between self-energy functions:

$$\Pi_e(Q^2 + \Pi_b) - Q^2 n^2 \Pi_c^2 = 0 \tag{5.137}$$

In the QED (Abelian) case, stronger relations are obtained from

$$0 = \frac{\delta_{\text{BRS}}}{\delta\zeta}\langle T(A_\mu(x)\bar{\eta}(y))\rangle$$

$$= \langle T(\partial_\mu\eta(x)\bar{\eta}(y))\rangle + \frac{1}{\xi}\langle T(A_\mu(x)f^\nu A_\nu(y))\rangle \tag{5.138}$$

and the fact that the ghost field is a free field. One obtains

$$f^\mu D_{\mu\nu} = f^\mu D^F_{\mu\nu} = \frac{-i\xi\, Q_\nu}{Q\cdot f} \tag{5.139}$$

From this relation one obtains $\Pi_c = \Pi_e = 0$ so that $\Pi_{\mu\nu}$ is transverse:

$$Q^\mu\Pi_{\mu\nu} = 0 \tag{5.140}$$

A simpler way to derive (5.140) is to use current conservation. Furthermore, we recall that in QED $\Pi_{\mu\nu}$ is gauge-fixing independent. Both (5.140) (except at $T=0$ in the Lorentz gauge) and the gauge independence of $\Pi_{\mu\nu}$ fail to hold in the non-Abelian case.

5.2.3 The gauge boson propagator at non-zero temperature

The preceding results are immediately generalized at non-zero temperature, provided one uses in the gauge condition only the four-vector $\tilde{n}_\mu$ which defines the heat bath rest frame. In what follows, we always make the choice $\tilde{n}_\mu = (1,\mathbf{0})$. If a vector other than $\tilde{n}_\mu$ is used to define the gauge condition, then the form of the propagator is more complicated. Our choice for $\tilde{n}_\mu$ includes the Lorentz gauge, the Coulomb gauge and the time axial gauge. Let us write the corresponding propagators first in the imaginary-time (Euclidean space) formalism:

(i) Covariant gauge $\partial_\mu A_\mu = 0$ (remember that $Q^2 = q^2 + q_4^2$):

$$D^F_{\mu\nu} = \frac{1}{Q^2}\left(\delta_{\mu\nu} - (1-\xi)\frac{Q_\mu Q_\nu}{Q^2}\right) \tag{5.141}$$

(ii) Coulomb gauge $\partial_i A_i = 0$:

$$D^F_{\mu\nu} = \frac{\delta_{4\mu}\delta_{4\nu}}{q^2} + \frac{1}{Q^2}P^T_{\mu\nu} + \frac{\xi Q^2}{q^4}\frac{Q_\mu Q_\nu}{Q^2} \tag{5.142}$$

The usual Coulomb gauge is recovered by setting $\xi = 0$.

(iii) Temporal axial gauge $A_4 = 0$:

$$D^F_{\mu\nu} = \frac{1}{Q^2}P^T_{\mu\nu} + \frac{1}{q_4^2}\hat{q}_i\hat{q}_j + \frac{\xi Q^2}{q_4^2}\frac{Q_\mu Q_\nu}{Q^2} \tag{5.143}$$

The usual temporal axial gauge is recovered by setting $\xi = 0$. At first sight the temporal axial gauge appears to be a nice gauge to work with, but there are problems due to the fact that the condition $A_4 = 0$ is not compatible with the periodicity of the field.

Let us now turn to the real-time formalism. A straightforward generalization of the results of chapter 3 allows us to write the gauge field propagator as a matrix:

$$D_{\mu\nu}(x) = \begin{pmatrix} \langle T(A_\mu(x)A_\nu(0))\rangle & \langle A_\mu(x^0 + i\sigma, \mathbf{x})A_\nu(0)\rangle \\ \langle A_\mu(x^0 - i\sigma, \mathbf{x})A_\nu(0)\rangle & \langle \tilde{T}(A_\mu(x)A_\nu(0))\rangle \end{pmatrix} \tag{5.144}$$

where σ is the time contour parameter of chapter 3 and $\tilde{T}$ the anti-time-ordering product. For the sake of simplicity, we shall write down all formulae in the Feynman gauge ($\xi = 1$ in (5.141)). In momentum space we have, from (3.93),

$$(D^F_{\mu\nu})_{11} = (D^F_{\mu\nu})^*_{22} = -g_{\mu\nu}\left[\frac{i}{Q^2 + i\eta}\right.$$
$$\left. + n(q_0)2\pi\delta(Q^2)\right] \tag{5.145a}$$

$$(D^F_{\mu\nu})_{12} = -g_{\mu\nu}e^{\sigma q_0}f(q_0)2\pi\,\varepsilon(q_0)\,\delta(Q^2) \tag{5.145b}$$

$$(D^F_{\mu\nu})_{21} = -g_{\mu\nu}e^{-\sigma q_0}(1 + f(q_0))\,2\pi\,\varepsilon(q_0)\,\delta(Q^2) \tag{5.145c}$$

where we recall that

$$f(q_0) = \frac{1}{e^{\beta q_0} - 1} \qquad n(q_0) = \frac{1}{e^{\beta|q_0|} - 1} \tag{5.146}$$

In a non-Abelian gauge theory one should also give the expression for the ghost propagator which is derived in a straightforward way from (3.93). In this approach, the unphysical components of the gauge fields, as well as the ghost propagator, acquire a thermal part. However, the thermal part of the propagator is proportional to a δ-function which puts the particle on shell: the thermal part represents the absorption of particles from the heat bath and the stimulated emission of particles into it. If one computes a partition function or an expectation value by taking the trace over physical states $|i>$:

$$Z = \sum_i < i|e^{-\beta H}|i > \qquad (5.147a)$$

$$\langle A \rangle = Z^{-1} \sum_i < i|e^{-\beta H}A|i > \qquad (5.147b)$$

the thermal part of the propagator is directly associated with the particles in the physical states $|i>$. Consequently, the physical degrees of freedom acquire a thermal part while the unphysical degrees of freedom never come to thermal equilibrium. This intuitive argument may be justified by going back to perturbation theory in the interaction picture and Wick's theorem. Thus, in the Feynman gauge and with the choice $\sigma = 0$, the free gauge field propagator (5.145) may be written by separating out a $T = 0$ part:

$$D^F_{\mu\nu}|_{T=0} = -g_{\mu\nu} \begin{pmatrix} i/(Q^2 + i\eta) & 2\pi\theta(-q_0)\delta(Q^2) \\ 2\pi\theta(q_0)\delta(Q^2) & -i/(Q^2 - i\eta) \end{pmatrix} \qquad (5.148a)$$

and a $T \neq 0$ part

$$D^F_{\mu\nu}|_{T\neq 0} = P^T_{\mu\nu} \, 2\pi \, \delta(Q^2) \, n(q_0) \begin{pmatrix} 1 & 1 \\ 1 & 1 \end{pmatrix} \qquad (5.148b)$$

In that case the ghost propagator of course has no thermal part. We emphasize that there is no contradiction between the conventional approach of (5.145) and the approach which is summarized in (5.148). In the conventional approach, the thermal contribution from the unphysical degrees of freedom is cancelled by the thermal part of the ghosts, as we saw in the calculation of the partition function. In the alternative approach, both contributions are set to zero from the beginning. This observation allows interesting simplifications in the calculation of Feynman diagrams.

5.3 Real photon and lepton pair production

In section 4.4 we studied the production of weakly interacting scalar particles from a heat bath. We now generalize these results to the realistic case of photon and lepton pair production from a quark–gluon plasma. The photons and leptons are assumed to interact weakly with the quarks of the heat bath: they are not thermalized and their interactions are treated to first order in the electron charge e.

5.3.1 Real photon production

The S-matrix element for the transition $(i) \rightarrow (f, \gamma)$ reads

$$S_{fi}^{(\lambda)}(Q) = -ie \int d^4x \, e^{iQ \cdot x} \varepsilon_\mu^{(\lambda)}(Q) < f|j^\mu(x)|i > \qquad (5.149)$$

where Q is the photon four-momentum, $\varepsilon_\mu^{(\lambda)}(Q)$ is its polarization four-vector and $j^\mu(x)$ the electromagnetic current. Taking the square of (5.149), summing over initial and final states and polarization indices we get

$$\frac{1}{\Omega} \frac{1}{Z(\beta)} \sum_{i,f,\lambda} e^{-\beta E_i} |S_{fi}^{(\lambda)}(Q)|^2$$

$$= -e^2 g^{\mu\nu} \int d^4x \, e^{iQ \cdot x} \langle j_\mu(0) j_\nu(x) \rangle_\beta \qquad (5.150)$$

where Ω is the space-time volume where the interaction takes place. The Fourier transform in (5.150) is denoted by $\tilde{\Pi}_{\mu\nu}^<(Q)$, where $\tilde{\Pi}_{\mu\nu}(Q)$ is the Fourier transform of the time-ordered current correlation function. To lowest order in e, but to all orders in g, $-ie^2 \tilde{\Pi}_{\mu\nu}(Q)$ is equal to the one-particle irreducible photon self-energy $\Pi_{\mu\nu}(Q)$. We obtain for the production rate per unit of time and volume

$$\Gamma = \frac{1}{\Omega} \frac{1}{Z(\beta)} \int \sum_{i,f,\lambda} e^{-\beta E_i} |S_{fi}^{(\lambda)}(Q)|^2 \frac{d^3q}{2q_0(2\pi)^3} \qquad (5.151)$$

so that the differential rate may be expressed in terms of the photon self-energy:

$$q_0 \frac{d\Gamma}{d^3q} = -\frac{g^{\mu\nu} \Pi_{\mu\nu}^<(q_0, \mathbf{q})}{2(2\pi)^3} = \frac{g^{\mu\nu} \text{Im} \, \Pi_{\mu\nu}(q_0, \mathbf{q})}{(2\pi)^3(e^{\beta q_0} - 1)} \qquad (5.152)$$

This result is valid to order e^2 and to all orders in g.

5.3.2 Lepton pair production

The lepton pair production rate is derived along the same lines. The S-matrix element for the transition $(i) \rightarrow (f, (l_1, l_2))$, where (l_1, l_2) represents the lepton pair, reads

$$S_{fi}(P_1, P_2) = -\frac{ie^2}{Q^2} \bar{u}(\mathbf{p}_1) \gamma^\mu v(\mathbf{p}_2) \int d^4x \, e^{iQ \cdot x} < f|j_\mu(x)|i > \qquad (5.153)$$

where $Q = P_1 + P_2$ is the pair four-momentum. This leads to the production rate

$$d\Gamma = \frac{e^4}{Q^4} \, l^{\mu\nu} \, \tilde{\Pi}^<_{\mu\nu}(Q) \frac{d^3 p_1}{2E_1(2\pi)^3} \frac{d^3 p_2}{2E_2(2\pi)^3} \qquad (5.154)$$

The leptonic tensor $l_{\mu\nu}$ is given in the limit of zero lepton masses by

$$l_{\mu\nu} = \sum_{\text{spins}} (\bar{u}(\mathbf{p}_1)\gamma_\mu v(\mathbf{p}_2))(\bar{u}(\mathbf{p}_1)\gamma_\nu v(\mathbf{p}_2))^*$$

$$= 4 [p_{1\mu} p_{2\nu} + p_{1\nu} p_{2\mu} - (p_1 \cdot p_2)g_{\mu\nu}] \qquad (5.155)$$

Using the identification between $e^2 \tilde{\Pi}_{\mu\nu}(Q)$ and the photon self-energy: $\Pi_{\mu\nu}(Q)$ and integrating over final lepton momenta yields for the rate

$$\Gamma = \frac{e^2}{Q^4} \int \frac{d^4 Q}{(2\pi)^4} \Pi^{\mu\nu}(Q) L_{\mu\nu}(Q) \qquad (5.156)$$

with

$$L_{\mu\nu}(Q) = \int \frac{d^3 p_1}{2E_1(2\pi)^3} \frac{d^3 p_2}{2E_2(2\pi)^3} (2\pi)^4 \delta^{(4)}(Q - P_1 - P_2) l_{\mu\nu}$$

$$= \frac{1}{6\pi}(Q_\mu Q_\nu - Q^2 g_{\mu\nu}) \qquad (5.157)$$

We finally express the differential rate for pair production in terms of the photon self-energy

$$\frac{d\Gamma}{d^4 Q} = -\frac{\alpha g^{\mu\nu} \Pi^<_{\mu\nu}(Q)}{24\pi^4 Q^2} = \frac{\alpha g^{\mu\nu} \operatorname{Im} \Pi_{\mu\nu}(Q)}{12\pi^4 Q^2(e^{\beta q_0} - 1)} \qquad (5.158)$$

Let us conclude with the following two remarks

(i) When the lepton masses are not neglected, (5.157) becomes

$$L_{\mu\nu}(Q) = \frac{1}{6\pi}\left(1 + \frac{2m^2}{Q^2}\right)\left(1 - \frac{4m^2}{Q^2}\right)^{1/2}(Q_\mu Q_\nu - Q^2 g_{\mu\nu}) \qquad (5.159)$$

(ii) We may try to work to all orders in e by using the relation between the photon self-energy Π and the current correlation function $\tilde{\Pi}$, which is proportional to the improper (one-particle reducible) photon self-energy

$$e^2 \tilde{\Pi} = i\Pi D D_F^{-1} \qquad (5.160)$$

where D is the photon propagator and D_F is the free propagator. From this relation one derives

$$e^2 \operatorname{Im} \tilde{\Pi}_{\mu\nu} = Q^4 \operatorname{Im} D_{\mu\nu} = \frac{1}{2} Q^4 \rho_{\mu\nu} \qquad (5.161)$$

where $\rho_{\mu\nu}$ is the spectral function of the photon propagator.

References and further reading

A clear introduction to path integrals for fermions can be found in Negele and Orland (1988), chap. 1 or in Brown (1992), chap. 2. The quantization of Dirac fields at $T = 0$ is explained in all textbooks on field theory: see, for example, Itzykson and Zuber (1980), chap. 3 or Le Bellac (1992), chap. 11. Our treatment of gauge field quantization in an axial gauge follows Bernard (1974) and Kapusta (1989), chap. 5, who also examines higher-order perturbative calculations of the partition function. A general discussion of linear gauges may be found in Kobes, Kunstatter and Rebhan (1991). For reviews of the temporal axial gauge, see James and Landshoff (1990) or Leibrandt and Staley (1993). Landshoff and Rebhan (1993a,b) pointed out that one could avoid the thermalization of unphysical degrees of freedom of gauge fields. The general derivation of lepton pair and photon production rates is due to McLerran and Toimela (1985), Weldon (1990) and Gale and Kapusta (1991). Weldon (1991a) proposed the use of (5.161) in order to measure the temperature of the plasma.

Exercises

5.1 Prove (5.13) by applying both sides of the equation on the kets $|0>$ and $|1>$. Prove (5.14) by writing

$$A = \sum_{i,j=0}^{1} A_{ij} |i><j|$$

5.2 Prove (5.42) by using the Fourier decomposition of the free Dirac field, the anticommutation relations (5.32), and (5.43).

5.3 Check (5.53) either by showing that the Fourier coefficients of $\tilde{\Delta}(\tau, E_p)$ are $\tilde{\Delta}(i\omega_n, E_p)$ or by computing (5.52) directly using a contour integration as in exercise 3.1. What is the value of $\tilde{\Delta}(\tau, E_p)$ in the interval $[-\beta, 0]$?

5.4 Compute the first line of (5.61) by using (5.59) and (5.60) and analytically continuing to real values of the external energy. Compare with the expression obtained from the second line of (5.61).

5.5 (a) Show that the inverse Fourier transform of

$$\frac{1}{(\omega_n - i\mu)^2 + E_p^2}$$

is

$$\tilde{\Delta}(\tau) = \frac{1}{2E_p}\left[(1 - \tilde{f}(E_p - \mu))e^{-(E_p - \mu)\tau} \right.$$
$$\left. - \tilde{f}(E_p + \mu)e^{(E_p + \mu)\tau}\right]$$

Use this formula to perform the frequency sum in (5.77).

(b) From the commutation relations (5.67) and the definition

$$S_{\alpha\beta}^{>}(x, x') = \frac{1}{Z(\beta, \mu)} \mathrm{Tr}\left(e^{-\beta(H-\mu Q)}\psi_\alpha(x)\overline{\psi}_\beta(x')\right)$$

show that

$$S_{\alpha\beta}^{>}(t, \mathbf{x}; t', \mathbf{x}') = -e^{-\beta\mu}S_{\alpha\beta}^{<}(t + i\beta, \mathbf{x}; t', \mathbf{x}')$$

and derive the relations

$$S_{\alpha\beta}^{>}(P) = (1 - \tilde{f}(p_0 - \mu))\rho_{\alpha\beta}(P)$$
$$S_{\alpha\beta}^{<}(P) = -\tilde{f}(p_0 - \mu)\rho_{\alpha\beta}(P)$$

(c) Using the representation (3.74) of the propagator suitably generalized to Dirac particles, show that the (11) matrix element of the real time propagator is, for non-zero chemical potential:

$$S_{11}^{F}(P) = (\not{P} + m)\left[\frac{i}{P^2 - m^2 + i\eta} - 2\pi\left(\theta(-p_0)\right.\right.$$
$$\left.\left. + \varepsilon(p_0)\tilde{f}(p_0 - \mu)\right)\delta(P^2 - m^2)\right]$$

(d) Show that the propagator $S_F'(\tau, \mathbf{x}) = e^{\mu\tau}S_F(\tau, \mathbf{x})$ is antiperiodic in imaginary time and that it obeys the correct partial differential equation which generalizes (5.65) to non-zero μ

$$(\gamma^0(\partial_\tau - \mu) - i\gamma \cdot \nabla + m)S_F'(\tau, \mathbf{x}) = \delta^{(4)}(x)$$

Obtain the Matsubara propagator by taking the Fourier transform of S_F'. Note that the boundary conditions are different in questions (a, d) and (b, c).

5.6 Check equation (5.119) by establishing a set of multiplication rules: $A^2 = A$, $A.B = 0$, etc. Check equations (5.122) and (5.123) for the propagator in the Coulomb and time-axial gauges.

5.7 Instead of using (5.57)–(5.60) for the frequency sums, one can use a generalization of (2.58) for the fermion propagator. Show that the free fermion propagator may be written, taking $m = \mu = 0$ for simplicity, as

$$S_F(i\omega_n, \mathbf{p}) = \int_{-\infty}^{\infty} \frac{dp_0}{2\pi} \frac{\not{P}\rho_F(P)}{p_0 - i\omega_n}$$

with

$$\rho_F(P) = 2\pi\varepsilon(p_0)\delta(P^2)$$

Use (5.72) to perform the frequency sums and show that in the boson–fermion case

$$T\sum_n \frac{1}{(p_0 - i\omega_n)(p_0' - i(\omega - \omega_n))} = -\frac{1 + f(p_0) - \tilde{f}(p_0')}{i\omega - p_0 - p_0'}$$

while in the fermion–antifermion case one has

$$T \sum_n \frac{1}{(p_0 - i\omega_n)(p'_0 - i(\omega_n - \omega))} = -\frac{\tilde{f}(p_0) - \tilde{f}(p'_0)}{i\omega - p_0 + p'_0}$$

Show that in QED the electron self-energy at one loop (see section 6.5) may be written with $\mathbf{q} = \mathbf{p} + \mathbf{k}$

$$\Sigma(i\omega, \mathbf{q}) = 2e^2 \int \frac{d^3k}{(2\pi)^3} \int \frac{dp_0}{2\pi} \int \frac{dk_0}{2\pi}$$

$$\times \mathcal{P}\rho_F(P)\rho_F(K)\frac{1 + f(k_0) - \tilde{f}(p_0)}{i\omega - p_0 - k_0}$$

and that the photon self-energy (see section 6.2) is at order one-loop ($\mathbf{q} = \mathbf{k} - \mathbf{k}'$)

$$\Pi_{\mu\nu}(i\omega, \mathbf{q}) = -4e^2 \int \frac{d^3k}{(2\pi)^3} \int \frac{dk_0}{2\pi} \int \frac{dk'_0}{2\pi}$$

$$\times \left(K_\mu K'_\nu + K'_\mu K_\nu - g_{\mu\nu} K \cdot K'\right)$$

$$\times \rho_F(K)\rho_F(K')\frac{\tilde{f}(k_0) - \tilde{f}(k'_0)}{i\omega - k_0 + k'_0}$$

6

Collective excitations in a plasma

It is well-known that the properties of elementary particles are modified when they propagate in a medium, as they become 'dressed' by their interactions: they acquire, for example, an effective mass which is different from the mass as measured in the vacuum. More generally, one speaks of the propagation of collective modes, or quasi-particles; in some cases, these quasi-particles can easily be identified with ordinary particles whose properties are only slightly modified by their interactions with the medium. In other cases collective modes bear little resemblance with particles in the vacuum.

Collective modes are characterized by a dispersion law $\omega(\mathbf{q})$ giving their energy ω as a function of their momenta $\mathbf{q}$. Their lifetime is not infinite, contrary to that of stable particles in the vacuum; thus another relevant quantity is the decay (or damping) rate $\gamma(\mathbf{q})$ of the collective modes.

In general, collective modes appear mathematically as poles of propagators with well-defined quantum numbers in the complex plane of the energy: the real part of the pole gives the dispersion law, while the imaginary part gives the damping rate; see, however, the remarks following (6.19). In the present chapter we shall study the propagation of gauge bosons and fermions in a plasma, and we shall compute explicitly the dispersion laws to a first approximation. The most important outcome of the calculations is that they reveal the fundamental role which is played by the energy scale gT, where g is the gauge coupling constant (the electron charge e in the QED case). Actually, in a gauge theory, a plasma of massless particles is characterized by two quantities only: the temperature T and the (dimensionless) coupling constant g, where, in the spirit of perturbation theory, we assume that g is small: $g \ll 1$; remember that $g(T) \sim 1/\ln(T/\Lambda)$, where Λ is the QCD coupling constant. Masses may be neglected if gT is much larger than all $T = 0$ masses.

114

The energy scale T is characteristic of individual particles: their average energy is $\sim T$ and the average distance between two neighbouring particles is $\bar{r} \sim T^{-1}$. The energy scale gT which appears in this chapter is associated with the collective motion of individual particles which takes place over distances $\sim 1/gT$, with frequencies $\sim gT$; in a field theoretic approach, this scale appears as the energy of the quasi-particles. In the following chapter, it will be shown that the decay rate of these collective excitations is of order $g^2 T$, so that the quasi-particles are well defined in the small g limit. It is of course possible to think of a whole hierarchy of energy scales: gT, $g^2 T$,... and perhaps $g^{3/2} T$, etc. As we shall see in chapter 10, the scale $g^2 T$ is associated with non-perturbative phenomena in the transverse part of the gauge boson propagator: this is the so-called 'magnetic scale', while the scale gT is the 'electric scale'. The fact that the typical wavelength $\lambda \sim gT$ of a collective excitation is much larger than the thermal wavelength $\lambda_T \sim 1/T$ suggests that these collective effects might be described semi-classically. Indeed, the results which we shall derive by field-theoretical methods can also be obtained from kinetic theory (see sections 6.4 and 7.4).

The general framework which is used in order to interpret the results of our calculations physically is linear response theory: one probes the medium with a weak perturbation and one studies the response of the medium in the linear approximation. We shall thus begin with a short review of linear response theory.

6.1 Linear response theory

We take a general quantum mechanical system described in the Schrödinger picture by a state vector $|\Psi_S(t) >$ whose time evolution is governed by a time-independent Hamiltonian H:

$$i\frac{\mathrm{d}}{\mathrm{d}t}|\Psi_S(t) >= H|\Psi_S(t) > \tag{6.1}$$

The state vector in the Heisenberg picture is chosen to coincide with that of the Schrödinger picture at $t = 0$:

$$|\Psi_H >= \mathrm{e}^{iHt}|\Psi_S(t) > \tag{6.2}$$

Let $V(t)$ be an external perturbation to our system; we assume that the perturbation vanishes for $t \le t_0$: $V(t) = 0$ if $t \le t_0$. The new state vector $|\overline{\Psi}_S(t) >$ can be written as

$$|\overline{\Psi}_S(t) >= \mathrm{e}^{-iHt} U(t, t_0)|\Psi_S(0) > \tag{6.3}$$

where the operator $U(t, t_0)$ obeys the following differential equation and boundary condition:

$$i\frac{dU(t, t_0)}{dt} = V_H(t) \, U(t, t_0) \qquad U(t, t_0) = 1 \text{ if } t \le t_0 \tag{6.4}$$

$$V_H(t) = e^{iHt} \, V(t) \, e^{-iHt} \tag{6.5}$$

$V_H(t)$ is nothing other than the perturbation in the Heisenberg picture (in elementary time-dependent perturbation theory, H is generally denoted by H_0 and (6.5) defines the perturbation in the interaction picture). Expanding U in powers of V, one finds in a standard way for $t > t_0$

$$U(t, t_0) = 1 - i \int_{t_0}^{t} dt' \, V_H(t') + \ldots \tag{6.6}$$

whence

$$|\overline{\Psi}_S(t) > = e^{-iHt} \, |\Psi_S(0) >$$
$$- i e^{-iHt} \int_{t_0}^{t} dt' V_H(t') |\Psi_S(0) > + \ldots \tag{6.7}$$

Let $\hat{O}(t)$ be any operator in the Schrödinger picture and let $\langle \hat{O}(t) \rangle$ be its expectation value

$$\langle \hat{O}(t) \rangle = < \overline{\Psi}_S(t) | \hat{O}(t) | \overline{\Psi}_S(t) > \tag{6.8}$$

We denote by $\delta \langle \hat{O}(t) \rangle$ the difference between the expectation value of $\langle \hat{O}(t) \rangle$ with $V(t)$ and without $V(t)$ (the 'induced expectation value')

$$\delta \langle \hat{O}(t) \rangle = < \overline{\Psi}_S(t) | \hat{O}(t) | \overline{\Psi}_S(t) > - < \Psi_S(t) | \hat{O}(t) | \Psi_S(t) > \tag{6.9}$$

From (6.7) we have, to first order in V,

$$\delta \langle \hat{O}(t) \rangle = -i \int_{t_0}^{\infty} dt' < \Psi_H | \theta(t - t') [\hat{O}_H(t), V_H(t')] | \Psi_H > \tag{6.10}$$

Equation (6.10) features the retarded commutator of the perturbation $V_H(t)$ and the operator $\hat{O}(t)$ in the Heisenberg picture. Note that we could write a similar equation for any matrix element of $\hat{O}_H(t)$, instead of an expectation value, and, more importantly, we could also write a similar equation for any kind of canonical (or grand canonical) average by taking for $|\Psi_H >$ eigenvectors $|n >$ of H:

$$\langle \ldots \rangle = \frac{1}{Z} \sum_n e^{-\beta E_n} < n | \ldots | n > \tag{6.11}$$

We now apply (6.10) to a particularly important case:

$$V(t') = \int d^3x' \, j(\mathbf{x}',t')\hat{\varphi}_S(\mathbf{x}') \qquad (6.12a)$$

$$\hat{O}(t) = \hat{\varphi}_S(\mathbf{x}) \qquad (6.12b)$$

where $j(x) = j(\mathbf{x},t)$ is a (c-number) external source and $\hat{\varphi}_S(\mathbf{x})$ the field operator of a scalar theory in the Schrödinger picture. We obtain from (6.10) and (6.12)

$$\delta\langle\hat{\varphi}(x)\rangle = -i \int d^4x' \, D_R(x-x')j(x') \qquad (6.13)$$

where $\langle\ldots\rangle$ represents from now on a thermal average and $D_R(x-x')$ is the retarded propagator

$$D_R(x-x') = \theta(t-t')\langle[\hat{\varphi}_H(x),\hat{\varphi}_H(x')]\rangle \qquad (6.14)$$

At zero temperature $\langle\ldots\rangle$ would of course represent a vacuum expectation value. In Fourier space (6.13) is translated into

$$\delta\langle\hat{\varphi}(q_0,\mathbf{q})\rangle = -ij(q_0,\mathbf{q}) \, D_R(q_0,\mathbf{q}) \qquad (6.15)$$

This equation will be used later on in order to study the behaviour of static potentials. It is also instructive to apply (6.13) to the case of an impulsive perturbation

$$j(\mathbf{x},t) = j(\mathbf{q})e^{i\mathbf{q}\cdot\mathbf{x}}\delta(t) \qquad (6.16)$$

Then we learn that

$$\delta\langle\hat{\varphi}(\mathbf{x},t)\rangle = -ij(\mathbf{q})e^{i\mathbf{q}\cdot\mathbf{x}} \int_{-\infty}^{\infty} \frac{dq_0}{2\pi} e^{-iq_0 t} D_R(q_0,\mathbf{q}) \qquad (6.17)$$

As explained in the introduction, we expect collective excitations to show up as poles in the propagators; we thus try the following form for D_R:

$$D_R(q_0,\mathbf{q}) = \frac{iR(q_0,\mathbf{q})}{q_0 - \omega(\mathbf{q}) + i\gamma(q_0,\mathbf{q})} \qquad (6.18)$$

Because of the factor $\theta(t-t')$ in (6.14) we know that $D_R(q_0,\mathbf{q})$ is analytic in the upper half complex q_0-plane Im $q_0 > 0$, and γ must be positive. The response to an impulsive perturbation is then a damped travelling wave with a dispersion law $q_0 = \omega(\mathbf{q})$ and a damping rate $\gamma = \gamma(\omega(\mathbf{q}),\mathbf{q})$:

$$\delta\langle\hat{\varphi}(\mathbf{x},t)\rangle = -ij(\mathbf{q}) \, R(\omega(\mathbf{q}),\mathbf{q}) \, \theta(t) \exp\left(i(\mathbf{q}\cdot\mathbf{x} - \omega(\mathbf{q})t) - \gamma t\right) \qquad (6.19)$$

For a free field we would have $\omega(\mathbf{q}) = (\mathbf{q}^2 + m^2)^{1/2}$ and $\gamma = 0$. The amplitude of the travelling wave is proportional to the residue at the pole of the retarded Green function; however, the analytic structure of (6.18) might well be too naïve and lead to difficulties in some cases: see section 10.4. Our program is now to use these results of linear response theory in the case of QED and QCD plasmas, and, for that purpose, we need the retarded propagators, at least to some approximation.

6.2 The photon propagator in a QED plasma

We wish to apply linear response theory first to the QED plasma, and for that purpose we need to compute the photon propagator: this will be done in the present section at the one-loop approximation and in the high-temperature limit, which means that the temperature is much larger than the electron mass $(T \gg m_e)$ and the external momenta. In all that follows, we shall neglect the fermion (electron or quark) mass and put it to zero, except when the contrary is stated explicitly. Our calculations are first performed in the imaginary-time formalism (Euclidean space), and later on continued analytically to Minkowski space.

6.2.1 General form of the photon propagator

From its definition (5.127) and the transversality property (5.140), the photon polarization tensor (or photon self-energy) $\Pi_{\mu\nu}$ may be written as (note that there is no factor of i in Euclidean space)

$$\Pi_{\mu\nu} = D^{-1}_{\mu\nu} - D^{-1}_{F,\mu\nu} = FP^L_{\mu\nu} + GP^T_{\mu\nu} \tag{6.20}$$

where the transverse and longitudinal projectors $P^T_{\mu\nu}$ and $P^L_{\mu\nu}$ are given in Euclidean space by $(Q_\mu = (q_4, \mathbf{q}) = (-\omega, \mathbf{q}), \ \hat{\mathbf{q}} = \mathbf{q}/q)$:

$$P^T_{44} = P^T_{4i} = 0 \quad P^T_{ij} = \delta_{ij} - \hat{\mathbf{q}}_i \hat{\mathbf{q}}_j$$

$$P^L_{\mu\nu} = \delta_{\mu\nu} - \frac{Q_\mu Q_\nu}{Q^2} - P^T_{\mu\nu} \tag{6.21}$$

These projectors obey the usual relations

$$(P^T)^2 = P^T \quad (P^L)^2 = P^L \quad P^T P^L = P^L P^T = 0 \tag{6.22}$$

In a covariant gauge the full propagator $D_{\mu\nu}$ reads

$$D_{\mu\nu} = \frac{1}{G+Q^2} P^T_{\mu\nu} + \frac{1}{F+Q^2} P^L_{\mu\nu} + \frac{\xi}{Q^2} \frac{Q_\mu Q_\nu}{Q^2} \tag{6.23}$$

Our task now is to evaluate F and G, or equivalently $\Pi_{\mu\nu}$, in the one-loop approximation (fig. 6.1). The explicit expression of $\Pi_{\mu\nu}$ to one loop is

$$\Pi_{\mu\nu} = e^2 \int \frac{d^4K}{(2\pi)^4} \text{Tr}[\gamma_\mu \not{K} \gamma_\nu (\not{K} - \not{Q})] \tilde{\Delta}(K) \tilde{\Delta}(K-Q) \tag{6.24}$$

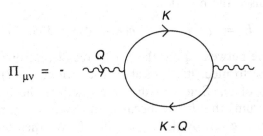

$$\Pi_{\mu\nu} = -$$

Fig. 6.1. One-loop approximation to the photon propagator. Wavy lines: photons. Solid lines: electrons.

where the overall sign is fixed by our definition (6.20), $\tilde{\Delta}(K)$ is defined in (5.56) and we have used the notational convention

$$T \sum_n \int \frac{\mathrm{d}^3k}{(2\pi)^3} = \int \frac{\mathrm{d}^4K}{(2\pi)^4} \qquad (6.25)$$

6.2.2 *The hard thermal loop approximation for $\Pi_{\mu\nu}$*

The complete calculation of $\Pi_{\mu\nu}$ is rather involved, even at one loop, and we shall limit ourselves to the high-temperature limit: in a terminology to be defined soon, we shall evaluate only the 'hard thermal loop' (HTL) part of $\Pi_{\mu\nu}$. The loop integral in (6.24) is quadratically divergent, and one expects its thermal part to be proportional to T^2, since the momentum integral is cut off at momenta of order T. This T^2-behaviour has already been seen in section 3.1.2, in the case of the φ^4-theory, where its origin is rather obvious. However, gauge theories are much richer than scalar theories, and there are other possible origins for T^2-behaviour, which we shall discover soon. Let us examine a typical loop integral appearing in the evaluation of $\Pi_{\mu\nu}$:

$$J = \int \frac{\mathrm{d}^4K}{(2\pi)^4} k^2 \tilde{\Delta}(K) \tilde{\Delta}(K - Q) \qquad (6.26)$$

The frequency sum is performed by using (5.59):

$$J = -\frac{1}{8\pi^2} \int \frac{k^2 \, \mathrm{d}k \, \mathrm{d}\Omega}{4\pi} \frac{k^2}{E_1 E_2} \left[(1 - \tilde{n}_1 - \tilde{n}_2) \right.$$
$$\times \left(\frac{1}{i\omega - E_1 - E_2} - \frac{1}{i\omega + E_1 + E_2} \right)$$
$$\left. - (\tilde{n}_1 - \tilde{n}_2) \left(\frac{1}{i\omega + E_1 - E_2} - \frac{1}{i\omega - E_1 + E_2} \right) \right] \qquad (6.27)$$

where we have used the notations

$$E_1 = k \quad E_2 = |\mathbf{q} - \mathbf{k}| \quad \tilde{n}_i = \tilde{n}(E_i) \quad \mathrm{d}\Omega = \mathrm{d}(\cos\theta)\mathrm{d}\varphi \qquad (6.28)$$

θ being the angle between $\mathbf{q}$ and $\mathbf{k}$. Let us recall that once the frequency sum has been performed, it is possible to continue (6.27) analytically to arbitrary values of ω. Since we are interested in the high-temperature limit, we may assume that $k \gg q$ because the leading contribution in T to the loop integral is given by the region $k \sim T$. We then use the following approximations:

$$E_2 \simeq k - q\cos\theta$$

$$\tilde{n}_2 \simeq \tilde{n}(k - q\cos\theta) \simeq \tilde{n}(k) - q\cos\theta\,\frac{\mathrm{d}\tilde{n}(k)}{\mathrm{d}k} \qquad (6.29)$$

In the large k-limit, the first term in the square bracket of (6.27) behaves as

$$\frac{1}{k}(1 - 2\tilde{n}(k)) \qquad (6.30)$$

The $1/k$ term leads to a quadratic divergence which is taken care of by renormalization; of course, because of gauge-invariance, this divergence is only logarithmic if one uses a gauge-invariant regularization. In the second term, $-2\tilde{n}(k)/k$, the Fermi–Dirac factor cuts off the integral at values of $k \sim T$, so that the $T = 0$ quadratic divergence is turned into T^2-behaviour. The exact value of the integral is deduced from

$$\int_0^\infty k\,\mathrm{d}k\,\tilde{n}(k) = \frac{\pi^2 T^2}{12} \qquad (6.31)$$

The leading (that is T^2) contribution in T comes from loop momenta of order T: following Braaten and Pisarski, these momenta will be called 'hard momenta' and the approximation we are working with will be called the 'hard thermal loop' (HTL) approximation. In what follows, another scale, of order eT, will turn out to play a crucial role, and momenta of order eT will be called 'soft momenta'.

A closer inspection of (6.27) reveals another possible origin for terms of order T^2; indeed,

$$i\omega \pm E_1 \mp E_2 \simeq i\omega \pm q\cos\theta \qquad (6.32)$$

while

$$\tilde{n}_1 - \tilde{n}_2 \simeq q\cos\theta\frac{\mathrm{d}\tilde{n}(k)}{\mathrm{d}k} \qquad (6.33)$$

The angular and k-integrations are now decoupled, and the k-integral again leads to a T^2-behaviour, which can easily be found by integrating by parts (6.31).

The other frequency sum which is needed for the evaluation of $\Pi_{\mu\nu}$ is

$$J_l = \int \frac{\mathrm{d}^4K}{(2\pi)^4}\, \omega_n\, k_l\, \tilde{\Delta}(K)\tilde{\Delta}(K - Q) \qquad (6.34)$$

The frequency sum is performed by using (5.60):

$$J_l = \frac{1}{8\pi^2} \int \frac{k^2\mathrm{d}k\mathrm{d}\Omega}{4\pi} \frac{ik_l}{E_2}\left[(1 - \tilde{n}_1 - \tilde{n}_2)\right.$$
$$\times \left(\frac{1}{i\omega - E_1 - E_2} + \frac{1}{i\omega + E_1 + E_2}\right)$$
$$\left. + (\tilde{n}_1 - \tilde{n}_2)\left(\frac{1}{i\omega + E_1 - E_2} + \frac{1}{i\omega - E_1 + E_2}\right)\right] \qquad (6.35)$$

In this case, the contribution from the first term in the square bracket is non-leading, while the second term leads to a T^2-behaviour.

6.2.3   Evaluation of $\Pi_{\mu\nu}$

Within the HTL approximation, we can clearly neglect $\not{Q}$ with respect to $\not{K}$ in the term between square brackets in (6.24); in all that follows, this kind of term will be called a term from the 'numerator', in contrast to $\tilde{\Delta}(K) = K^{-2}$, which is a term from the 'denominator'. Thus,

$$\Pi_{\mu\nu} \simeq e^2 \int \frac{\mathrm{d}^4K}{(2\pi)^4}(8K_\mu K_\nu - 4K^2\delta_{\mu\nu})\tilde{\Delta}(K)\tilde{\Delta}(K - Q)$$
$$= I_{\mu\nu} - 4e^2\delta_{\mu\nu} \int \frac{\mathrm{d}^4K}{(2\pi)^4}\tilde{\Delta}(K - Q) \qquad (6.36)$$

The second term in (6.36) is easily evaluated since

$$\int \frac{\mathrm{d}^4K}{(2\pi)^4}\tilde{\Delta}(K - Q) = \int \frac{\mathrm{d}^4K}{(2\pi)^4}\tilde{\Delta}(K)$$
$$= \int \frac{\mathrm{d}^3k}{(2\pi)^3}\frac{1}{2k}(1 - 2\tilde{n}(k)) \simeq -\frac{T^2}{24} \qquad (6.37)$$

The quadratic divergence has been absorbed in the zero-temperature renormalization. Finite renormalizations are non-leading in T, and we have retained only the leading behaviour: $\simeq$ in (6.37) means that we keep only the T^2-part of the result. The evaluation of $I_{\mu\nu}$ is slightly more involved; it will be convenient to define the light-like four-vectors $\hat{K}$ and $\hat{K}'$:

$$\hat{K} = (-i, \hat{\mathbf{k}}) \quad \hat{K}' = (-i, -\hat{\mathbf{k}}) \qquad (6.38)$$

Within our approximations the denominators in the second term of the square bracket of (6.27) and (6.35) can be written ($\omega = -q_4$)

$$i\omega + E_1 - E_2 \simeq i\omega + \mathbf{q} \cdot \hat{\mathbf{k}} = -iq_4 + \mathbf{q} \cdot \hat{\mathbf{k}} = Q \cdot \hat{K}$$
$$i\omega - E_1 + E_2 \simeq i\omega - \mathbf{q} \cdot \hat{\mathbf{k}} = -iq_4 - \mathbf{q} \cdot \hat{\mathbf{k}} = Q \cdot \hat{K}' \qquad (6.39)$$

Using (6.27), (6.39) and simple symmetries of the integrand, we write I_{ij} as

$$
\begin{aligned}
I_{ij} &\simeq -\frac{2e^2}{\pi^2} \int \frac{k^2 dk d\Omega}{4\pi} \hat{k}_i \hat{k}_j \left[\frac{\tilde{n}(k)}{k} - \frac{d\tilde{n}(k)}{dk} + \frac{d\tilde{n}(k)}{dk} \frac{i\omega}{Q \cdot \hat{K}} \right] \\
&= -\frac{e^2 T^2}{6} \delta_{ij} + \frac{e^2 T^2}{3} \int \frac{d\Omega}{4\pi} \frac{i\omega}{Q \cdot \hat{K}} \hat{k}_i \hat{k}_j \qquad (6.40)
\end{aligned}
$$

As noted before, the k- and Ω-integrals are decoupled, and the k-integration has been performed explicitly when going from the first to the second line of (6.40). We immediately deduce Π_{ij}:

$$\Pi_{ij} = \frac{e^2 T^2}{3} \int \frac{d\Omega}{4\pi} \frac{i\omega}{Q \cdot \hat{K}} \hat{k}_i \hat{k}_j \qquad (6.41a)$$

The calculation of Π_{4i} makes use of (6.35):

$$\Pi_{4i} = \frac{e^2 T^2}{3} \int \frac{d\Omega}{4\pi} \frac{\omega \hat{k}_i}{Q \cdot \hat{K}} \qquad (6.41b)$$

while, writing $\omega_n^2 = K^2 - k^2$, we obtain for Π_{44}

$$\Pi_{44} = 4e^2 \int \frac{d^4 K}{(2\pi)^4} \tilde{\Delta}(K) - 8e^2 \int \frac{d^4 K}{(2\pi)^4} k^2 \tilde{\Delta}(K) \tilde{\Delta}(K - Q) \qquad (6.42)$$

The last integral in (6.43) is nothing other than $\delta_{ij} I_{ij}$ and

$$\Pi_{44} = \frac{e^2 T^2}{3} \left(1 - \int \frac{d\Omega}{4\pi} \frac{i\omega}{Q \cdot \hat{K}} \right) \qquad (6.41c)$$

The results (6.41a,b,c) can be summarized by

$$\Pi_{\mu\nu} = 2m^2 \int \frac{d\Omega}{4\pi} \left(\frac{i\omega \hat{K}_\mu \hat{K}_\nu}{Q \cdot \hat{K}} + \delta_{\mu 4} \delta_{\nu 4} \right) \qquad (6.43)$$

where we have defined the photon thermal mass m by

$$m^2 = e^2 T^2 / 6 \qquad (6.44)$$

One may check that, despite their superficial appearance, all integrals in (6.43) are real. $\Pi_{\mu\nu}$ in (6.43) obeys the Ward identity

$$Q_\mu \Pi_{\mu\nu}(Q) = 0 \qquad (6.45)$$

in agreement with (5.140). This property, in addition to rotational invariance, guarantees that the decomposition (6.20) for $\Pi_{\mu\nu}$ is indeed valid.

One could have derived the Ward identity directly from (6.24) by noting that, in the HTL approximation,

$$Q \cdot K = \frac{1}{2} \left[K^2 - (Q - K)^2 \right] \tag{6.46}$$

6.2.4   *Transverse and longitudinal photons*

Before we extract the physics from the preceding results, we must identify the coefficients F and G of the transverse and longitudinal projectors $P^T_{\mu\nu}$ and $P^L_{\mu\nu}$ (6.20). Taking $\mathbf{q}$ parallel to the z-axis, we have for F and G

$$F = \frac{Q^2}{\omega q} \Pi_{4z} \qquad G = \Pi_{xx} \tag{6.47}$$

We are left with the evaluation of the angular integrals in (6.43). Let us begin with F:

$$
\begin{aligned}
F &= \frac{2m^2 Q^2}{\omega q} \int \frac{d\Omega}{4\pi} \frac{\omega \cos\theta}{i\omega + q \cos\theta} \\
&= \frac{2m^2 Q^2}{q^2} \left(1 - \frac{i\omega}{2q} \ln \frac{i\omega + q}{i\omega - q} \right)
\end{aligned} \tag{6.48a}
$$

The results can be expressed in terms of Legendre functions of the second kind:

$$Q_0(x) = \frac{1}{2} \ln \frac{x+1}{x-1} \qquad Q_1(x) = x Q_0(x) - 1 \tag{6.49}$$

and one may write F as

$$F = \frac{2m^2 Q^2}{q^2} \left(1 - \frac{i\omega}{q} Q_0 \left(\frac{i\omega}{q} \right) \right) = -\frac{2m^2 Q^2}{q^2} Q_1 \left(\frac{i\omega}{q} \right) \tag{6.48b}$$

Let us now turn to G:

$$
\begin{aligned}
G = \Pi_{xx} &= 2m^2 \int \frac{d\Omega}{4\pi} \frac{i\omega \sin^2\theta \cos^2\varphi}{i\omega + q \cos\theta} \\
&= m^2 \left(\frac{i\omega}{q} \right) \left[\left(1 - \left(\frac{i\omega}{q} \right)^2 \right) Q_0 \left(\frac{i\omega}{q} \right) + \frac{i\omega}{q} \right]
\end{aligned} \tag{6.50}
$$

Lastly we analytically continue (6.48) and (6.50) from Euclidean to Minkowski space. As explained in section 5.1.2, this continuation is performed by making the substitutions $i\omega \to q_0 + i\eta$ and $Q^2 \to -Q^2$, with q_0 real, as we are interested in the retarded propagator. The retarded photon propagator reads

$$D^R_{\mu\nu} = \frac{i}{Q^2 - G} P^T_{\mu\nu} + \frac{i}{Q^2 - F} P^L_{\mu\nu} - i \frac{\xi}{Q^2} \frac{Q_\mu Q_\nu}{Q^2} \tag{6.51}$$

where

$$P^L_{\mu\nu} = -g_{\mu\nu} + \frac{Q_\mu Q_\nu}{Q^2} - P^T_{\mu\nu} \qquad (6.52)$$

and

$$F = -\frac{2m^2 Q^2}{q^2}\left(1 - \frac{q_0}{q}Q_0\left(\frac{q_0}{q}\right)\right) \qquad (6.53a)$$

$$G = m^2\left(\frac{q_0}{q}\right)\left[\left(1 - \left(\frac{q_0}{q}\right)^2\right)Q_0\left(\frac{q_0}{q}\right) + \frac{q_0}{q}\right]$$

$$= \frac{1}{2}(2m^2 - F) \qquad (6.53b)$$

The Legendre function $Q_0(q_0/q)$ is defined in the complex q_0-plane cut from $-q$ to $+q$; it is real for q_0 real and $|q_0|$ larger than q, namely, for time-like Q^2 ($Q^2 > 0$). It is complex for space-like Q^2:

$$Q_0\left(\frac{q_0}{q}\right) = \frac{1}{2}\ln\left|\frac{q_0 + q}{q_0 - q}\right| - \frac{i\pi}{2} \quad (Q^2 < 0) \qquad (6.54a)$$

for $i\omega = q_0 + i\eta, \eta \to 0^+$ so that the general form reads

$$Q_0\left(\frac{q_0}{q}\right) = \frac{1}{2}\ln\left|\frac{q_0 + q}{q_0 - q}\right| - \frac{i\pi}{2}\theta(q^2 - q_0^2) \qquad (6.54b)$$

It is useful to comment on the origin of these properties: at zero temperature, $\Pi_{\mu\nu}$ is complex for $Q^2 > 0$. This stems from the fact that the virtual photon may then decay into physical final states, and the imaginary part of $\Pi_{\mu\nu}$ is directly related to the decay rate. The leading contribution in T has exactly the opposite property, since $\Pi_{\mu\nu}$ is real for $Q^2 > 0$ and complex for $Q^2 < 0$: here the imaginary part stems from the scattering of electrons and positrons with momenta of order T on the low-momentum photon. In plasma physics, this phenomenon is known as Landau damping and will be discussed in section 6.4.

6.3 Debye screening and plasma oscillations

6.3.1 Linear response and Debye screening

Having computed the photon propagator, we can now exploit linear response theory. Let us apply a (weak) external electromagnetic field $(\mathbf{E}_{cl}, \mathbf{B}_{cl})$ to a QED plasma. We call $j^\mu_{cl} = (\rho_{cl}, \mathbf{j}_{cl})$ the classical source of $(\mathbf{E}_{cl}, \mathbf{B}_{cl})$, which is related to the fields through Maxwell's equations and obeys the conservation condition $\partial_\mu j_{cl}^\mu = 0$. The perturbation V of section 6.1 reads*

$$V = \int d^3x\, j_{cl}^\mu(x)\hat{A}_\mu(x) \tag{6.55}$$

where $\hat{A}_\mu$ is the quantized electromagnetic field. The average field $\delta\langle\hat{A}_\mu\rangle$, denoted simply by A_μ in what follows, as $\langle\hat{A}_\mu\rangle_{eq} = 0$, is given by a slight generalization of (6.13):

$$A_\mu = -i\int d^4x'\,\theta(t-t')\langle[\hat{A}_\mu(x),\hat{A}_\nu(x')]\rangle_\beta j_{cl}^\nu(x') \tag{6.56}$$

Equation (6.56) involves the retarded commutator of the components of the electromagnetic field. We write (6.56) in Fourier space, using the photon propagator (6.51) in a covariant gauge and remembering the condition for current conservation $Q_\mu j_{cl}^\mu(Q) = 0$

$$A^0 = -\frac{\rho_{cl}}{Q^2 - F} \qquad \mathbf{A} = -\left(\frac{\mathbf{j}_{cl}^L}{Q^2 - F} + \frac{\mathbf{j}_{cl}^T}{Q^2 - G}\right) \tag{6.57}$$

where $\mathbf{j}_{cl}^T$ and $\mathbf{j}_{cl}^L$ are the transverse and longitudinal components of $\mathbf{j}_{cl}$ (with respect to $\mathbf{q}$). From (6.57) we derive the electric and magnetic fields

$$\mathbf{E} = i(q_0\mathbf{A} - \mathbf{q}A_0)$$

$$= -i\left(\frac{\rho_{cl}}{Q^2 - F}\frac{Q^2}{q^2}\mathbf{q} + \frac{q_0\,\mathbf{j}_{cl}^T}{Q^2 - G}\right) \tag{6.58a}$$

$$\mathbf{B} = i\mathbf{q}\times\mathbf{A} = -i\frac{\mathbf{q}\times\mathbf{j}_{cl}^T}{Q^2 - G} \tag{6.58b}$$

Using Maxwell's equations we express (6.58) in terms of the external fields

$$\mathbf{E} = \frac{Q^2}{Q^2 - F}\mathbf{E}_{cl}^L + \frac{q_0}{Q^2 - G}\left(\mathbf{q}\times\mathbf{B}_{cl} + q_0\mathbf{E}_{cl}^T\right) \tag{6.59a}$$

$$\mathbf{B} = \frac{Q^2}{Q^2 - G}\mathbf{B}_{cl} \tag{6.59b}$$

where $\mathbf{E}_{cl}^T$ and $\mathbf{E}_{cl}^L$ are the transverse and longitudinal components of the external electric field. Note that in the absence of interactions ($F = G = 0$), one recovers $\mathbf{E} = \mathbf{E}_{cl}, \mathbf{B} = \mathbf{B}_{cl}$, and, as is of course obvious from (6.57), $\mathbf{A} = \mathbf{A}_{cl}$. It is important to understand that $\mathbf{E}$ and $\mathbf{B}$, which are the average values of the quantized $\hat{\mathbf{E}}$ and $\hat{\mathbf{B}}$, represent the total electromagnetic field, and A_μ in (6.56) is the total electromagnetic potential. It is straightforward, but instructive, to check that the same results are obtained in the Coulomb

---

* In the non-Abelian case, the straightforward generalization of (6.55) may run into problems.

and time-axial gauges, although the time and longitudinal components of A_μ are of course different (exercise 6.2).

Let us also derive Kubo's formula, which relates the induced current, namely, the difference between the total current and the classical current, to the electromagnetic field. Dropping Lorentz indices and the R subscript, we write

$$A = -iD j_{\text{cl}} = DD_F^{-1} A_{\text{cl}} \tag{6.60}$$

where D_F is the free propagator (for example, $D_F^{-1} = -i\partial_\mu \partial^\mu$ in the Feynman gauge), and

$$
\begin{aligned}
j_{\text{ind}} = j_{\text{tot}} - j_{\text{cl}} &= iD_F^{-1} A - iD_F^{-1} A_{\text{cl}} \\
&= i(D_F^{-1} - D^{-1})A = \Pi A
\end{aligned} \tag{6.61a}
$$

We have used (6.20), with a factor of i in Minkowski space. A more explicit form of (6.61a) is

$$j_{\text{ind}}^\mu(x) = \int \mathrm{d}^4 x' \Pi_R^{\mu\nu}(x - x') A_\nu(x') \tag{6.61b}$$

This equation allows us to compute the current induced in the plasma by the total field $A_\mu(x)$ from the retarded polarization tensor $\Pi_R^{\mu\nu}$. Note that from (5.140) the induced current is conserved and gauge-independent.

Let us now specialize our discussion to the case of a static external electric field $\mathbf{E}_{\text{cl}}(\mathbf{x})$. From (6.59) we have in Fourier space

$$\mathbf{E}(\mathbf{q}) = \frac{\mathbf{q}\,(\mathbf{q} \cdot \mathbf{E}_{\text{cl}}(\mathbf{q}))}{\mathbf{q}^2 + F(q_0 = 0, \mathbf{q})} = -i\frac{\mathbf{q}\,\rho_{\text{cl}}(\mathbf{q})}{\mathbf{q}^2 + F(q_0 = 0, \mathbf{q})} \tag{6.62}$$

As $\mathbf{E}(\mathbf{q}) = -i\mathbf{q}V(\mathbf{q})$, where $V(\mathbf{q})$ is the static potential (not to be confused with V in (6.55)!), we may compute from (6.62) the potential of a static charge Q at the origin:

$$V(r) = Q \int \frac{\mathrm{d}^3 q}{(2\pi)^3} \frac{e^{i\mathbf{q}\cdot\mathbf{r}}}{\mathbf{q}^2 + F(q_0 = 0, \mathbf{q})} \tag{6.63}$$

The large-distance behaviour of the static potential is governed by the zeros of the denominator in (6.63). It can be shown from general field theoretical arguments that the position of the poles of two-point functions is gauge-fixing independent; of course in the present case we know that in QED, $\Pi_{\mu\nu}$ itself is gauge-fixing independent, and *a fortiori* the position of the pole in (6.63) shares the same property. This position is called the electric (or Debye) mass m_{el} (or m_{D}). Its inverse is the Debye radius $r_{\text{D}} = m_{\text{D}}^{-1}$ and the potential $V(r)$ of a static charge Q is screened with a characteristic length r_{D}

$$V(r) = \frac{Q}{r} \exp(-r/r_{\text{D}}) \tag{6.64}$$

Since within our HTL approximation F is independent of q when $q_0 = 0$, we simply have

$$m_{\rm el}^2 = F(q_0 = 0, q \to 0)$$

$$= -\Pi_{00}(q_0 = 0, q \to 0) = 2m^2 = \frac{1}{3}e^2 T^2 \tag{6.65}$$

In conclusion: static electric fields are screened with a characteristic length $r_{\rm D} = (\sqrt{2}\, m)^{-1}$. On the other hand one immediately checks that $G(q_0 = 0, q \to 0)$ vanishes: static magnetic fields are not screened, and this result holds to all orders of perturbation theory (see exercise 6.7). We shall see in chapter 8 that low-frequency magnetic fields may nevertheless be screened: this is the dynamical screening phenomenon.

6.3.2 Plasma oscillations

From linear response theory, we know that the poles of the transverse and longitudinal parts of the photon propagator give the dispersion laws for transverse and longitudinal damped travelling waves in the plasma; these are the collective excitations or quasi-particles. Let us look first at the transverse modes; the dispersion law $\omega_T(q)$ and the damping rate $\gamma_T(q)$ ($\gamma \ll \omega$) are given by

$$\omega_T^2(q) = q^2 + \mathrm{Re}\, G(q_0 = \omega_T(q), q) \tag{6.66a}$$

$$\gamma_T(q) = -\frac{1}{2\omega_T} \mathrm{Im}\, G(q_0 = \omega_T(q), q) \tag{6.66b}$$

From the explicit expression (6.50) of G, the quasi-particle poles in (6.66a) are given by the equation

$$x^2 - 1 - \frac{m^2}{q^2}\left[x^2 + \frac{x(1-x^2)}{2} \ln \frac{x+1}{x-1}\right] = 0 \tag{6.67}$$

where we have defined x (not to be confused with the position four-vector x!) by $x = q_0/q = \omega_T/q$. This equation has to be solved numerically; the solution $\omega_T(q)$ is plotted as a function of q in fig. 6.2. It is possible to find approximate analytical solutions for small or large values of q:

$$q \ll m \quad \omega_T^2 \simeq \frac{2}{3}m^2 + \frac{6}{5}q^2 = \omega_P^2 + \frac{6}{5}q^2 \tag{6.68}$$

where we have defined the plasma frequency ω_P:

$$\omega_P = \sqrt{\frac{2}{3}}\, m = \frac{1}{3}eT \tag{6.69}$$

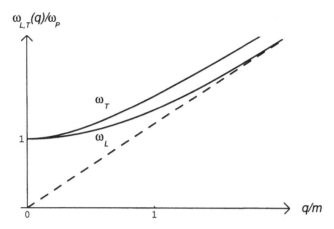

Fig. 6.2   Dispersion laws for transverse and longitudinal photons.

The plasma frequency gives the oscillation frequency for vanishing wave vectors, namely, spatially uniform oscillations. For large values of q, namely, for $eT \ll q \ll T$ we obtain

$$q \gg m : \quad \omega_T^2 \simeq q^2 + m^2 \tag{6.70}$$

Thus m acts as an effective, gauge-invariant mass for a photon propagating in the plasma. Since $\omega_T(q) > q$ in all cases, the imaginary part of G is zero at order T^2, and the damping rate is zero at this order in T.

The dispersion law for the collective longitudinal modes is given by

$$\omega_L^2(q) = q^2 + \operatorname{Re} F(\omega_L(q), q) \tag{6.71}$$

From the explicit expression (6.53a) of F, ω_L is given by the equation $(x = \omega_L/q)$:

$$1 + \frac{2m^2}{q^2}\left(1 - \frac{x}{2}\ln\frac{x+1}{x-1}\right) = 0 \tag{6.72}$$

The solution of (6.72) is plotted in fig. 6.2. Again, one can obtain the approximate form of $\omega_L(q)$ for small and large values of q:

$$q \ll m : \quad \omega_L^2(q) \simeq \omega_P^2 + \frac{3}{5}q^2$$

$$q \gg m : \quad \omega_L \simeq q\left(1 + 2\exp\left(-\frac{q^2 + m^2}{m^2}\right)\right) \tag{6.73}$$

For $q = 0$ one cannot distinguish between transverse and longitudinal excitations, and one finds that, indeed, $\omega_L(0) = \omega_T(0) = \omega_P$. This property can also be deduced from the fact that $\Pi_{\mu\nu}$ is non-singular for small values of q. Finally, a comment about terminology might be useful: we call our quasi-particles transverse and longitudinal photons, while some authors refer to photons as transverse excitations and to plasmons as longitudinal excitations.

6.3.3 Sum rules and residues

We are now going to exploit the analytic properties of the propagator in order to derive some results which will turn out to be very useful in the physical interpretation of our collective excitations, as well as in the calculations of the following chapters. It is convenient to define transverse and longitudinal photon propagators Δ_T and Δ_L through (note that Δ_T is the free transverse photon propagator for $m = 0$)

$$
\begin{aligned}
\Delta_T(q_0, q) &= \frac{-1}{Q^2 - G} \\
&= \frac{-1}{q_0^2 - q^2 - m^2 \left[x^2 + \frac{x(1-x^2)}{2} \ln \frac{x+1}{x-1} \right]}
\end{aligned}
\tag{6.74}
$$

where, as above, $x = q_0/q$, and

$$
\Delta_L(q_0, q) = \frac{Q^2}{q^2} \frac{-1}{Q^2 - F} = \frac{-1}{q^2 + 2m^2(1 - \frac{x}{2} \ln \frac{x+1}{x-1})}
\tag{6.75}
$$

The functions Δ_L and Δ_T are analytic in the complex q_0-plane cut from $-q$ to $+q$; in addition they have poles at $q_0 = \pm \omega_{T,L}(q)$. Using the contour Γ of fig. 6.3 we write Cauchy's theorem for $\Delta = \Delta_T$ or Δ_L:

$$
\begin{aligned}
\Delta(q_0, q) &= \oint_\Gamma \frac{dz}{2i\pi} \frac{\Delta(z, q)}{z - q_0} \\
&= \int_{-\infty}^{\infty} \frac{dq_0'}{2i\pi} \frac{\Delta(q_0' + i\eta, q) - \Delta(q_0' - i\eta, q)}{q_0' - q_0} \\
&\quad + \oint_{\Gamma'} \frac{dz}{2i\pi} \frac{\Delta(z, q)}{z}
\end{aligned}
\tag{6.76}
$$

where Γ' is a circle whose radius tends to infinity. Equation (6.76) can be rewritten in terms of the spectral density $\rho(q_0, q)$, which contains both the discontinuities across the cuts and the residues at the poles:

$$
\rho(q_0, q) = 2 \operatorname{Im} \Delta(q_0 + i\eta, q)
\tag{6.77}
$$

as

$$
\Delta(q_0, q) = \int_{-\infty}^{\infty} \frac{dq_0'}{2\pi} \frac{\rho(q_0', q)}{q_0' - q_0} + \oint_{\Gamma'} \frac{dz}{2i\pi} \frac{\Delta(z, q)}{z}
\tag{6.78}
$$

As $\Delta_T(z, q) \sim 1/z^2$ for $z \to \infty$, there is no contribution from Γ' in this case. However, $\Delta_L(z, q) \sim 1/q^2$ for $z \to \infty$ and the contribution from Γ' is

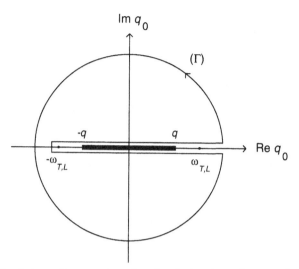

Fig. 6.3   Integration contour for the derivation of sum rules.

q^{-2}. Two sum rules are obtained by setting $q_0 = 0$ in the above equation; we obtain for Δ_L

$$\int_{-\infty}^{\infty} \frac{dq_0}{2\pi} \frac{\rho_L(q_0, q)}{q_0} = \frac{1}{q^2} + \Delta_L(0, q) = \frac{2m^2}{q^2(q^2 + 2m^2)} \qquad (6.79)$$

and for Δ_T

$$\int_{-\infty}^{\infty} \frac{dq_0}{2\pi} \frac{\rho_T(q_0, q)}{q_0} = \Delta_T(0, q) = \frac{1}{q^2} \qquad (6.80)$$

Other sum rules are obtained by examining the asymptotic behaviour when $q_0 \to \infty$; since the support of $\rho_L(q_0, q)$ is bounded, we may write for $q_0 \to \infty$

$$\Delta_L(q_0, q) = -\frac{1}{q^2} - \frac{1}{q_0} \sum_{n=0}^{\infty} \int_{-\infty}^{\infty} \frac{dq_0'}{2\pi} \left(\frac{q_0'}{q_0}\right)^{2n+1} \rho_L(q_0', q) \qquad (6.81)$$

where we have used the parity property

$$\rho_L(q_0, q) = -\rho_L(-q_0, q) \qquad (6.82)$$

On the other hand, we may expand (6.75) in powers of q_0^{-1}; comparing with (6.81) we obtain for $n = 0$

$$\int_{-\infty}^{\infty} \frac{dq_0}{2\pi} q_0 \rho_L(q_0, q) = \frac{2}{3} \frac{m^2}{q^2} = \frac{\omega_P^2}{q^2} \qquad (6.83)$$

and for $n = 1$

$$\int_{-\infty}^{\infty} \frac{dq_0}{2\pi} q_0^3 \rho_L(q_0, q) = \frac{2}{5} m^2 + \frac{4}{9} \frac{m^4}{q^2} = \frac{3}{5} \omega_P^2 + \frac{\omega_P^4}{q^2} \qquad (6.84)$$

etc. An equation that is analogous to (6.81) can be written for Δ_T, but without the $1/q^2$ factor. Expanding Δ_T in (6.74) in powers of q_0^{-1}, and comparing with the analogue of (6.81) yields

$$\int_{-\infty}^{\infty} \frac{dq_0}{2\pi} q_0 \rho_T(q_0, q) = 1 \tag{6.85}$$

$$\int_{-\infty}^{\infty} \frac{dq_0}{2\pi} q_0^3 \rho_T(q_0, q) = \omega_P^2 + q^2 \tag{6.86}$$

The RHS of (6.85) is independent of q: indeed this particular sum rule may be shown to be a consequence of the canonical commutation relations, independently of the HTL approximation.

Let us now turn to the problem of the residues of Δ_L and Δ_T at the poles. The residues for the longitudinal excitations are given by

$$Z_L(q) = - \left(\left[\partial \Delta_L^{-1}(q_0, q) / \partial q_0 \right]_{q_0 = \omega_L(q)} \right)^{-1}$$
$$= \frac{\omega_L(\omega_L^2 - q^2)}{q^2(q^2 + 2m^2 - \omega_L^2)} \tag{6.87}$$

For $q \ll m$ we find

$$Z_L(q) \simeq \frac{\omega_P}{2q^2} \left(1 - \frac{3}{10} \frac{q^2}{\omega_P^2} \right) \tag{6.88}$$

while for $q \gg m$ we have

$$Z_L(q) \simeq \frac{2q}{m^2} \exp\left(-\frac{q^2 + m^2}{m^2} \right) \tag{6.89}$$

The corresponding expressions for $Z_T(q)$ are

$$Z_T(q) = \frac{\omega_T(\omega_T^2 - q^2)}{3\omega_P^2 \omega_T^2 - (\omega_T^2 - q^2)^2} \tag{6.90}$$

with the following limits for small and large q:

$$q \ll m \quad Z_T(q) \simeq \frac{1}{2\omega_P} \left(1 - \frac{4}{5} \frac{q^2}{\omega_P^2} \right) \tag{6.91}$$

$$q \gg m \quad Z_T(q) \simeq \frac{1}{2q} \tag{6.92}$$

The complete expression for the spectral function reads, in the longitudinal case, with $x = q_0/q$,

$$(2\pi)^{-1} \rho_L(q_0, q) = Z_L(q)[\delta(q_0 - \omega_L(q))$$
$$- \delta(q_0 + \omega_L(q))] + \beta_L(q_0, q) \tag{6.93}$$

with

$$\beta_L(q_0, q) = \frac{m^2 x\ \theta(1 - x^2)}{[q^2 + 2m^2(1 - \frac{x}{2}\ln|\frac{x+1}{x-1}|)]^2 + \pi^2 m^4 x^2} \tag{6.94}$$

while the corresponding expression for the transverse spectral function is

$$\frac{1}{(2\pi)}\rho_T(q_0, q) = Z_T(q)[\delta(q_0 - \omega_T(q))$$
$$- \delta(q_0 + \omega_T(q))] + \beta_T(q_0, q) \tag{6.95}$$

with

$$\beta_T(q_0, q) = \left[m^2 x(1 - x^2)\theta(1 - x^2)/2 \right]$$
$$\times \left[\left(q^2(x^2 - 1) - m^2 \left[x^2 + \frac{x(1 - x^2)}{2} \right. \right. \right.$$
$$\left. \left. \times \ln\left|\frac{x + 1}{x - 1}\right| \right) \right)^2 + \pi^2 m^4 x^2 (1 - x^2)^2/4 \right]^{-1} \tag{6.96}$$

One notes that the quasi-particle poles almost saturate the sum rules for small values of q (remember that $q_0\rho_{L,T} \geq 0$). For large values of q, the transverse spectral function is dominated by the quasi-particle poles; indeed, within corrections of order m^2/q^2, one can interpret the sum rules for $q \gg m$ by writing

$$\rho_T(Q) = 2\pi\varepsilon(q_0)\delta(Q^2) \tag{6.97}$$

which is nothing other than the spectral function of the free theory. On the other hand, the longitudinal excitations decouple for large values of q: see (6.89).

Let us conclude by summarizing the physical interpretation of our results: at $T = 0$ the only physical modes are the transverse modes, and these are the only modes which propagate for large values of q at non-zero T. However, transverse photons acquire a gauge-invariant mass of order eT. The longitudinal excitations decouple at large values of q, but at low values of q ($q \sim m$), transverse and longitudinal excitations are equally important. It must be understood that there are three different masses, all of order eT for dimensional reasons, but which have quite different physical interpretations as they are obtained by taking different limits:

(i) the electric mass: $m_{\text{el}}^2 = \frac{1}{3}e^2 T^2$ ($q_0 = 0$, $q \to 0$);
(ii) the T-dependent mass: $m^2 = \frac{1}{6}e^2 T^2$ ($q_0, q \gg eT$);
(iii) the plasma frequency: $\omega_P^2 = \frac{1}{9}e^2 T^2$ ($q_0 \sim eT$, $q = 0$).

6.3.4 Non-zero chemical potential

It is not very difficult to generalize the preceding discussion to the case of non-zero chemical potential. Let μ be the electron chemical potential, that of the positrons being of course $-\mu$. The overall electrical neutrality of the plasma is ensured by ions, whose contribution to the photon self-energy could be taken into account, if necessary, but which will be neglected in what follows. In order to compute the integral in (6.26), we use equation (5.77) to perform the frequency sum; (6.27) becomes

$$
J = -\frac{1}{8\pi^2} \int \frac{k^2\,dk\,d\Omega}{4\pi}\, \frac{k^2}{E_1 E_2} \left[\frac{1 - \tilde{n}_+(E_1) - \tilde{n}_-(E_2)}{i\omega - E_1 - E_2} \right.
$$
$$
- \frac{\tilde{n}_-(E_1) - \tilde{n}_-(E_2)}{i\omega + E_1 - E_2} + \frac{\tilde{n}_+(E_1) - \tilde{n}_+(E_2)}{i\omega - E_1 + E_2}
$$
$$
\left. - \frac{1 - \tilde{n}_-(E_1) - \tilde{n}_+(E_2)}{i\omega + E_1 + E_2} \right] \tag{6.98}
$$

The analogue of (6.35) is written by inspection. We recall that in (6.98) $\tilde{n}_+(E)$ $(\tilde{n}_-(E))$ is the Fermi–Dirac distribution function for electrons (positrons):

$$
\tilde{n}_\pm(E) = \frac{1}{e^{\beta(E \mp \mu)} + 1} \tag{6.99}
$$

Equation (6.98) is exact and can be used in order to compute $\Pi_{\mu\nu}$ to various approximations. We limit ourselves to the case where T and/or μ are much larger than the electron mass and the external momenta, so that we can use all our HTL machinery. The first and last terms in (6.98) give contributions that are proportional to

$$
\int_0^\infty [\tilde{n}_-(k) + \tilde{n}_+(k)]k\,dk = \frac{\mu^2}{2} + \frac{\pi^2 T^2}{6} \tag{6.100}
$$

For the second and last terms we use (see (6.29))

$$
\tilde{n}_\pm(E_1) - \tilde{n}_\pm(E_2) \simeq -q\cos\theta \frac{d\tilde{n}_\pm(k)}{dk} \tag{6.101}
$$

and

$$
\int_0^\infty k^2 dk \left(\frac{d\tilde{n}_+(k)}{dk} + \frac{d\tilde{n}_-(k)}{dk} \right) = -\left(\mu^2 + \frac{\pi^2 T^2}{3} \right) \tag{6.102}
$$

It is clear that all our HTL strategy remains unchanged: the only quantity which must be modified is the thermal mass, which becomes an effective mass:

$$
m^2 = \frac{e^2}{2\pi^2} \left(\mu^2 + \frac{\pi^2 T^2}{3} \right) \tag{6.103}
$$

Actually, all the results which follow in the present chapter and in the next one are generalized to the case of non-zero chemical potential, provided one uses the definition (6.103) of the effective mass.

6.4 Kinetic derivation of the photon self-energy

Historically, the preceding results were derived in a simple and intuitive way from kinetic theory. Let us consider a plasma and call $f^{(0)}(\mathbf{k})$ the equilibrium distribution function of charged particles with charge e; at equilibrium, this distribution function is of course position-independent. Assume that the plasma interacts with a weak space-time dependent electromagnetic field; the distribution function $f(\mathbf{x}, \mathbf{k}, t)$ obeys the Boltzmann equation:

$$Df(\mathbf{x}, \mathbf{k}, t) \equiv \frac{\partial f}{\partial t} + \mathbf{v} \cdot \frac{\partial f}{\partial \mathbf{x}} + \dot{\mathbf{k}} \cdot \frac{\partial f}{\partial \mathbf{k}} = C[f]$$

$$\dot{\mathbf{k}} = e(\mathbf{E} + \mathbf{v} \times \mathbf{B})$$

(6.104)

where D is the total derivative along a trajectory in phase space and $\mathbf{v} = \mathbf{k}/E_k$ is the velocity. In a rarefied plasma, the collision term $C[f]$ may be neglected; in our case this approximation is valid provided that $T \gg e^2/\bar{r}$, where $\bar{r} \sim T^{-1}$ is the average distance between neighbouring particles. This condition means that the kinetic energy of the particles is much larger than the potential energy of two neighbouring particles, and this is satisfied in the weak coupling limit $e^2 \ll 1$. The fields $\mathbf{E}$ and $\mathbf{B}$ are macroscopic averages, which cannot vary significantly over the microscopic scale $\bar{r}$: they must describe slow space-time variations of the plasma, associated to the collective motion of charged particles. For a weak electromagnetic field, the deviation $f^{(1)}$ with respect to equilibrium:

$$f(\mathbf{x}, \mathbf{k}, t) = f^{(0)}(\mathbf{k}) + f^{(1)}(\mathbf{x}, \mathbf{k}, t)$$

(6.105a)

obeys the linearized equation

$$\frac{\partial}{\partial t} f^{(1)} + \mathbf{v} \cdot \frac{\partial}{\partial \mathbf{x}} f^{(1)} = -e(\mathbf{E} + \mathbf{v} \times \mathbf{B}) \cdot \frac{\partial f^{(0)}}{\partial \mathbf{k}}$$

(6.105b)

Since the equilibrium distribution is isotropic, the magnetic field $\mathbf{B}$ does not contribute to the RHS of (6.105b). The induced charge density $\rho_{\text{ind}}(t, \mathbf{x})$ and current density $\mathbf{j}_{\text{ind}}(t, \mathbf{x})$ due to the deviation from equilibrium are

$$\rho_{\text{ind}}(t, \mathbf{x}) = e \int \frac{d^3k}{(2\pi)^3} f^{(1)}(\mathbf{x}, \mathbf{k}, t)$$

(6.106a)

$$\mathbf{j}_{\text{ind}}(t, \mathbf{x}) = e \int \frac{d^3k}{(2\pi)^3} \mathbf{v} f^{(1)}(\mathbf{x}, \mathbf{k}, t)$$

(6.106b)

This current is conserved from (6.105b), as the integral over $\mathbf{k}$ of the RHS of (6.105b) is zero for an isotropic distribution. In (6.106) we should of course sum over all kinds of charged particle; we shall take this remark into account later on. At equilibrium the plasma is electrically neutral and $\rho_{eq} = j_{eq} = 0$.

Assume that we apply a weak, slowly varying, external electromagnetic field to the plasma; the source of this external field is denoted by $j^{\mu}_{cl} = (\rho_{cl}, j_{cl})$, and it is assumed to be switched on adiabatically: at $t = -\infty$ the plasma is at equilibrium. We wish to determine the response of the plasma from kinetic theory, in parallel with what was done in section 6.3 from field theory. The source of the total electromagnetic field in (6.105) is obtained by adding to the classical current (ρ_{cl}, j_{cl}) the induced current (ρ_{ind}, j_{ind}) of (6.106). Let us define the polarization $\mathbf{P}$ by

$$\frac{\partial \mathbf{P}}{\partial t} = j_{ind} \qquad \nabla \cdot \mathbf{P} = -\rho_{ind} \qquad (6.107)$$

Because the current (ρ_{ind}, j_{ind}) is conserved, and because we switch on the external current adiabatically, the two equations in (6.107) give equivalent expressions of the polarization. If we now define the displacement $\mathbf{D}$ by $\mathbf{D} = \mathbf{E} + \mathbf{P}$, Maxwell's equations read

$$\nabla \times \mathbf{E} = -\frac{\partial \mathbf{B}}{\partial t} \qquad \nabla \cdot \mathbf{B} = 0$$

$$\nabla \times \mathbf{B} = j_{cl} + \frac{\partial \mathbf{D}}{\partial t} \qquad \nabla \cdot \mathbf{D} = \rho_{cl} \qquad (6.108a)$$

or in Fourier space

$$\mathbf{q} \times \mathbf{E} = q_0 \mathbf{B} \qquad i\mathbf{q} \cdot \mathbf{B} = 0$$

$$\mathbf{q} \times \mathbf{B} = -i j_{cl} - q_0 \mathbf{D} \qquad \mathbf{q} \cdot \mathbf{D} = -i\rho_{cl} \qquad (6.108b)$$

Equations (6.105), (6.106) and (6.108) form a closed system of equations, the Vlasov equations, which may be solved self-consistently, given the external sources (ρ_{cl}, j_{cl}). Taking the isotropy of $f^{(0)}(k)$ into account, we solve (6.105b) for $f^{(1)}(\mathbf{x}, \mathbf{k}, t)$ as

$$f^{(1)}(\mathbf{x}, \mathbf{k}, t) = -e \frac{df^{(0)}}{dk} \int_{-\infty}^{t} dt' \, \hat{\mathbf{k}} \cdot \mathbf{E}(t', \mathbf{x} - \mathbf{v}(t - t')) e^{\eta t'} \qquad (6.109)$$

where the factor $\exp(\eta t'), \eta \to 0^+$ takes care of the adiabatic switching of the external sources. Then we obtain the induced charge and current from (6.106). Let us write, for example, the induced current:

$$j_{ind}(t, \mathbf{x}) = -e^2 \int \frac{d^3 k}{(2\pi)^3} \frac{df^{(0)}}{dk} \mathbf{v} \int_0^{\infty} d\tau \, \hat{\mathbf{k}} \cdot \mathbf{E}(t - \tau, \mathbf{x} - \mathbf{v}\tau) e^{-\eta \tau} \qquad (6.110)$$

In the case of an ultrarelativistic plasma $\mathbf{v} = \hat{\mathbf{k}}$, so that the k and angular (Ω) integrations decouple. Note that this is not the general case, as the argument of $\mathbf{E}$ depends on $\mathbf{v}$, which is a function of k ($\mathbf{v} = \mathbf{k}/(k^2 + m_e^2)^{1/2}$), except when $m_e = 0$.

Specializing from now on to the case of an ultrarelativistic plasma, we take the Fourier transform of (6.110):

$$\mathbf{j}_{\text{ind}}(q_0, \mathbf{q}) = e^2 \int \frac{k^2 dk}{2\pi^2} \frac{df^{(0)}}{dk} \int \frac{d\Omega}{4\pi} \frac{\hat{k}\hat{k}_m E_m(q_0, \mathbf{q})}{i(q_0 - \mathbf{q} \cdot \hat{\mathbf{k}} + i\eta)} \qquad (6.111a)$$

Taking for $f^{(0)}(k)$ the Fermi–Dirac distribution $\tilde{n}_{\pm}(k)$ of massless electrons and positrons in a QED plasma, taking the helicity degrees of freedom into account and performing the k-integral thanks to (6.102) yields

$$\mathbf{j}_{\text{ind}}(q_0, \mathbf{q}) = 2m^2 i \int \frac{d\Omega}{4\pi} \frac{\hat{k}\hat{k}_m E_m(q_0, \mathbf{q})}{q_0 - \mathbf{q} \cdot \hat{\mathbf{k}} + i\eta} \qquad (6.111b)$$

with m^2 given in (6.103). This equation is now cast into an expression for the permittivity tensor ε_{lm} defined (in Fourier space) by

$$D_l = \varepsilon_{lm} E_m = E_l + P_l = E_l + \frac{i}{q_0} j_l^{\text{ind}} \qquad (6.112)$$

One finds

$$\varepsilon_{lm} = \delta_{lm} - \frac{2m^2}{q_0} \int \frac{d\Omega}{4\pi} \frac{\hat{k}_l \hat{k}_m}{q_0 - \mathbf{q} \cdot \hat{\mathbf{k}} + i\eta} \qquad (6.113)$$

It is convenient to decompose the dielectric tensor into transverse (ε_T)- and longitudinal (ε_L)-components:

$$\varepsilon_{lm} = (\delta_{lm} - \hat{q}_l \hat{q}_m) \varepsilon_T + \hat{q}_l \hat{q}_m \varepsilon_L \qquad (6.114)$$

These transverse and longitudinal components are given by, with as usual, $x = q_0/q$,

$$\varepsilon_T = 1 - \frac{m^2}{q_0^2} \left[x^2 + \frac{x(1 - x^2)}{2} \ln \frac{x+1}{x-1} \right] \qquad (6.115a)$$

$$\varepsilon_L = 1 - \frac{2m^2}{q^2} \left[\frac{x}{2} \ln \frac{x+1}{x-1} - 1 \right] \qquad (6.115b)$$

We are now in a position to display electromagnetic waves which are solutions of Maxwell's equations (6.108) in the absence of external sources ($\rho_{\text{cl}} = \mathbf{j}_{\text{cl}} = 0$), and thus represent waves propagating in the plasma. Longitudinal waves $\mathbf{E} \times \mathbf{q} = 0$ give a solution if $\mathbf{D} = \mathbf{B} = 0$; they are obtained from (6.110) if $\varepsilon_L = 0$. Transverse waves $\mathbf{E} \cdot \mathbf{q} = 0$ give a solution if

$$D_l = \frac{q^2}{q_0^2} (\delta_{lm} - \hat{q}_l \hat{q}_m) E_m \qquad (6.116)$$

with a transverse magnetic field; they are obtained if $\varepsilon_T = q^2/q_0^2$. From (6.74), (6.75) and (6.114) one checks immediately that $\varepsilon_L = 0$ is equivalent to $Q^2 - F = 0$ while $\varepsilon_T = q^2/q_0^2$ is equivalent to $Q^2 - G = 0$. Thus one recovers the dispersion laws for the longitudinal and transverse collective modes from the kinetic approach. One may also note the expressions for ε_L and ε_T as functions of F and G

$$\varepsilon_L = 1 - \frac{F}{Q^2} \qquad \varepsilon_T = 1 - \frac{G}{q_0^2} \tag{6.117}$$

The dielectric constants are complex for $|x| \le 1$; from (6.54) and (6.114) we find

$$\operatorname{Im} \varepsilon_T = \frac{\pi m^2}{2q_0^2} x(1 - x^2)\theta(1 - x^2) \tag{6.118a}$$

$$\operatorname{Im} \varepsilon_L = \frac{\pi m^2}{q^2} x\theta(1 - x^2) \tag{6.118b}$$

Note that $\operatorname{Im} \varepsilon_T$ and $\operatorname{Im} \varepsilon_L$ are both positive quantities for $x > 0$. Another way of computing $\operatorname{Im} \varepsilon_{lm}$ is to take the imaginary part of (6.113) directly: then the integrand involves the delta-function $\delta(\mathbf{q} \cdot \hat{\mathbf{k}} - q_0)$, which shows that the electrons contributing to $\operatorname{Im} \varepsilon_{lm}$ are those whose velocity component along the wave vector $\mathbf{q}$ is equal to the phase velocity of the external electromagnetic field: $v_z = \hat{\mathbf{q}} \cdot \hat{\mathbf{k}} = \cos\theta = q_0/q$, if one takes the z-axis parallel to $\hat{\mathbf{q}}$. This is the reason why $\operatorname{Im} \varepsilon_{lm}$ is non-zero only for $q_0^2 \le q^2$. Furthermore one can compute the amount of energy transferred to the plasma from the electromagnetic field. A standard formula of classical electrodynamics yields the energy $\langle dW/dt \rangle$ dissipated in the plasma by unit of time and volume for a periodic electric field $\mathbf{E}(t, \mathbf{x}) = \mathbf{E}(\mathbf{x}) \cos q_0 t$:

$$\left\langle \frac{dW}{dt} \right\rangle = \frac{1}{2} \int \frac{d^3 q}{(2\pi)^3} q_0 E_l(-\mathbf{q}) \left[\operatorname{Im} \varepsilon_{lm}(q_0, \mathbf{q}) \right] E_m(\mathbf{q}) \tag{6.119a}$$

This is the energy transferred from the gauge field to the plasma constituents, in the form of mechanical work done by the electric field on the charged particles. Using the explicit expression (6.113) for $\varepsilon_{lm}(q_0, \mathbf{q})$ yields

$$\left\langle \frac{dW}{dt} \right\rangle = \pi m^2 \int \frac{d^3 q}{(2\pi)^3} \int \frac{d\Omega}{4\pi} \delta(q_0 - \mathbf{q} \cdot \hat{\mathbf{k}})|\hat{\mathbf{k}} \cdot \mathbf{E}(\mathbf{q})|^2 \tag{6.119b}$$

$\langle dW/dt \rangle$ can also be expressed in terms of the external current j_μ^{cl}, which ultimately supplies the energy given to the field: see exercise 6.8.

This dissipation of energy in the plasma, or Landau damping, occurs in a collisionless plasma and is not linked to an increase in entropy; the energy transfer between the field and the particles is a reversible process. An

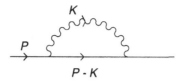

Fig. 6.4. One-loop approximation to the electron propagator. Solid lines: electrons. Wavy lines: photons.

easy calculation shows that it is proportional to $-\mathrm{d}f^{(0)}(k_z)/\mathrm{d}k_z|_{v_z} = q_0/q$; energy dissipation comes from the electrons whose velocity component along the wave vector $\mathbf{q}$ is close to q_0/q. Electrons travelling with a velocity slightly larger than q_0/q yield energy to the plasma, while electrons travelling with a velocity slightly smaller than q_0/q receive energy from the plasma. For an isotropic distribution, $-\mathrm{d}f^{(0)}(k_z)/\mathrm{d}k_z$ is always positive, so that $\langle \mathrm{d}W/\mathrm{d}t \rangle > 0$. However, one may well imagine situations where $-\mathrm{d}f^{(0)}(k_z)/\mathrm{d}k_z$ is negative, and where the field receives energy from the plasma. In this latter case the plasma is unstable; thus the positivity of $\mathrm{Im}\,\varepsilon_T$ and $\mathrm{Im}\,\varepsilon_L$ is directly linked to the plasma stability.

6.5 The electron propagator in a QED plasma

We now want to examine the electron propagator within the HTL approximation. We must thus evaluate the graph of fig. 6.4, which also defines the kinematics. The minus sign in the definition of the electron self-energy Σ is chosen in such a way that the full electron propagator $S(P)$ reads, in Euclidean space $(P_\mu = (p_4, \mathbf{p}) = (-\omega, \mathbf{p}))$,

$$S(P) = \frac{1}{\not{P} + m + \Sigma} \qquad (6.120)$$

6.5.1 Evaluation of Σ

The expression of Σ at one loop is, in the Feynman gauge,

$$\Sigma(P) = -e^2 \int \frac{\mathrm{d}^4 K}{(2\pi)^4} \gamma_\mu (\not{K} - \not{P}) \gamma_\mu \Delta(K) \tilde{\Delta}(P - K) \qquad (6.121)$$

In the HTL approximation we may neglect $\not{P}$ with respect to $\not{K}$ in (6.121) and

$$\Sigma(P) \simeq -2e^2 \int \frac{\mathrm{d}^4 K}{(2\pi)^4} \not{K} \Delta(K) \tilde{\Delta}(P - K) \qquad (6.122)$$

The frequency sum is performed by using (5.57) and (5.58)

$$\int \frac{d^4K}{(2\pi)^4} \Delta(K)\tilde{\Delta}(P-K) = -\frac{1}{8\pi^2} \int \frac{k^2\, dk\, d\Omega}{4\pi} \frac{1}{E_1 E_2} \Big[(1 + n_1 - \tilde{n}_2)$$
$$\times \left(\frac{1}{i\omega - E_1 - E_2} - \frac{1}{i\omega + E_1 + E_2} \right) + (n_1 + \tilde{n}_2)$$
$$\times \left(\frac{1}{i\omega + E_1 - E_2} - \frac{1}{i\omega - E_1 + E_2} \right) \Big] \qquad (6.123a)$$

$$\int \frac{d^4K}{(2\pi)^4} \omega_n \Delta(K)\tilde{\Delta}(P-K) = \frac{1}{8\pi^2} \int \frac{k^2\, dk\, d\Omega}{4\pi} \frac{i}{E_2} \Big[(1 + n_1 - \tilde{n}_2)$$
$$\times \left(\frac{1}{i\omega - E_1 - E_2} + \frac{1}{i\omega + E_1 + E_2} \right) - (n_1 + \tilde{n}_2)$$
$$\times \left(\frac{1}{i\omega + E_1 - E_2} + \frac{1}{i\omega - E_1 + E_2} \right) \Big] \qquad (6.123b)$$

where

$$n_1 = n(E_1) = n(k) \qquad n_2 = \tilde{n}(E_2) = \tilde{n}(|\mathbf{k} - \mathbf{p}|) \qquad (6.124)$$

The first term in the square bracket of (6.123a) leads to a behaviour that is linear in T, and the corresponding term in (6.123b) is even more convergent; thus these terms are non-leading in T. Only the second term in (6.123) contributes to order T^2 and this term involves the sum of the statistical distributions for two different statistics. The situation is simpler than in the case of the photon propagator, and this simplification stems of course from the fact that the electron propagator is only linearly divergent at zero temperature. Following the same strategy as given in section 6.2 and using

$$\int_0^\infty k\, dk\, n(k) = 2 \int_0^\infty k\, dk\, \tilde{n}(k) = \frac{\pi^2 T^2}{6} \qquad (6.125)$$

we easily obtain

$$\int \frac{d^4K}{(2\pi)^4} k_i \Delta(K)\tilde{\Delta}(P-K) = -\frac{T^2}{16} \int \frac{d\Omega}{4\pi} \frac{\hat{k}_i}{P \cdot \hat{K}} \qquad (6.126a)$$

$$\int \frac{d^4K}{(2\pi)^4} \omega_n \Delta(K)\tilde{\Delta}(P-K) = -\frac{iT^2}{16} \int \frac{d\Omega}{4\pi} \frac{1}{P \cdot \hat{K}} \qquad (6.126b)$$

The expression for Σ follows:

$$\Sigma(P) = \frac{e^2 T^2}{8} \int \frac{d\Omega}{4\pi} \frac{\hat{K}}{P \cdot \hat{K}} \qquad \hat{K} = (-i, \hat{\mathbf{k}}) \qquad (6.127)$$

with $\gamma_4 = i\gamma_0$ and $P \cdot \hat{K} = i\omega + \mathbf{p} \cdot \hat{\mathbf{k}} = -ip_4 + \mathbf{p} \cdot \hat{\mathbf{k}}$. Explicit calculations in various gauges suggest that this result, obtained in the Feynman gauge, is in fact gauge-independent; a general field theoretical proof of this result can be given, but the argument is somewhat technical and we refer the reader to the literature.

Let us finally make the continuation to Minkowski space; the analytic continuation of $\hat{K}$ is in principle $\hat{K} = (-1, \hat{\mathbf{k}})$, but because of symmetries of the integrand, one can use $\hat{K} = (1, \hat{\mathbf{k}})$ as well, while other continuations are $\rlap{/}{A} \rightarrow -\rlap{/}{A}, A \cdot B \rightarrow -A \cdot B$ (see (5.51)). Thus in Minkowski space Σ has the same formal expression as in (6.127):

$$\Sigma(P) = \frac{e^2 T^2}{8} \int \frac{d\Omega}{4\pi} \frac{\hat{\rlap{/}{K}}}{P \cdot \hat{K}} \qquad \hat{K} = (1, \hat{\mathbf{k}}) \qquad (6.128)$$

The retarded dressed electron propagator reads

$$S(P) = \frac{i}{\rlap{/}{P} - \Sigma} \qquad (6.129)$$

6.5.2   Quasi-particles

The explicit evaluation of the expression of Σ in (6.128) is straightforward. One finds

$$\Sigma = \frac{m_f^2}{p} \gamma_0 Q_0\left(\frac{p_0}{p}\right) + \frac{m_f^2}{p} \gamma \cdot \hat{\mathbf{p}}\left(1 - \frac{p_0}{p} Q_0\left(\frac{p_0}{p}\right)\right) \qquad (6.130)$$

where we have introduced the characteristic mass scale, or fermion thermal mass m_f, through

$$m_f^2 = \frac{1}{8} e^2 T^2 \qquad (6.131)$$

Let us rewrite

$$iS^{-1}(P) = \rlap{/}{P} - \Sigma = A_0 \gamma_0 - A_S \gamma \cdot \hat{\mathbf{p}} \qquad (6.132)$$

with

$$A_0 = p_0 - \frac{m_f^2}{p} Q_0\left(\frac{p_0}{p}\right) \qquad A_S = p + \frac{m_f^2}{p}\left(1 - \frac{p_0}{p} Q_0\left(\frac{p_0}{p}\right)\right) \qquad (6.133)$$

The physical interpretation of our calculations is particularly transparent if we rewrite

$$-iS(P) = \frac{1}{2}\Delta_+(P)(\gamma_0 - \gamma \cdot \hat{\mathbf{p}}) + \frac{1}{2}\Delta_-(P)(\gamma_0 + \gamma \cdot \hat{\mathbf{p}}) \qquad (6.134)$$

where

$$\Delta_\pm(P) = (A_0 \mp A_S)^{-1}$$

$$= \left(p_0 \mp p - \frac{m_f^2}{2p}\left[\left(1 \mp \frac{p_0}{p}\right) \ln \frac{p_0 + p}{p_0 - p} \pm 2\right]\right)^{-1} \qquad (6.135)$$

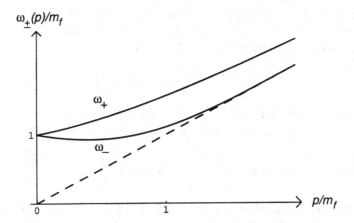

Fig. 6.5   Dispersion laws for the fermionic excitations.

Equation (6.134) is easily checked by multiplying both sides by $iS^{-1}(P)$. The denominator in (6.134) has an imaginary part for space-like values of P ($p_0^2 < p^2$), as in the case of the photon propagator. It is useful to note the parity properties:

$$\text{Re } \Delta_+(p_0, p) = -\text{Re } \Delta_-(-p_0, p) \qquad (6.136a)$$

$$\text{Im } \Delta_+(p_0, p) = \text{Im } \Delta_-(-p_0, p) \qquad (6.136b)$$

The zeros of the denominators in (6.134) give the position of the quasi-particle poles. One finds that the equation $A_0 - A_S = 0$ has two solutions:

$$p_0 = \omega_+(p) \qquad p_0 = -\omega_-(p) \qquad (6.137)$$

and, from the parity properties (6.136), the equation $A_0 + A_S = 0$ has also two solutions:

$$p_0 = \omega_-(p) \qquad p_0 = -\omega_+(p) \qquad (6.138)$$

where $\omega_\pm(p)$ is chosen to be positive. At $T = 0$, the theory is chirally invariant, since the fermions are massless. Chiral invariance is not broken by the temperature: the self-energy $\Sigma(P)$ (6.130) and the propagator $S(P)$ (6.134) anticommute with γ_5. If we look at the pole at $p_0 = \omega_+(p)$, the Dirac spinors associated with it are eigenstates of $(\gamma_0 - \gamma \cdot \hat{\mathbf{p}})$, and they have a positive helicity over chirality ratio: $\chi = +1$. On the other hand, if $p_0 = \omega_-(p)$, the Dirac spinors which are eigenstates of $(\gamma_0 + \gamma \cdot \hat{\mathbf{p}})$ have a negative helicity over chirality ratio: $\chi = -1$. In the vacuum, positive energy fermions have $\chi = +1$; introducing the heat bath allows the emergence of collective fermionic modes having $\chi = -1$. These collective modes are sometimes called antiquark holes, or plasminos.

The dispersion laws $\omega_\pm(p)$ are illustrated in fig. 6.5. One notices the negative derivative of $\omega_-(p)$ at $p = 0$ and the minimum at $p/m \simeq 0.41$. For small values of p the dispersion laws are

$$p \ll m_f : \quad \omega_\pm(p) \simeq m_f \pm \frac{1}{3}p \tag{6.139}$$

Since helicity is not defined for $p = 0$, one finds of course that $\omega_+(0) = \omega_-(0)$. For large values of p ($m_f \ll p \ll T$) we have

$$p \gg m_f : \quad \omega_+(p) \simeq p + \frac{m_f^2}{p} \quad \omega_-(p) \simeq p + \frac{2p}{e}\exp\left(-\frac{2p^2}{m_f^2}\right) \tag{6.140}$$

As in the case of the photon propagator, we define the spectral functions $\rho_\pm(p_0, p)$ corresponding to the propagators $\Delta_\pm(p)$ (note the minus sign in the definition of $\rho_\pm$):

$$
\begin{aligned}
\rho_\pm(p_0, p) = &- 2\mathrm{Im}\,\Delta_\pm(p_0, p) = 2\pi\left[Z_\pm(p)\delta(p_0 - \omega_\pm(p))\right. \\
&\left. + Z_\mp(p)\delta(p_0 + \omega_\mp(p))\right] + \frac{\pi}{p}m_f^2(1 \mp x)\theta(1 - x^2) \\
&\times \left[\left(p(1 \mp x) \pm \frac{m_f^2}{2p}\left[(1 \mp x)\right.\right.\right. \\
&\times \left.\left.\left. \ln\left|\frac{x+1}{x-1}\right| \pm 2\right]\right)^2 + \frac{\pi^2 m_f^4}{4p^2}(1 \mp x)^2\right]^{-1}
\end{aligned}
\tag{6.141}
$$

The residues at the quasi-particle poles have very simple expressions:

$$Z_\pm(p) = \frac{\omega_\pm^2(p) - p^2}{2m_f^2} \tag{6.142}$$

For small values of p one obtains the approximate expressions

$$p \ll m_f : \quad Z_\pm(p) \simeq \frac{1}{2} \pm \frac{p}{3m_f} \tag{6.143}$$

while for large values of p one has

$$p \gg m_f : \quad Z_+(p) \simeq 1 + \frac{m_f^2}{2p^2}\left(1 - \ln\left(\frac{2p^2}{m_f^2}\right)\right) \tag{6.144a}$$

$$Z_-(p) \simeq \frac{2p^2}{em_f^2}\exp\left(-\frac{2p^2}{m_f^2}\right) \tag{6.144b}$$

Before we comment on these equations, let us derive sum rules analogous to those obtained for transverse and longitudinal photons. Since $\Delta_+(p_0, p)$ and $\Delta_-(p_0, p)$ are meromorphic functions in the complex p_0-plane cut from $-p$ to $+p$, which decrease as $1/p_0$ when $|p_0| \to \infty$, we may write the dispersion relations:

$$\Delta_\pm(p_0 + i\eta, p) = -\int_{-\infty}^{\infty} \frac{dp_0'}{2\pi} \frac{\rho_\pm(p_0', p)}{p_0' - p_0 - i\eta} \tag{6.145}$$

Following the same strategy as in section (6.3.3) we derive the sum rules:

$$\int_{-\infty}^{\infty} \frac{dp_0}{2\pi} \rho_\pm(p_0, p) = 1 \tag{6.146}$$

$$\int_{-\infty}^{\infty} \frac{dp_0}{2\pi} p_0 \, \rho_\pm(p_0, p) = \pm p \tag{6.147}$$

$$\int_{-\infty}^{\infty} \frac{dp_0}{2\pi} p_0^2 \, \rho_\pm(p_0, p) = p^2 + m_f^2 \tag{6.148}$$

etc. From the parity properties (6.136) of $\rho_\pm$, one sees that non-trivial sum rules are obtained for the combination $\rho_+ + \rho_-$ in (6.146) and (6.148), and for the combination $\rho_+ - \rho_-$ in (6.147). The RHS of (6.146) is independent of p, because this sum rule is a consequence of the canonical anticommutation relations:

$$\left\{ \psi(\mathbf{x}, t), \psi^\dagger(\mathbf{x}', t) \right\} = \delta^{(3)}(\mathbf{x} - \mathbf{x}') \tag{6.149}$$

as is shown in (5.41). One also notices from (6.143) and (6.144) that the sum rule (6.146) is almost exhausted by the quasi-particle poles for $p \ll m_f$ as well as for $p \gg m_f$, as this sum rule can be written

$$Z_+(p) + Z_-(p) + \int_{-p}^{+p} \frac{dp_0}{2\pi} \rho_\pm(p_0, p) = 1 \tag{6.150}$$

For $p \ll m_f$, both kinds of quasiparticle give a similar contribution to the sum rules, but for $p \gg m_f$ the $\chi = -1$ excitation decouples and the $\chi = +1$ excitation almost saturates the sum rule; since a zero-temperature massless fermion has $\chi = +1$, one can see that 'normal' zero-temperature fermions are recovered at large values of p. For intermediate values of p one finds numerically that

$$0.8 \le Z_+(p) + Z_-(p) \le 1$$

so that the contribution from the continuum to the sum rule is always small.

Let us conclude with three remarks.

(i) One should note the close similarity between the behaviour of the photon and fermion collective modes. At large (or hard) momentum, these collective modes resemble the particles in the vacuum closely, namely, transverse photons and fermions with $\chi = +1$, while the longitudinal modes, as well as the fermionic modes with $\chi = -1$, decouple from the plasma. The difference between $T = 0$ and $T \ne 0$ is that photons and massless fermions acquire a thermal mass of order eT. However, this does not break the symmetries of the $T = 0$ theory: gauge invariance and chiral symmetry are not broken. At low (= soft) momenta, transverse and longitudinal photon modes are equally important, and the same property is true of $\chi = +1$ and $\chi = -1$ fermionic excitations.

(ii) It is interesting to consider the case of massive fermions (at $T = 0$); let m_0 be the mass in the vacuum. If $m_0 \ll eT$, one recovers the case of massless fermions. When m_0 increases, the $\chi = -1$ branch decouples progressively, and completely disappears for $m_0 \gg eT$. The only effect of temperature is to add a thermal correction to m_0.

(iii) We have completely ignored finite renormalizations, because they are non-leading in T: they are negligible with respect to the T^2-term at large T. However, Σ is infrared divergent, and care must be taken of the infrared divergences, which should cancel in physical quantities. This point will be examined more fully in chapter 10.

6.6 Elementary excitations in a QCD plasma

It is remarkable that the results of the preceding sections can be generalized to QCD with only minor modifications. Contrary to the photon self-energy in QED, the gluon self-energy is not gauge-fixing independent and it is indeed very surprising to discover that its HTL approximation does not depend on the choice of gauge. This property has been checked by explicit computations in different gauges (covariant, Coulomb, temporal axial), and, as in the case of the fermion propagator, it is possible to give a general field theoretical proof of this result.

We first compute the gluon self-energy in Euclidean space and in the Feynman gauge; four diagrams (fig. 6.6) contribute at the one-loop order. The first one (tadpole) is Q-independent:

$$\Pi_{\mu\nu}^{(a)}(Q) = 3 \, C_A \, g^2 \delta_{\mu\nu} \int \frac{\mathrm{d}^4 K}{(2\pi)^4} \Delta(K) \tag{6.151}$$

where g is the QCD coupling constant and C_A ($C_A = 3$) is the group factor. The calculation of the second graph is slightly simplified in the HTL approximation, if one notes that the three-gluon vertex can be written to this approximation, with all momenta incoming:

$$\Gamma_{\mu\nu\rho}^{abc} = ig f_{abc} [2\delta_{\nu\rho} K_\mu - \delta_{\mu\nu} K_\rho - \delta_{\mu\rho} K_\nu] \tag{6.152}$$

where a, b, c are colour indices, μ, ν, ρ are Lorentz indices and K is the (hard) loop-momentum in fig. 6.6 (see Appendix A for details of our

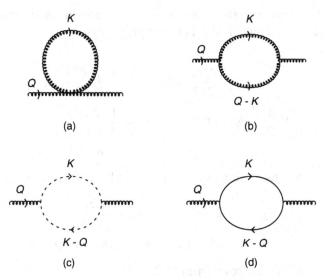

Fig. 6.6.   Feynman graphs for the gluon self-energy. Curly lines: gluons. Dashed lines: ghosts. Solid lines: quarks.

conventions). We find for $\Pi_{\mu\nu}^{(b)}$ (gluon loop)

$$\Pi_{\mu\nu}^{(b)}(Q) = -g^2 C_A \int \frac{d^4 K}{(2\pi)^4} (K^2 \delta_{\mu\nu} + 5K_\mu K_\nu)\Delta(K)\Delta(Q-K) \qquad (6.153)$$

and for $\Pi_{\mu\nu}^{(c)}$ (ghost loop)

$$\Pi_{\mu\nu}^{(c)}(Q) = g^2 C_A \int \frac{d^4 K}{(2\pi)^4} K_\mu K_\nu \Delta(K)\Delta(Q-K) \qquad (6.154)$$

Adding the gluon and ghost loops gives

$$\Pi_{\mu\nu}^{(a+b+c)}(Q) = -g^2 C_A \int \frac{d^4 K}{(2\pi)^4}$$
$$\times (4K_\mu K_\nu - 2K^2 \delta_{\mu\nu})\Delta(K)\Delta(Q-K) \qquad (6.155)$$

Finally, the quark–antiquark loop gives, within a factor $N_f/2$, where N_f is the number of flavours, the same expression as in QED:

$$\Pi_{\mu\nu}^{(d)}(Q) = \frac{g^2}{2}N_f \int \frac{d^4 K}{(2\pi)^4} (8K_\mu K_\nu - 4K^2 \delta_{\mu\nu})\tilde{\Delta}(K)\tilde{\Delta}(Q-K) \qquad (6.156)$$

As has already been noted, in performing the frequency sums, one goes from a bosonic ($\Pi_{\mu\nu}^{(b)}$ or $\Pi_{\mu\nu}^{(c)}$) to a fermionic ($\Pi_{\mu\nu}^{(d)}$) loop by the substitution $n(E) \to -\tilde{n}(E)$. Furthermore, one notes that

$$\int_0^\infty k\, dk\, \tilde{n}(k) = \frac{1}{2}\int_0^\infty k\, dk\, n(k) \qquad (6.157)$$

and that a similar equation holds for the derivatives of the statistical factors. Thus we obtain in the HTL approximation

$$\int \frac{\mathrm{d}^4 K}{(2\pi)^4} \tilde{\Delta}(K) = -\frac{1}{2} \int \frac{\mathrm{d}^4 K}{(2\pi)^4} \Delta(K) \tag{6.158a}$$

$$\int \frac{\mathrm{d}^4 K}{(2\pi)^4} \tilde{\Delta}(K) \tilde{\Delta}(Q - K) = -\frac{1}{2} \int \frac{\mathrm{d}^4 K}{(2\pi)^4}$$
$$\times \Delta(K) \Delta(Q - K) \tag{6.158b}$$

Taking all these results into account, we find, for the gluon propagator,

$$\Pi_{\mu\nu}(Q) = -g^2 \left(C_A + \frac{1}{2} N_f \right) \int \frac{\mathrm{d}^4 K}{(2\pi)^4}$$
$$\times (4 K_\mu K_\nu - 2 K^2 \delta_{\mu\nu}) \Delta(K) \Delta(Q - K) \tag{6.159}$$

The final result can be read from (6.43):

$$\Pi_{\mu\nu}(Q) = \frac{1}{3} g^2 T^2 \left(C_A + \frac{1}{2} N_f \right) \int \frac{\mathrm{d}\Omega}{4\pi} \left(\frac{i\omega \hat{K}_\mu \hat{K}_\nu}{Q \cdot \hat{K}} + \delta_{\mu 4} \delta_{\nu 4} \right) \tag{6.160}$$

The result involves the thermal gluon mass squared m_g^2, or simply m^2:

$$m^2 = \frac{1}{6} g^2 T^2 \left(C_A + \frac{1}{2} N_f \right) \tag{6.161}$$

The only modification to the QED result (6.43) is a change of the thermal mass which becomes (6.161) instead of (6.44), and $\Pi_{\mu\nu}$ again obeys the Ward identity (6.45). In the case of the fermion self-energy, going from QED to QCD is even simpler: the result (6.127) is only modified by a group factor C_F ($C_F = 4/3$). The quark self-energy reads

$$\Sigma(P) = \frac{1}{8} g^2 T^2 C_F \int \frac{\mathrm{d}\Omega}{4\pi} \frac{\hat{K}}{P \cdot \hat{K}} \tag{6.162}$$

The result involves the fermion thermal mass squared:

$$m_f^2 = \frac{1}{8} g^2 T^2 C_F \tag{6.163}$$

Due to the close connection between QED and QCD results, all our preceding discussions on collective excitations, sum rules, etc. can readily be transposed to QCD, the only change being for gluons:

$$e^2 T^2 \rightarrow g^2 T^2 (C_A + \frac{1}{2} N_f) \tag{6.164}$$

and for quarks:

$$e^2 T^2 \rightarrow g^2 T^2 C_F \tag{6.165}$$

There is a third self-energy graph, the ghost self-energy, which we have ignored up to now. However, due to the fact that one power of momentum is associated with an external line, there are not enough powers of K to produce a hard thermal loop: the ghost self-energy does not exhibit a T^2-behaviour.

References and further reading

Linear response theory is reviewed, for example, in Fetter and Walecka (1971), chap. 5, or in Kapusta (1989), chap. 6. The photon hard thermal loop was first computed by Silin (1960), using well-known methods of plasma physics, and then by Fradkin (1965) using field theoretical methods. An early discussion of collective excitations in the QCD plasma may be found in Weldon (1982a). The importance of hard thermal loops was fully appreciated in the fundamental work of Braaten and Pisarski (1990a,b) and Frenkel and Taylor (1990); see also Barton (1990). The kinetic derivation of the photon hard thermal loop in section 6.4 was inspired by Lifschitz and Pitaevskii (1980), chap. 3, who also examine in detail the Landau damping mechanism. The electron hard thermal loop was first discussed by Klimov (1981), who discovered the 'plasmino', and by Weldon (1982b); see also Weldon (1989). The case of massive fermions is examined by Petitgirard (1992) and by Baym, Blaizot and Svetitsky (1992). The gluon hard thermal loop first appears in Kalashnikov and Klimov (1981) and in Weldon (1982a). The gauge independence of hard thermal loops and of dispersion laws was proved by Kobes, Kunstatter and Rebhan (1991). The magnetic mass in QED and scalar QED is studied by Blaizot, Iancu and Parwani (1995).

Exercises

6.1 Using $(\mathbf{E})_i = \partial_0 A_i - \partial_i A_0 = F_{0i}$ and the canonical commutation relations

$$[A_i(x), \dot{A}_j(x')]\delta(t - t') = i\delta_{ij}\delta^{(4)}(x - x')$$

show that

$$\theta(t - t')\langle[E_i(x), E_j(x')]\rangle = \partial_i\partial'_j D^R_{00}(x - x')$$
$$- \partial_i\partial_0 D^R_{0j}(x - x') - \partial_0\partial'_j D^R_{i0}(x - x')$$
$$+ \partial_0\partial'_0 D^R_{ij}(x - x') - i\delta_{ij}\delta^{(4)}(x - x')$$

6.2 Obtain (6.59) by using the retarded photon propagator in the Coulomb gauge

$$D^R_{\mu\nu} = \frac{i}{Q^2 - G}P^T_{\mu\nu} + \frac{Q^2}{q^2}\frac{i\delta_{\mu 0}\delta_{\nu 0}}{Q^2 - F}$$

and in the time axial gauge

$$D^R_{ij} = \frac{i}{Q^2 - G} P^T_{ij} + \frac{Q^2}{q_0^2} \frac{i}{Q^2 - F} \hat{q}_i \hat{q}_j$$

$$D^R_{00} = D^R_{i0} = 0$$

6.3 Obtain Kubo's formula (6.61) by applying linear response theory to the perturbation

$$V = \int d^3x \, A^{cl}_\mu(t, \mathbf{x}) \hat{j}^\mu(t, \mathbf{x})$$

with

$$\hat{j}^\mu = e \overline{\psi} \gamma^\mu \psi$$

Compute $j^{ind}_\mu = \delta \langle \hat{j}_\mu \rangle$ from the retarded commutator $\langle [\hat{j}_\mu(x), \hat{j}_\nu(x')]_R \rangle$ and use (5.160) to recover Kubo's formula.

6.4 Derive (6.43) from the formula for $\Pi_{\mu\nu}$ obtained in exercise 5.7.

6.5 Check equations (6.87)–(6.92).

6.6 Show that for a non-zero chemical potential the quark thermal mass becomes, in QCD,

$$m_f^2 = \frac{1}{8} g^2 C_F \left(T^2 + \frac{\mu^2}{\pi^2} \right)$$

6.7 (a) Starting from the Schwinger–Dyson equation for the photon propagator in QED, written in imaginary time:

$$\Pi_{\mu\nu}(Q) = e^2 \int \frac{d^4P}{(2\pi)^4} \, \text{Tr} \left[\gamma_\mu S(P) \Gamma_\nu(P, P + Q) S(P + Q) \right]$$

where S is the (full) electron propagator and Γ_ν the (full) electron–photon vertex, and using the general QED Ward identity in the form

$$\Gamma_\nu(P, P) = \frac{\partial S^{-1}(P)}{\partial P_\nu}$$

show that

$$\Pi_{\mu\nu}(0, q \to 0) = e^2 \int \frac{d^4P}{(2\pi)^4} \, \text{Tr} \left(\gamma_\mu \frac{\partial S(P)}{\partial P_\nu} \right)$$

After analytical continuation to real time, S is a function of $p_0 - \mu$, where μ is the chemical potential, so that one may trade the p_0 derivative for a μ-derivative. From this observation show that

$$\Pi_{00}(0, q \to 0) = -e^2 \left(\frac{\partial n}{\partial \mu} \right)_T = -e^2 \left(\frac{\partial^2 P(\mu, T)}{\partial \mu^2} \right)$$

where n is the particle density $\mathrm{Tr}\,(\hat{\psi}^{\dagger}(0,0)\hat{\psi}(0,0))$ and P is the pressure. Show that one can obtain the Debye mass to lowest order from (1.13), namely, from the ideal gas result!

(b) Show from the above results that $\Pi_{ii}(0, q \to 0)$ vanishes to all orders in perturbation theory: the magnetic mass is necessarily zero in QED. This result is formal in the sense that it could be spoiled by infrared divergences. However, a detailed analysis shows that these divergences are absent.

6.8 Use Maxwell's equations to write (6.119b) in terms of the external current

$$\left\langle \frac{dW}{dt} \right\rangle = -\frac{1}{2} \int \frac{d^3 q}{(2\pi)^3}\, q_0 \left[\frac{|\rho^{\mathrm{cl}}(\mathbf{q})|^2}{q^2} \mathrm{Im}\, \frac{1}{\varepsilon_L(q_0, \mathbf{q})} \right.$$
$$\left. + \frac{|\mathbf{j}_T^{\mathrm{cl}}(\mathbf{q})|^2}{q_0^2} \mathrm{Im}\, \frac{1}{\varepsilon_T(q_0, \mathbf{q}) - q^2/q_0^2} \right]$$

7

Hard thermal loops and resummation

In the preceding chapter we have learned that the leading behaviour in temperature of the gauge particle and fermion self-energy is proportional to T^2, and that this behaviour is obtained without too much effort in the HTL approximation. In the present chapter we generalize these results to N-point functions, computed at the one-loop approximation. We shall show that some (but not all!) N-point functions also behave as T^2, and that these N-point functions have rather simple expressions in the HTL approximation. This situation should be constrasted with that of the φ^4-theory, where only the two-point function behaves as T^2: once more, gauge theories are much richer than scalar theories.

We shall also discover that these N-point functions obey remarkable Ward identities, and that there are again striking similarities between QED and QCD. All our results can be expressed in a compact way by writing an effective Lagrangian. Most importantly, we shall show how to correct naïve perturbation theory, which breaks down for soft external momenta, by using a resummed (or effective) perturbative expansion. Some applications to physical processes will be given in section 7.3, and more will be given in the following chapters. We conclude with a kinetic derivation of hard thermal loops, which generalizes the results of section 6.4.

7.1 Hard thermal loops in QED

In order to begin with the simplest case, we first examine QED. We compute in turn the three-point function (photon–electron proper vertex) and the four-point functions.

7.1.1 The three-point function in QED

The one-loop approximation for the photon–electron vertex is illustrated in fig. 7.1, which also defines the momenta. The one-loop expression for

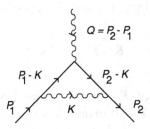

Fig. 7.1. Kinematics of the electron–photon vertex. Solid lines: electrons. Wavy lines: photons.

the vertex Γ_μ, including a factor of e in the definition, is, in the Feynman gauge and in Euclidean space,

$$e\Gamma_\mu(P_1, P_2) = e^3 \int \frac{\mathrm{d}^4 K}{(2\pi)^4} [\gamma_\alpha(\slashed{K} - \slashed{P}_2)\gamma_\mu(\slashed{K} - \slashed{P}_1)\gamma_\alpha]$$
$$\times \Delta(K)\tilde\Delta(P_2 - K)\tilde\Delta(P_1 - K) \tag{7.1}$$

We wish to compute Γ_μ within the HTL approximation, which holds, after frequency sums have been performed, if the external momenta P_1 and P_2 are much smaller than T: $\omega_1, p_1 \ll T$, $\omega_2, p_2 \ll T$. Since the main contribution from the loop integration comes from momenta $\sim T$, our first rule in HTL calculations will be

Rule 1: neglect all external momenta in 'numerators', that is, in the term between square brackets in (7.1), and analogous terms in further calculations.

Using this rule Γ_μ becomes

$$\Gamma_\mu(P_1, P_2) \simeq e^2 \int \frac{\mathrm{d}^4 K}{(2\pi)^4} \gamma_\alpha \left[-2K_\mu\slashed{K} + \gamma_\mu K^2\right]\gamma_\alpha$$
$$\times \Delta(K)\tilde\Delta(P_2 - K)\tilde\Delta(P_1 - K) \tag{7.2}$$

where, from now on, $\simeq$ means that we are working in the framework of the HTL approximation. Then we note that

$$\int \frac{\mathrm{d}^4 K}{(2\pi)^4} K^2 \Delta(K)\tilde\Delta(P_2 - K)\tilde\Delta(P_1 - K)$$
$$\simeq \int \frac{\mathrm{d}^4 K}{(2\pi)^4} \tilde\Delta(K)\tilde\Delta(P_2 - P_1 - K) \tag{7.3}$$

The last integral has already been encountered in (6.26), except that in (6.26) we found an additional k^2 factor which led to the T^2-behaviour. In fact (7.3) is quite similar to an integral which occurs in the one-loop correction to the vertex in φ^4-theory, and can easily be shown to behave as $\ln T$ (exercise 4.2). Thus, in the HTL approximation, we may neglect K^2 in (7.2) and write

$$\mathcal{K}\gamma_\mu\mathcal{K} \simeq -2K_\mu\mathcal{K} \qquad (7.4)$$

This a particular case of
Rule 2: ignore factors of K^2 in the numerators of one-loop integrals, except when the $T = 0$ expression is quadratically divergent. This divergence occurs only in the case of the gauge particle self-energy.
Our starting point for Γ_μ is then $(P_1 = (-\omega_1, \mathbf{p}_1), \; P_2 = (-\omega_2, \mathbf{p}_2))$:

$$\Gamma_\mu(P_1, P_2) = -4\, e^2\gamma_\alpha \int \frac{\mathrm{d}^4K}{(2\pi)^4} K_\mu K_\alpha \Delta(K)\tilde{\Delta}(P_2 - K)\tilde{\Delta}(P_1 - K) \qquad (7.5)$$

Let us denote by X_0 the frequency sum

$$X_0 = T\sum_n \Delta(K)\tilde{\Delta}(P_2 - K)\tilde{\Delta}(P_1 - K) \qquad (7.6)$$

Using (4.43) with $f(s_i E_i) \to -\tilde{f}(s_i E_i)$ we obtain

$$X_0 = -\sum_{s,s_1,s_2} \frac{ss_1s_2}{8EE_1E_2} \frac{1}{i(\omega_1 - \omega_2) - s_1E_1 + s_2E_2}$$

$$\times \left[\frac{1 + f(sE) - \tilde{f}(s_1E_1)}{i\omega_1 - sE - s_1E_1} - \frac{1 + f(sE) - \tilde{f}(s_2E_2)}{i\omega_2 - sE - s_2E_2}\right] \qquad (7.7)$$

where s, s_1, $s_2 = \pm 1$ and

$$E = k \quad E_1 = |\mathbf{p}_1 - \mathbf{k}| \quad E_2 = |\mathbf{p}_2 - \mathbf{k}| \qquad (7.8)$$

Counting powers of k, it is easy to see that the only possibility of obtaining a T^2-behaviour is to have

$$s = -s_1 = -s_2$$

Thus,

$$X_0 \simeq -\frac{1}{8EE_1E_2}\left(\frac{1}{i(\omega_1 - \omega_2) + E_1 - E_2}\left[\frac{n + \tilde{n}_1}{i\omega_1 - E + E_1}\right.\right.$$

$$\left.- \frac{n + \tilde{n}_2}{i\omega_2 - E + E_2}\right] + \frac{1}{i(\omega_1 - \omega_2) - E_1 + E_2}$$

$$\left.\times \left[\frac{n + \tilde{n}_1}{i\omega_1 + E - E_1} - \frac{n + \tilde{n}_2}{i\omega_2 + E - E_2}\right]\right) \qquad (7.9)$$

In the HTL approximation we may use

$$E = k \quad E_1 \simeq k - \mathbf{p}_1 \cdot \hat{\mathbf{k}} \quad E_2 \simeq k - \mathbf{p}_2 \cdot \hat{\mathbf{k}} \qquad (7.10)$$

and $\tilde{n}_1 \simeq \tilde{n}_2 \simeq \tilde{n}(k)$ in order to simplify X_0:

$$X_0 \simeq \frac{n(k) + \tilde{n}(k)}{8k^3} \left[\frac{1}{(i\omega_1 - E + E_1)(i\omega_2 - E + E_2)} \right.$$
$$\left. + \frac{1}{(i\omega_1 + E - E_1)(i\omega_2 + E - E_2)} \right] \qquad (7.11)$$

Introducing the light-like four-vectors

$$\hat{K} = (-i, \hat{\mathbf{k}}) \quad \hat{K}' = (-i, -\hat{\mathbf{k}}) \qquad (7.12)$$

and remembering that $\omega_1 = -p_{14}$, $\omega_2 = -p_{24}$, X_0 can be rewritten as

$$X_0 = \frac{n(k) + \tilde{n}(k)}{8k^3} \left[\frac{1}{(P_1 \cdot \hat{K}')(P_2 \cdot \hat{K}')} + \frac{1}{(P_1 \cdot \hat{K})(P_2 \cdot \hat{K})} \right] \qquad (7.13)$$

We need two other frequency sums: X_1, which is obtained from (7.9) by using $\omega_n \rightarrow -isE$ (see (4.26)):

$$X_1 = T \sum_n \omega_n \Delta(K) \tilde{\Delta}(P_1 - K) \tilde{\Delta}(P_2 - K)$$
$$\simeq -i \frac{n(k) + \tilde{n}(k)}{8k^2} \left[\frac{1}{(P_1 \cdot \hat{K}')(P_2 \cdot \hat{K}')} - \frac{1}{(P_1 \cdot \hat{K})(P_2 \cdot \hat{K})} \right] \qquad (7.14)$$

and X_2, for which we use rule 2:

$$X_2 = T \sum_n \omega_n^2 \Delta(K) \tilde{\Delta}(P_1 - K) \tilde{\Delta}(P_2 - K) \simeq -k^2 X_0 \qquad (7.15)$$

With all these results in hand we revert to (7.5) and complete our calculation, with m_f^2 defined in (6.131):

$$\Gamma_\mu(P_1, P_2) = -m_f^2 \int \frac{d\Omega}{4\pi} \frac{\hat{K}_\mu \hat{K}}{(P_1 \cdot \hat{K})(P_2 \cdot \hat{K})} \qquad (7.16)$$

In QED we know that Γ_μ obeys the Ward identity

$$(P_1 - P_2)_\mu \Gamma_\mu(P_1, P_2) = \Sigma(P_1) - \Sigma(P_2) \qquad (7.17)$$

which must also hold for the leading term in T for Γ and Σ; it is straightforward to check from (7.16) that (7.17) is indeed obeyed if one uses the result (6.127) for Σ.

In QED, Ward identities for HTL such as (7.17) are of course expected; however, in order to understand their generalization to QCD, where the

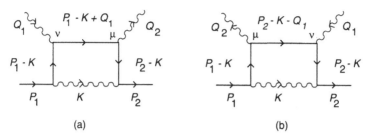

Fig. 7.2   Kinematics of the two-electron–two photon-vertex.

existence of similar identities is highly non-trivial, let us prove it directly
from our initial expression (7.5); we may write

$$(P_1 - P_2) \cdot K \simeq \frac{1}{2}(P_2 - K)^2 - \frac{1}{2}(P_1 - K)^2 \qquad (7.18)$$

Thus,

$$(P_1 - P_2)_\mu \Gamma_\mu(P_1, P_2) \simeq \int \frac{\mathrm{d}^4 K}{(2\pi)^4} \not{K} \left(\Delta(K)\tilde{\Delta}(P_1 - K) \right.$$
$$\left. -\Delta(K)\tilde{\Delta}(P_2 - K) \right) = \Sigma(P_1) - \Sigma(P_2) \qquad (7.19)$$

by comparison with (6.122). The other three-point function in QED is
the three-photon vertex which is of course zero from Furry's theorem or
charge conjugation.

7.1.2   *The four-point functions in QED*

Let us first consider the two-electron–two-photon vertex function. The
two one-loop graphs are illustrated in fig. 7.2, from which one can read
the kinematics. The contribution $\Gamma^{(a)}_{\mu\nu}$ of the first graph is written in the
Feynman gauge and in Euclidean space, within a factor of e^2:

$$\Gamma^{(a)}_{\mu\nu}(P_1, P_2, Q_1) \simeq 8e^2 \int \frac{\mathrm{d}^4 K}{(2\pi)^4} K_\mu K_\nu \not{K} \Delta(K)$$
$$\times \tilde{\Delta}(P_2 - K)\tilde{\Delta}(P_1 - K + Q_1)\tilde{\Delta}(P_1 - K) \qquad (7.20)$$

where we have repeatedly used (7.4), while $\Gamma^{(b)}_{\mu\nu}$ reads

$$\Gamma^{(b)}_{\mu\nu}(P_1, P_2, Q_1) \simeq 8e^2 \int \frac{\mathrm{d}^4 K}{(2\pi)^4} K_\mu K_\nu \not{K} \Delta(K)$$
$$\times \tilde{\Delta}(P_2 - K)\tilde{\Delta}(P_2 - K - Q_1)\tilde{\Delta}(P_1 - K) \qquad (7.21)$$

The expression for $\Gamma_{\mu\nu} = \Gamma^{(a)}_{\mu\nu} + \Gamma^{(b)}_{\mu\nu}$ could be worked out by generalizing
the method used for the vertex. Here we shall limit ourselves to proving

the Ward identities. As in the preceding section we use

$$Q_1 \cdot K \simeq \frac{1}{2}(K - P_1)^2 - \frac{1}{2}(K - Q_1 - P_1)^2 \qquad (7.22)$$

so that, comparing with (7.5),

$$Q_{1\mu}\Gamma_{\mu\nu}^{(a)}(P_1, P_2, Q_1) \simeq \Gamma_\nu(P_1, P_2) - \Gamma_\nu(P_1 + Q_1, P_2) \qquad (7.23)$$

Similarly, we obtain for $\Gamma_{\mu\nu}^{(b)}$

$$Q_{1\mu}\Gamma_{\mu\nu}^{(b)}(P_1, P_2, Q_1) \simeq \Gamma_\nu(P_1, P_2 - Q_1) - \Gamma_\nu(P_1, P_2) \qquad (7.24)$$

Adding (7.23) and (7.24) gives the Ward identity

$$Q_{1\mu}\Gamma_{\mu\nu}(P_1, P_2, Q_1) = \Gamma_\nu(P_1, P_2 - Q_1) - \Gamma_\nu(P_1 + Q_1, P_2) \qquad (7.25)$$

This is of course an identity which is generally valid in QED. However, we have in addition (rule 2)

$$\Gamma_{\mu\mu} \simeq 0 \qquad (7.26)$$

The explicit computation of $\Gamma_{\mu\nu}$ gives

$$\Gamma_{\mu\nu}(P_1, P_2; Q_1) = -m_f^2 \int \frac{d\Omega}{4\pi} \frac{\hat{K}_\mu \hat{K}_\nu \hat{K}}{[(P_1 + Q_1) \cdot \hat{K}][(P_2 - Q_1) \cdot \hat{K}]}$$
$$\times \left[\frac{1}{P_1 \cdot \hat{K}} + \frac{1}{P_2 \cdot \hat{K}} \right] \qquad (7.27)$$

It is easy to check that $\Gamma_{\mu\nu}$ as given by (7.27) obeys (7.25).

Let us now turn to the other four-point functions. The four-photon vertex $\Gamma_{\mu\nu\rho\sigma}$ is traceless in any pair of indices (rule 2) and obeys Ward identities of the form (7.25) involving the three-photon vertex on the RHS. However, the three-photon vertex is zero: these properties are sufficient to prove that the four-photon vertex is zero in the HTL approximation; details are given in section 7.2.3. Finally, the loop integral in the calculation of the four-electron vertex has only two powers of K in the numerator: thus it vanishes in the HTL approximation. It is not very difficult to show in general that the only non-zero proper vertices are in the HTL approximation:

- the photon self-energy $\Pi_{\mu\nu}$;
- the electron self-energy Σ;
- the N-photon–two-electron proper vertex.

7.1.3 The effective Lagrangian in QED

The preceding results can be summarized in a compact way by writing an effective Lagrangian:

$$\delta\mathscr{L} = \mathscr{L}_\gamma + \mathscr{L}_f \tag{7.28}$$

where $\mathscr{L}_\gamma$ and $\mathscr{L}_f$ are effective Lagrangians which generate (with some caveats, to be discused after (7.32)) hard thermal loops with zero and two electron external lines respectively; since hard thermal loops obey tree-like Ward identities and since there exists a field theoretical proof that they are gauge-fixing independent, one is led to look for a gauge-invariant $\delta\mathscr{L}$. Let us begin with $\mathscr{L}_f$, which is easier to derive. We work in Euclidean space, where Fourier transforms imply that $\partial_\mu \to ip_\mu$.

Equation (6.127) for the electron propagator suggests an effective Lagrangian $\mathscr{L}'_f$ of the form

$$\mathscr{L}'_f = m_f^2\, \overline{\psi}(p) \int \frac{d\Omega}{4\pi} \frac{\hat{K}}{P \cdot \hat{K}}\, \psi(-p) \tag{7.29}$$

which reads in x-space

$$\mathscr{L}'_f = im_f^2\, \overline{\psi}(x) \int \frac{d\Omega}{4\pi} \frac{\hat{K}}{\hat{K} \cdot \partial}\, \psi(x) \tag{7.30}$$

By construction, $\mathscr{L}'_f$ generates the correct electron propagator; however, it is not gauge-invariant. In order to make it gauge-invariant, it is sufficient to substitute the covariant derivative D_μ for the ordinary derivative ∂_μ:

$$D_\mu = \partial_\mu - ieA_\mu \tag{7.31}$$

Thus we obtain the gauge-invariant effective Lagrangian

$$\mathscr{L}_f = im_f^2\, \overline{\psi}(x) \int \frac{d\Omega}{4\pi} \frac{\hat{K}}{\hat{K} \cdot D}\, \psi(x) \tag{7.32}$$

By computing $\delta\mathscr{L}_f/\delta\overline{\psi}\delta\psi\delta A_\mu$, it is straightforward to recover the electron–photon vertex (7.16); it is also clear from (7.32) that there are no HTLs with more than two electron lines. Since $\hat{K} \cdot D$ appears in the denominator of (7.32), this effective Lagrangian is non-local. If one performs an analytical continuation to Minkowski space, the denominator in (7.32) may vanish, and the effective Lagrangian is only defined on fields $\psi(x)$ such that $\hat{K} \cdot D\psi(x)$ does not vanish. In Fourier space, and in the case of the electron self-energy, this means that $P \cdot \hat{K}$ must not vanish, which is certainly true if P is time-like: in this region the electron self-energy has no imaginary part.

Let us now turn to the photon Lagrangian $\mathscr{L}_\gamma$. From (6.43) one may easily guess the following form for $\mathscr{L}_\gamma$:

$$\mathscr{L}_\gamma = \frac{1}{2} A_\mu \Pi_{\mu\nu} A_\nu = m^2 \left[A_4^2 + \int \frac{d\Omega}{4\pi} (\hat{K} \cdot A) \frac{i\omega}{\hat{K} \cdot Q} (A \cdot \hat{K}) \right] \qquad (7.33)$$

or, in x-space,

$$\mathscr{L}_\gamma = m^2 \left[A_4^2 - \int \frac{d\Omega}{4\pi} (\hat{K} \cdot A) \frac{i\partial_4}{\hat{K} \cdot \partial} (A \cdot \hat{K}) \right] \qquad (7.34)$$

The A_4^2-term in (7.33) or (7.34) is reminiscent of a mass term; indeed, in a purely static situation, this term gives rise to the Debye mass $m_D = \sqrt{2}m$. This term is clearly gauge-invariant for purely static gauge transformations

$$A_\mu \to A_\mu - \partial_\mu \chi \qquad (7.35)$$

where χ is x_4-independent. Although this property is not obvious at first sight, $\mathscr{L}_\gamma$ is gauge-invariant under the transformation (7.35) for a general function $\chi(x)$, which gives the following variations:

$$\delta A_4^2 = -2A_4 \partial_4 \chi \qquad (7.36a)$$

$$\delta \left[(\hat{K} \cdot A) \frac{-i\partial_4}{\hat{K} \cdot \partial} (A \cdot \hat{K}) \right] = 2(A \cdot \hat{K}) i \partial_4 \chi \qquad (7.36b)$$

Integrating (7.36b) over Ω gives $2A_4 \partial_4 \chi$. Equation (7.33) for $\mathscr{L}_\gamma$ gives the simplest expression for the effective Lagrangian in the case of QED. However, for easy generalization to QCD, it is useful to write $\mathscr{L}_\gamma$ in terms of the field strength tensor $F_{\mu\alpha} = \partial_\mu A_\alpha - \partial_\alpha A_\mu$; then $\mathscr{L}_\gamma$ reads

$$\mathscr{L}_\gamma = -\frac{m^2}{2} \int \frac{d\Omega}{4\pi} F_{\mu\alpha} \frac{\hat{K}_\alpha \hat{K}_\beta}{(\hat{K} \cdot \partial)^2} F_{\beta\mu} \qquad (7.37)$$

The equivalence between (7.34) and (7.37) is shown by working in q-space and using the following identities:

$$\int \frac{d\Omega}{4\pi} \frac{1}{(\hat{K} \cdot Q)^2} = -\frac{1}{Q^2} \qquad (7.38a)$$

$$\int \frac{d\Omega}{4\pi} \frac{\hat{k}_i}{(\hat{K} \cdot Q)^2} = \hat{q}_i \left[\frac{1}{q^2} Q_0 \left(\frac{i\omega}{q} \right) + \frac{i\omega}{qQ^2} \right] \qquad (7.38b)$$

$$\int \frac{d\Omega}{4\pi} \frac{\hat{k}_i \hat{k}_j}{(\hat{K} \cdot Q)^2} = \frac{\delta_{ij}}{q^2} Q_1 \left(\frac{i\omega}{q} \right)$$

$$- \hat{q}_i \hat{q}_j \left[\frac{3}{q^2} Q_1 \left(\frac{i\omega}{q} \right) + \frac{1}{Q^2} \right] \qquad (7.38c)$$

Expanding (7.37) and performing the angular integrations explicitly allows one to show the equivalence of (7.37) with (7.34). As in the case of the electron Lagrangian $\mathscr{L}_f$, one obtains a non-local photon Lagrangian $\mathscr{L}_\gamma$ and one must keep in mind the same warnings as in the electron case, namely, that $\mathscr{L}_\gamma$ is well-defined only on the space of fields such as $(\hat{K} \cdot \partial)$ does not vanish. If the fields are Fourier analyzed, this means that they must be functions of time-like momenta only, or, from (6.119b), that they propagate without dissipation. Finally, it is clear from either (7.34) or (7.37), since both equations are bilinear in A_μ, that Green's functions with N external photon lines and no electron external lines are zero in the HTL approximation for $N \geq 3$.

7.2 The QCD plasma

7.2.1 *The three-point functions in QCD*

The QCD case is slightly more complicated than the QED case, but the properties of HTLs are more unexpected and more remarkable. We shall discover a close relationship between the two cases and, in particular, the Ward identities for HTLs will turn out to be quite similar in both cases, in contradistinction to the $T = 0$ results. Let us begin with the three-gluon vertex: there are contributions from the gluon loop, the ghost loop and the quark loop (fig. 7.3 (a), (b) and (c)). The graph with a four-gluon vertex (fig. 7.3 (d)) does not contribute to the HTL approximation in the Feynman gauge, because it is of the self-energy type, but with one power of K less in the numerator. In the HTL approximation, the following result:

$$
\begin{aligned}
&\left[2\delta_{\lambda\sigma}K_v - \delta_{v\lambda}K_\sigma - \delta_{v\sigma}K_\lambda\right]\left[2\delta_{\sigma\tau}K_\rho - \delta_{\sigma\rho}K_\tau - \delta_{\tau\rho}K_\sigma\right] \\
&\simeq 4\delta_{\lambda\tau}K_vK_\rho - 2\delta_{\lambda\rho}K_vK_\tau - 2\delta_{v\tau}K_\lambda K_\rho \\
&\quad - \delta_{\tau\rho}K_\lambda K_v - \delta_{v\lambda}K_\rho K_\tau + \delta_{v\rho}K_\lambda K_\tau
\end{aligned} \tag{7.39}
$$

allows one to compute the contribution from the gluon loop easily:

$$
\begin{aligned}
\Gamma^{(\text{gl})}(P,Q,R) = ig^3 C_A f_{abc} \int \frac{\mathrm{d}^4 K}{(2\pi)^4}\, 9 K_\mu K_v K_\rho \\
\times \Delta(K)\Delta(R-K)\Delta(Q+K)
\end{aligned} \tag{7.40}
$$

while the ghost contribution reads

$$
\begin{aligned}
\Gamma^{(\text{gh})}(P,Q,R) = -ig^3 C_A f_{abc} \int \frac{\mathrm{d}^4 K}{(2\pi)^4}\, K_\mu K_v K_\rho \\
\times \Delta(K)\Delta(R-K)\Delta(Q+K)
\end{aligned} \tag{7.41}
$$

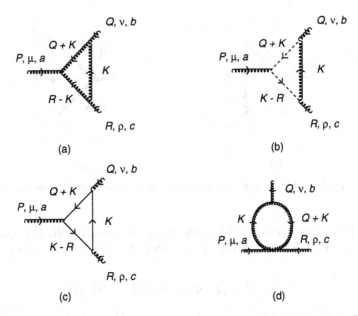

Fig. 7.3.   Feynman graphs for the three-gluon vertex. Curly lines: gluons. Dashed lines: ghosts. Solid lines: quarks.

Factoring out igf_{abc} which is common to the bare vertex, we obtain

$$\Gamma_{\mu\nu\rho}(P,Q,R) = g^2 C_A \int \frac{d^4K}{(2\pi)^4} \, 8K_\mu K_\nu K_\rho$$
$$\times \Delta(K)\Delta(R-K)\Delta(Q+K) \qquad (7.42)$$

This expression allows us to derive a QED-like Ward identity by writing

$$P \cdot K \simeq \frac{1}{2}(R-K)^2 - \frac{1}{2}(Q+K)^2 \qquad (7.43)$$

Comparing with (6.155) we derive

$$P_\mu \Gamma_{\mu\nu\rho} = \Pi_{\nu\rho}(R) - \Pi_{\nu\rho}(Q) \qquad (7.44)$$

where on the RHS we have taken into account the gluon and ghost contributions to $\Pi_{\mu\nu}$. It is worth noting that the Ward identity does not hold separately for the gluon and ghost loops (fig. 6.6); only the sum obeys (7.44). The quark loop gives

$$\Gamma^{(f)}_{\mu\nu\rho}(P,Q,R) = 4g^2 N_f \int \frac{d^4K}{(2\pi)^4} \, K_\nu K_\rho \mathrm{Tr}(\not{K}\gamma_\mu)$$
$$\times \tilde{\Delta}(K)\tilde{\Delta}(K-R)\tilde{\Delta}(Q+K) \qquad (7.45)$$

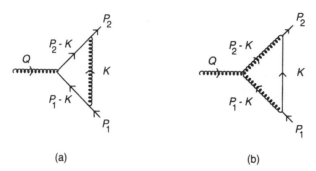

Fig. 7.4   The quark–gluon vertex. Same definitions as in fig. 7.3.

Using (7.43) once more and comparing the result with the quark contribution $\Pi_{\mu\nu}^{(f)}$ (6.156) of the quark loop to the gluon self-energy, one
finds

$$P_\mu \Gamma_{\mu\nu\rho}^{(f)}(P,Q,R) = \Pi_{\nu\rho}^{(f)}(R) - \Pi_{\nu\rho}^{(f)}(Q) \qquad (7.46)$$

It remains to evaluate the loop integrals in (7.42) and (7.45); this evaluation
follows closely that performed in the case of the electron–photon vertex,
and we give the final result directly, which includes gluon, ghost and quark
loops:

$$\Gamma_{\mu\nu\rho}(P,Q,R) = 2m^2 \int \frac{d\Omega}{4\pi} \hat{K}_\mu \hat{K}_\nu \hat{K}_\rho$$
$$\times \left[\frac{i\omega_r}{(P\cdot\hat{K})(R\cdot\hat{K})} - \frac{i\omega_q}{(P\cdot\hat{K})(Q\cdot\hat{K})} \right] \qquad (7.47)$$

and which obeys the Ward identity (7.44) with $\Pi_{\nu\rho}$ as in (6.160).
   We now evaluate the quark–gluon vertex. There are two graphs: the
first one (fig. 7.4 (a)) is QED like, and the second one (fig. 7.4 (b)) is
typical of QCD. The group factor for graph (a) is $(C_F - C_A/2)t_a$, where t_a
is a colour matrix of the fundamental representation, and its expression is
obtained from (7.16) by multiplying the QED result by this group factor.
Factoring out gt_a, which is common to the bare vertex, one can write the
expression for graph (b) as

$$\Gamma_\mu^{(b)}(P_1,P_2) = 2g^2 C_A \int \frac{d^4K}{(2\pi)^4} K_\mu \tilde{K}\tilde{\Delta}(K)\Delta(P_2-K)\Delta(P_1-K) \qquad (7.48)$$

In performing the summation over ω_n, one goes from graph (a) to graph
(b) with the usual substitution $f \to -\tilde{f}$:

$$\frac{1+f(sE)-\tilde{f}(s_1E_1)}{i\omega_1 - sE - s_1E_1} \to \frac{1-\tilde{f}(sE)+f(s_1E_1)}{i\omega_1 - sE - s_1E_1} \qquad (7.49)$$

and in the HTL approximation this substitution leads to a change of sign. Thus, upon adding graphs (a) and (b), the terms proportional to C_A cancel out and it is only necessary to change the thermal mass in the QED result (7.16):

$$\Gamma_\mu(P_1, P_2) = -m_f^2 \int \frac{d\Omega}{4\pi} \frac{\hat{K}_\mu \hat{K}}{(P_1 \cdot \hat{K})(P_2 \cdot \hat{K})} \qquad (7.50)$$

where m_f is given in (6.163). In general, graphs (a) and (b) are not related, and it is quite remarkable that they are proportional in the HTL approximation. As in the case of the two-point functions, the proper vertices with external ghosts do not have hard thermal loops, as there are not enough powers of K in the numerators.

7.2.2 *General remarks on hard thermal loops*

From our experience with specific cases, we are now ready to understand the various mechanisms which may lead to hard thermal loops by using simple power-counting arguments in imaginary time. Let N be the number of external legs of a proper vertex. We first consider $N = 2$.

(a) $N = 2$.
The frequency sum leads schematically to

$$\int \frac{d^4K}{(2\pi)^4} \Delta(K)\Delta(P - K) \sim \int dk\, X \qquad (7.51)$$

One must distinguish between three different cases, with $p \ll T$, $\omega \ll T$, and p generic for ω or p.

(i) X involves a sum of statistical factors of the same kind (both Bose–Einstein or both Fermi–Dirac)

$$X = \frac{n_1 + n_2}{i\omega + E_1 + E_2} \rightarrow \frac{n(k)}{k} \qquad (7.52)$$

In order to obtain a HTL, the numerator must contain two powers of k. In the case of the gluon self-energy, these powers of k come either from the fermion and antifermion propagators in the case of the quark loop, or from the vertices in the case of the gluon or ghost loops. This case is the only one which is not associated with Landau damping; it is associated with the $T = 0$ quadratic divergence. In fact, the complete numerator in the LHS of (7.52) is $(1 + n_1 + n_2)$, where the factor of one corresponds to the $T = 0$ limit.

(ii) X is associated with Landau damping and involves a difference of statistical factors when particles 1 and 2 obey identical statistics:

$$X = \frac{n_1 - n_2}{i\omega + E_1 - E_2} \rightarrow \frac{dn}{dk} \qquad (7.53)$$

The numerator must then contain two powers of k: this is the case of the gauge particle self-energy.

(iii) X is again associated with Landau damping, but involves a sum of statistical factors when particles 1 and 2 obey different statistics:

$$X = \frac{n_1 + \tilde{n}_2}{i\omega + E_1 - E_2} \rightarrow \frac{n(k) + \tilde{n}(k)}{p} \tag{7.54}$$

Then the numerator must contain one power of k; this is the case of the fermion self-energy.

(b) $N = 3$.

Using partial fractioning (see (4.42)), the frequency sum gives a leading behaviour of the form

$$\int \frac{d^4K}{(2\pi)^4} \Delta(K)\Delta(P_1 - K)\Delta(P_2 - K) \sim \int \frac{dk}{k} \frac{1}{p} X \tag{7.55}$$

As in the $N = 2$ case, one encounters three different possibilities (i)–(iii). Case (i) may occur for the three-gluon vertex, and one could expect a T^2-term not associated with Landau damping, since there are three powers of k in the numerator, stemming, for example, from the three vertices. However, it is easy to check by an explicit calculation that these terms cancel out: this cancellation must occur, because case (i) is associated with the $T = 0$ divergence, and the three-gluon proper vertex is only linearly, and not quadratically, divergent. Then HTLs can only be obtained from case (ii), with three powers of k in the numerator (three-gluon proper vertex (7.45)) and from case (iii), with two powers of k in the numerator (quark–gluon vertex (7.50)).

(c) $N \geq 4$.

The case $N = 3$ already belongs to the general case, and we may examine $N \geq 4$ only briefly. Using partial fractioning we write the leading contribution after performing the frequency sum:

$$\int \frac{d^4K}{(2\pi)^4} \Delta(K) \dots \Delta(P_{N-1} - K) \sim \int \frac{dk}{k^{N-2}} \frac{1}{p^{N-2}} X \tag{7.56}$$

As for $N = 3$, case (i) does not lead to HTLs, because the $T = 0$ integral is only logarithmically divergent ($N = 4$), or even convergent ($N > 4$). Case (ii) applies to the N-gluon vertex, where the numerator contains N powers of k, and case (iii) applies to the $(N - 2)$-gluon–two-quark vertex, where the numerator contains $(N - 1)$ powers of k. These are the only N-point proper vertices which have HTLs. All other N-point proper vertices, and in particular all proper vertices with external ghosts, have a non-leading behaviour in T.

One may recover from (7.56) the power-counting rules first established by Braaten and Pisarski. We have, without taking the numerator into account, and with $X \sim 1/T$ (case (ii)) or $X \sim 1/p$ (case (iii)),

$$\int \frac{dk}{k^{N-2}} \frac{1}{p^{N-2}} X \sim T^{-N+3} \, p^{-N+2} \, X \tag{7.57}$$

and this power counting is equivalent to

(a) $\int d^3k \to T^3$.
(b) 1st propagator: $T \sum_n \to T^{-1}$.
(c) Each additional propagator $\to (pT)^{-1}$.
(d) The statistical factor is p/T for lines with identical statistics and 1 for lines with different statistics.
(e) Each power of k in the numerator $\to T$.

Either from (7.57) or from the power-counting rules (a)–(e) we obtain for the N-gluon vertex (case (ii)) with N powers of k in the numerator

$$g^N \, T^{-N+3} \, p^{-N+2} \, T^N \, T^{-1} = g^N \, T^2 \, p^{-N+2} \tag{7.58}$$

while for the two-quark–$(N-2)$-gluon vertex, with $(N-1)$ powers of k in the numerator (case (iii)) we have ($p \ll T$)

$$g^N \, T^{-N+3} \, p^{-N+2} \, T^{N-1} \, p^{-1} = g^N \, T^2 \, p^{-N+1} \tag{7.59}$$

If now $p \sim gT$, we find from (7.58) that the N-gluon HTL is of order $g^2 T^{4-N}$, while the two-quark–$(N-2)$-gluon vertex is of order gT^{3-N}. Thus, for external momenta of order gT, the three-gluon HTL is of order $g^2 T = g(gT)$, which is of the same order of magnitude as the bare vertex, while the quark–gluon vertex is of order g, which is again of the same order of magnitude as the bare quark–gluon vertex, hence the necessity to reorganize perturbation theory.

7.2.3 *Hard thermal loops from Ward identities*

From our experience with two- and three-point functions and the results obtained in section 7.2.2, we are now in a position to derive the general form of HTLs; the derivation will rely heavily on Ward identities. Let us start from the typical integral appearing in an N-gluon proper vertex, with incoming momenta $P_1, \ldots, P_N$:

$$T_{\mu_1 \ldots \mu_N}(P_1, \ldots, P_N) = \int \frac{d^4K}{(2\pi)^4} K_{\mu_1} \ldots K_{\mu_N}$$
$$\times \Delta(K)\Delta(K - R_1) \ldots \Delta(K - R_{N-1}) \tag{7.60}$$

where we have defined R_l to be the sum

$$R_l = P_1 + P_2 + \ldots + P_l \quad (R_N = 0) \tag{7.61}$$

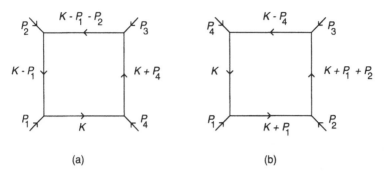

(a)   (b)

Fig. 7.5   Symmetry of the four-point function.

Note that $T_{\mu_1 \dots \mu_N}$ is invariant under any circular permutation of $P_1, \dots, P_N$ in the HTL approximation and that

$$T_{\mu_1 \dots \mu_N}(P_1, \dots, P_N) = (-)^N \, T_{\mu_1 \dots \mu_N}(P_N, \dots, P_1) \qquad (7.62)$$

as the permutation $(P_1, \dots, P_N) \to (P_N, \dots, P_1)$ corresponds to a change of sign of the loop momentum K (fig. 7.5). We also define $R_{l,m}$ by

$$R_{l,m} = -R_{m,l} = (R_l - R_m) \cdot \hat{K} = i(\omega_l - \omega_m) + (\mathbf{r}_l - \mathbf{r}_m) \cdot \hat{\mathbf{k}} \qquad (7.63)$$

In the HTL approximation we immediately derive for $N \geq 3$ a Ward identity from

$$P_2 \cdot K \simeq \frac{1}{2} \left[(K - R_1)^2 - (K - R_2)^2 \right] \qquad (7.64)$$

obtaining

$$P_{2\mu_1} T_{\mu_1 \dots \mu_N}(P_1, \dots, P_N) = \frac{1}{2} \Big(T_{\mu_2 \dots \mu_N}(P_1 + P_2, P_3, \dots, P_N)$$
$$- T_{\mu_2 \dots \mu_N}(P_1, P_2 + P_3, \dots, P_N) \Big) \qquad (7.65)$$

One should remark that Ward identities would also be obtained from the saturation with P_2 of any of the indices $\mu_1, \dots, \mu_N$. From our experience with the cases $N = 2$ and $N = 3$, our guess for $T_{\mu_1 \dots \mu_N}$ is

$N = 2$:

$$T_{\mu\nu}(P_1) = -\frac{T^2}{12} \int \frac{d\Omega}{4\pi} \left[\frac{i\omega_1 \hat{K}_\mu \hat{K}_\nu}{(P_1 \cdot \hat{K})} + \delta_{\mu 4} \delta_{\nu 4} - \frac{1}{2} \delta_{\mu\nu} \right] \qquad (7.66a)$$

$N \geq 3$:

$$T_{\mu_1 \dots \mu_N}(P_1, \dots, P_N) = -\frac{T^2}{3.2^N} \int \frac{d\Omega}{4\pi} \hat{K}_{\mu_1} \dots \hat{K}_{\mu_N}$$

$$\times \left(\frac{i\omega_1}{R_{1,2} R_{1,3} \dots R_{1,N}} + \frac{i\omega_2}{R_{2,1} R_{2,3} \dots R_{2,N}} \right.$$

$$\left. + \dots + \frac{i\omega_{N-1}}{R_{N-1,1} R_{N-1,2} \dots R_{N-1,N}} \right) \tag{7.66b}$$

As already explained in section 7.2.2, the case $N = 2$ is special, as some terms are not of the Landau damping type.

Let us first show that $T_{\mu_1 \dots \mu_N}$ obeys (7.65). We denote by R_l' the R-momenta for $T_{\mu_2 \dots \mu_N}(P_1 + P_2, P_3, \dots, P_N)$:

$$R_l' = R_l \quad R_{l,2}' = R_{l,2} \quad \text{if } l \geq 2 \tag{7.67}$$

and by R_l'' the R-momenta for $T_{\mu_2 \dots \mu_N}(P_1, P_2 + P_3, \dots, P_N)$

$$\begin{aligned} R_2'' &= R_1 & R_l'' &= R_l \quad \text{if } l \geq 3 \\ R_{l,2}'' &= R_{l,1} & R_{l,m}'' &= R_{l,m} \quad \text{if } l \geq 3 \end{aligned} \tag{7.68}$$

Furthermore, we remark that

$$P_2 \cdot \hat{K} = R_{2,1} = R_{l,1} - R_{l,2} \tag{7.69}$$

From these equations we obtain, for $l \geq 3$

$$\frac{(P_2 \cdot \hat{K}) \omega_l}{R_{l,1} R_{l,2} \dots R_{l,N}} = \frac{\omega_l'}{R_{l,2}' \dots R_{l,N}'} - \frac{\omega_l''}{R_{l,2}'' \dots R_{l,N}''} \tag{7.70}$$

which is clearly the combination to be expected on the RHS of (7.65). Working out the case $l \leq 2$ separately allows us to complete the proof of (7.65).

We next show that $T_{\mu_1 \dots \mu_N}$ is uniquely determined by the Ward identities for $N \geq 4$. Let us begin with the $N = 4$ case, and assume that there exists a completely symmetric tensor $H_{\mu\nu\rho\sigma}(P, Q, R, S)$, traceless in any two pair of indices, and which obeys

$$P_\mu H_{\mu\nu\rho\sigma} = Q_\mu H_{\mu\nu\rho\sigma} = R_\mu H_{\mu\nu\rho\sigma} = 0 \tag{7.71}$$

Since P, Q, R are linearly independent, (7.71) means that H has only one independent component. But this component is zero because the tensor is traceless. We thus conclude that the four-point vertex $T_{\mu\nu\rho\sigma}$ is uniquely determined from the knowledge of the three-point vertex $T_{\mu\nu\rho}$ and the Ward identities (7.65). Finally, the explicit calculation of section 7.2.1 shows that $T_{\mu\nu\rho}$ is of the form (7.66). Since $T_{\mu\nu\rho\sigma}$, as determined by (7.66), does obey the Ward identities (7.65), one may conclude that (7.66) is indeed the correct form of $T_{\mu_1 \ldots \mu_N}$ for $N = 4$. At this point, it should be obvious that (7.66) is correct for any N.

Let us now turn to the colour structure of the proper vertices. We shall limit ourselves to the cases $N = 3$ and $N = 4$, which are the only cases which will be needed later on. From section 7.2.1 (displaying the colour indices now) we may write the gluon and ghost contributions to the three-point vertex in the form

$$\Gamma^{abc}_{\mu\nu\rho}(P, Q, R) = (2g)^3 \mathrm{Tr}\left(T_a[T_b, T_c]\right) T_{\mu\nu\rho}(P, Q, R) \qquad (7.72)$$

where T_a is a colour matrix of the adjoint representation; this equation is identical to (7.47) when igf_{abc} has been factored out. For the four-point function we need the identity

$$T_{\mu\nu\rho\sigma}(P, Q, R, S) + T_{\mu\nu\rho\sigma}(P, R, Q, S) + T_{\mu\nu\rho\sigma}(P, R, S, Q) = 0 \qquad (7.73)$$

which is easily derived from (7.66). Then we have

$$\Gamma^{abcd}_{\mu\nu\rho\sigma}(P, Q, R, S) = (2g)^4 \left[\mathrm{Tr}\left(T_a[[T_b, T_c], T_d]\right) T_{\mu\nu\rho\sigma}(P, Q, R, S) \right.$$

$$\left. + \mathrm{Tr}\left(T_a[[T_b, T_d], T_c]\right) T_{\mu\nu\rho\sigma}(P, Q, S, R) \right] \qquad (7.74)$$

In this form, the four-gluon vertex is not explicitly symmetric under permutations of external legs, and one must use (7.73) to check Bose statistics.

In order to give simple formulae, we shall limit ourselves to the only case that we shall need explicitly: we sum over two adjacent colour indices and take $R = -Q$, $S = -P$:

$$\Gamma^{abbd}_{\mu\nu\rho\sigma}(P, Q, -Q, -P) = \frac{2}{3} g^4 T^2 C_A^2 \delta_{ad}$$

$$\times \int \frac{d\Omega}{4\pi} \frac{\hat{K}_\mu \hat{K}_\nu \hat{K}_\rho \hat{K}_\sigma}{[(P+Q)\cdot\hat{K}][(P-Q)\cdot\hat{K}]} \left[\frac{i\omega_q}{Q\cdot\hat{K}} - \frac{i\omega_p}{P\cdot\hat{K}} \right] \qquad (7.75)$$

Adding the fermion loops is very easy; one has to take the following into account:

- a factor of -1 because of the fermionic character of the loop;
- a factor of 2 because there are two possible orientations of the fermion loop;
- a factor of $-1/2$ because the Bose–Einstein distribution $n(k)$ has been replaced by a Fermi–Dirac distribution $\tilde{n}(k)$ (compare (6.31) and (6.125)).

On the other hand, the structure of the colour matrices is the same, but the matrices t_a now belong to the fundamental representation where

$$\operatorname{Tr} t_a t_b = \frac{1}{2} \delta_{ab} \tag{7.76}$$

The relevant combination of colour factors is, as in the cases $N = 2$ and $N = 3$, $C_A + N_f/2$. Finally, the two-quark–two-gluon vertex is given by (7.27), where the QCD expression (6.163) for the fermion thermal m_f mass should be used instead of the QED expression. The expressions for the two-quark–$(N-2)$-gluon proper vertices can be found in the literature.

At this point it might be useful to summarize the general properties of HTL which we have discovered up to now:

  (i) HTLs exist only for Green's functions with N external gluon lines or with $(N-2)$ external gluon and two external quark lines; the only group factors which appear are $C_A + N_f/2$ in the first case, C_F in the second case. There are no HTLs with external ghost lines.
 (ii) HTLs are totally symmetric and traceless with respect to external gluon indices.
(iii) HTLs do not involve the four-gluon vertex in the Feynman and Coulomb gauges, except in the case of the gluon propagator (contribution from the tadpole).
 (iv) HTLs are gauge-fixing-independent: it has been shown by explicit computation that they are the same in general covariant, Coulomb and axial gauges; we have already mentioned that there exists a general field theoretical proof of this result.
  (v) HTLs obey tree-like Ward identities. From now on we denote by $\delta\Pi$, etc., the HTL approximation. Ignoring the colour factors for notational simplicity we have:

$$\text{propagator}: Q_\mu \delta\Pi_{\mu\nu}(Q) = 0 \tag{7.77}$$

$$\text{three-gluon vertex}: P_\mu \delta\Gamma_{\mu\nu\rho}(P,Q,R)$$
$$= \delta\Pi_{\nu\rho}(Q) - \delta\Pi_{\nu\rho}(R) \tag{7.78}$$

$$\text{quark–gluon vertex}: Q_\mu \delta\tilde\Gamma_\mu(P_1,P_2;Q)$$
$$= \delta\Sigma(P_1) - \delta\Sigma(P_2) \tag{7.79}$$

In general, P_μ dotted into an N-point HTL gives a combination of $(N-1)$-point HTLs.

As already explained in the preceding chapter, all HTL results can be generalized to the case of a non-zero chemical potential: it suffices to replace the gauge particles and fermion thermal masses by effective masses depending on the temperature T and the chemical potential μ. We have in QED (see (6.103))

$$m^2 = \frac{1}{6}e^2\left(T^2 + \frac{3}{\pi^2}\mu^2\right) \tag{7.80}$$

$$m_f^2 = \frac{1}{8}e^2\left(T^2 + \frac{\mu^2}{\pi^2}\right) \tag{7.81}$$

and in QCD

$$m^2 = \frac{1}{6}C_A g^2 T^2 + \frac{1}{12}g^2 C_F\left(T^2 + \frac{3}{\pi^2}\mu^2\right) \tag{7.82}$$

$$m_f^2 = \frac{1}{8}g^2 C_F\left(T^2 + \frac{\mu^2}{\pi^2}\right) \tag{7.83}$$

As in the QED case, it is possible to write an effective Lagrangian which generates hard thermal loops. The fermion Lagrangian is identical to (7.32), provided that one uses the quark thermal mass (7.83) and the covariant derivative ($A^\mu = A_a^\mu t_a$):

$$D_\mu = \partial_\mu + igA_\mu \tag{7.84}$$

The effective gluon Lagrangian can be shown to be

$$\mathcal{L}_g = -m^2 \mathrm{Tr} \int \frac{\mathrm{d}\Omega}{4\pi} \, F_{\mu\alpha} \frac{\hat{K}_\alpha \hat{K}_\beta}{(\hat{K}\cdot D)^2} F_{\beta\mu} \tag{7.85}$$

where the trace is taken over colour indices, and $F^{\mu\nu} = F_a^{\mu\nu} T_a$ is the QCD field strength tensor:

$$F_{\mu\nu}^a = \partial_\mu A_\nu^a - \partial_\nu A_\mu^a + gf_{abc}A_\mu^b A_\nu^c \tag{7.86}$$

This Lagrangian is manifestly gauge invariant. An explicit, but tedious, computation shows that it generates the two- and three-point functions correctly, which is sufficient to ensure that it is also correct for all N-point functions.

7.3 The effective expansion

7.3.1 General theory

We have shown previously that the gluon and quark self-energies $\delta\Pi_{\mu\nu}$ and $\delta\Sigma$, as well as the three- and four-point vertices, are of the same order of magnitude as the bare inverse propagators and the bare vertices when all external momenta are of order gT. This observation has important consequences if one wishes to establish a consistent perturbative expansion. Let us look, for example, at the one-loop correction to the gluon propagator for soft values of the external momentum. If we wish to evaluate the contribution of the integration over soft loop momenta, it is clear that we must use dressed (or effective) propagators and vertices in the calculation, since they are of the same order of magnitude as the bare ones.

Let us recall that we denoted HTLs for propagators and vertices by $\delta\Pi$, $\delta\Sigma$, $\delta\Gamma$, which have been computed previously and are known to be gauge-fixing independent. The effective propagators and vertices will be denoted by $^*D_{\mu\nu}$, $^*S$, $^*\Gamma$. Below we summarize the main ingredients which are needed in the effective expansion, all results being given in Euclidean space.

(i) Gluon propagator. In a covariant gauge, the effective gluon propagator $^*D_{\mu\nu}$ is given by (6.23), the functions F and G being given by (6.49) and (6.50). Let us also write the expression of $^*D_{\mu\nu}$ in the Coulomb gauge:

$$^*D_{\mu\nu}(Q) = \frac{\delta_{\mu 4}\delta_{\nu 4}}{q^2 + \delta\Pi_L} + \frac{P^T_{\mu\nu}}{Q^2 + \delta\Pi_T} \tag{7.87}$$

where $P^T_{\mu\nu}$ is given in (6.21); furthermore,

$$\delta\Pi_L = \frac{q^2}{Q^2}F = -2m^2 Q_1\left(\frac{i\omega}{q}\right) \tag{7.88}$$

$$\delta\Pi_T = G = m^2\left(\frac{i\omega}{q}\right)\left[\left(1 - \left(\frac{i\omega}{q}\right)^2\right)Q_0\left(\frac{i\omega}{q}\right) + \frac{i\omega}{q}\right] \tag{7.89}$$

(ii) Effective quark propagator. The effective quark propagator $^*S(P)$ reads ($\gamma_4 = i\gamma_0$)

$$^*S(P) = \frac{1}{2}\Delta_+(P)(i\gamma_4 + \gamma\cdot\hat{\mathbf{p}}) + \frac{1}{2}\Delta_-(P)(i\gamma_4 - \gamma\cdot\hat{\mathbf{p}})$$

$$= \frac{1}{2}\Delta_+(P)\not{P}_+ + \frac{1}{2}\Delta_-(P)\not{P}_- \tag{7.90}$$

with $P_\pm = (i, \pm\hat{\mathbf{p}})$ and

$$\Delta_\pm(P) = \left(i\omega \mp p - \frac{m_f^2}{p}\left[\left(1 \mp \frac{i\omega}{p}\right)Q_0\left(\frac{i\omega}{p}\right) \pm 1\right]\right)^{-1} \tag{7.91}$$

(iii) Effective three-gluon vertex. The effective three-gluon vertex is obtained by adding the HTL loop (7.47) to the bare vertex:

$$^*\Gamma^{abc}_{\mu\nu\rho}(P,Q,R) = igf_{abc}\,^*\Gamma_{\mu\nu\rho}(P,Q,R)$$

$$= igf_{abc}\Big(\delta_{\mu\nu}(P-Q)_\rho + \delta_{\nu\rho}(Q-R)_\mu$$

$$+ \delta_{\rho\mu}(R-P)_\nu + 2m^2\int\frac{d\Omega}{4\pi}\,\hat{K}_\mu\hat{K}_\nu\hat{K}_\rho$$

$$\times\left[\frac{i\omega_r}{(P\cdot\hat{K})(R\cdot\hat{K})} - \frac{i\omega_q}{(P\cdot\hat{K})(Q\cdot\hat{K})}\right]\Big) \tag{7.92}$$

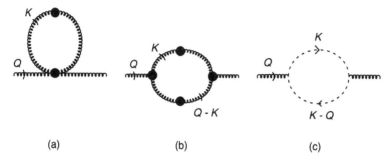

Fig. 7.6. The gluon self-energy in the effective expansion. Heavy dots indicate resummed propagators and vertices.

(iv) Effective quark–gluon vertex. Similarly, we write for the effective quark–gluon vertex

$$*\Gamma_a^\mu(P_1, P_2) = gt_a \, {}^*\Gamma^\mu(P_1, P_2)$$

$$= gt_a\left(\gamma^\mu - m_f^2 \int \frac{d\Omega}{4\pi} \frac{\hat{K}^\mu \hat{\not{K}}}{(P_1 \cdot \hat{K})(P_2 \cdot \hat{K})}\right) \tag{7.93}$$

(v) Effective two-gluon–two-quark and four-gluon vertices. Since there is no bare two-gluon–two-quark vertex, the effective vertex is nothing other than the HTL (7.27). Finally, we quote the four-gluon vertex summed over two colour indices, in a particular momentum configuration:

$$*\Gamma_{\mu\nu\rho\sigma}^{abbd}(P, Q, -Q, -P) = g^2 C_A \delta_{ad} \, {}^*\Gamma_{\mu\nu\rho\sigma}(P, Q, -Q, -P)$$

$$= g^2 C_A \delta_{ad}\left(2\delta_{\mu\sigma}\delta_{\nu\rho} - \delta_{\mu\nu}\delta_{\rho\sigma} - \delta_{\mu\rho}\delta_{\nu\sigma} + 4m^2 \int \frac{d\Omega}{4\pi}\right.$$

$$\left. \times \frac{\hat{K}_\mu \hat{K}_\nu \hat{K}_\rho \hat{K}_\sigma}{[(P+Q)\cdot\hat{K}][(P-Q)\cdot\hat{K}]}\left[\frac{i\omega_q}{Q\cdot\hat{K}} - \frac{i\omega_p}{P\cdot\hat{K}}\right]\right) \tag{7.94}$$

We write the QCD Lagrangian as

$$\mathcal{L} = (\mathcal{L}_{QCD} + \delta\mathcal{L}) - \delta\mathcal{L} = \mathcal{L}_{\text{eff}} - \delta\mathcal{L} \tag{7.95}$$

where $\delta\mathcal{L} = \mathcal{L}_g + \mathcal{L}_f$ is the sum of the effective gluon and quark Lagrangians obtained in the preceding section. The term $-\delta\mathcal{L}$ is treated as a counter-term, so as to avoid any double counting; the effective expansion follows from the Lagrangian (7.95). Effective propagators and vertices will be represented graphically by heavy dots.

For the sake of definiteness, let us examine the gluon self-energy, taking as always soft external momenta $(q_0, q \sim gT)$ for time-like values of Q

after analytical continuation. We have (fig. 7.6)

$$
\Pi_{\mu\nu}(Q) = \frac{g^2 C_A}{2} \int \frac{\mathrm{d}^4 K}{(2\pi)^4} \Big[\, {}^*\Gamma_{\mu\lambda\sigma}(Q,-K,K-Q)
$$
$$
\times \, {}^*D_{\lambda\lambda'}(K) \, {}^*D_{\sigma\sigma'}(Q-K) \, {}^*\Gamma_{\nu\lambda'\sigma'}(-Q,K,Q-K)
$$
$$
+ \, {}^*\Gamma_{\mu\lambda\sigma\nu}(Q,K,-K,-Q){}^*D_{\lambda\sigma}(K) \Big] + \text{ghost loop} \qquad (7.96)
$$

Let us examine (7.96) from a power-counting point of view. In the loop integration, it is convenient to distinguish between the soft region ($K \sim gT$) and the hard region ($K \sim T$). We already know the hard contribution, since, for hard values of K, ${}^*D$ and ${}^*\Gamma$ may be replaced by their bare counterparts, leading to $\mathrm{Re}\,\Pi_{\mu\nu} \sim g^2 T^2$. We also know that $\mathrm{Im}\,\Pi_{\mu\nu}$ is zero at this order because Q is assumed to be time-like, and we know that HTLs have imaginary parts for $Q^2 < 0$ only. In order to evaluate the contribution from the soft region (loop momentum $\sim gT$), we use the same power-counting as in section 7.2.2, with the substitution $T \to gT$ following the same order (a)–(e) as in section 7.2.2

$$
g^2 \int_{\text{soft}} \frac{\mathrm{d}^4 K}{(2\pi)^4} \cdots \sim g^2 (gT)^3 \frac{1}{gT} \frac{1}{g^2 T^2} \frac{1}{g} g^2 T^2 \sim g^3 T^2 = g(gT)^2 \qquad (7.97)
$$

This result can also be read from (6.27). The factor $1/g$ comes from the Bose–Einstein distribution (for a Fermi–Dirac distribution we would have 1 instead of $1/g$: this is the reason why we do not consider the quark loop in fig. 7.6). The main conclusion to be drawn from (7.97) is that the contribution from the integration over soft loop momenta is g times the leading order: such corrections will be called order g corrections. However, there are two other possible origins for terms of order $g(gT)^2$, which we examine now.

(i) Corrections to the HTL approximations in one-loop calculations. In the preceding analysis, we have systematically neglected terms of order q/k, for example when neglecting $\not{Q}$ with respect to $\not{K}$ in (6.24). Since q/k is of order g, we find corrections of the order of g; however, these corrections occur only in the real part of Π.

(ii) Two-loop terms. Let us, for example, consider the two-loop graph of fig. 7.7, with hard internal momenta. Since $\Pi_{\mu\nu}(Q) \sim g^2 T^2$, power counting leads to

$$
g^2 T^3 \frac{1}{T} \Big(\frac{1}{qT} \Big)^2 \frac{q}{T} T^2 (gT)^2 = \frac{g^4 T^3}{q} \sim g^3 T^2 \qquad (7.98)
$$

Once more, this correction of order g contributes only to the real part of Π because the sum of the three internal momenta must add up to a soft momentum, and phase space is then very limited.

Fig. 7.7   Two-loop graph for the gluon self-energy.

7.3.2   Production of soft dileptons in a quark–gluon plasma

Our first application of the effective expansion will be the computation of the production rate of dileptons from a quark–gluon plasma. These dileptons could be a signal of plasma formation in heavy ion collisions because, once they are produced, they escape without further interactions. Unfortunately, the signal will probably be buried in the background of dileptons of a completely different origin. Nevertheless, it is worth computing the production rate, assuming that the plasma can be taken at equilibrium with temperature T. We shall limit ourselves to the case where the virtual photon which decays into a lepton pair is at rest in the plasma rest frame: $Q = (q_0, \mathbf{0})$, $q_0 > 0$. Calculations with $\mathbf{q} \neq 0$ have been carried out, but they are more complicated by an order of magnitude. If q_0 is hard, $q_0 \sim T$, we may apply ordinary (or bare) perturbation theory; this will be done in chapter 10. Here we assume that q_0 is soft, $q_0 \sim gT$, and we must use the effective expansion. The differential rate for producing pairs of massless leptons is related to the photon self-energy thanks to (5.158):

$$\frac{d\Gamma}{dq_0 d^3 q} = \frac{\alpha}{12\pi^4} \frac{1}{Q^2} \frac{1}{e^{\beta q_0} - 1} \, \mathrm{Im} \, \Pi^\mu_\mu(q_0 + i\eta, \mathbf{q}) \tag{7.99}$$

In the bare expansion, the one-loop, or Drell–Yan, approximation (fig. 7.8(a)) has a very simple expression when $\mathbf{q} = 0$:

$$\left. \frac{d\Gamma}{dq_0 d^3 q} \right|_{\mathbf{q}=0} = \frac{5\alpha^2}{36\pi^4} \, \tilde{n}\left(\frac{q_0}{2}\right) \tag{7.100}$$

In the effective expansion, the one-loop approximation (fig. 7.8(b)) for $\Pi_{\mu\mu}$ is written in Euclidean space using the photon–quark effective vertex $^*\Gamma_\mu$ and the quark effective propagator $^*S$:

$$\Pi_{\mu\mu} = \frac{5}{3} e^2 \int \frac{d^4 P}{(2\pi)^4} \mathrm{Tr}\left[\, ^*S(P) \, ^*\Gamma_\mu(P, P - Q) \right.$$
$$\left. \times \, ^*S(P - Q) \, ^*\Gamma_\mu(P - Q, P) \right] \tag{7.101}$$

where the trace is taken over Dirac indices; when continuing from Euclidean to Minskowski space one should not forget that $\delta_{\mu\nu} \to -g_{\mu\nu}$. The

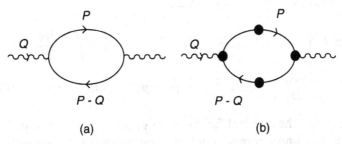

P

Q

P - Q

P

Q

P - Q

(a)   (b)

Fig. 7.8. Bare and resummed one-loop diagrams for dilepton production. Solid lines: quarks. Wavy lines: photons. Heavy dots indicate resummed propagators and vertices.

diagram built from the effective two-photon–two-quark vertex does not contribute as $^*\Gamma_{\mu\mu} = 0$ (see (7.26)), although it is essential in order that the Ward identity

$$Q_\mu \Pi_{\mu\nu}(Q) = 0 \qquad (7.102)$$

be satisfied. The effective photon–quark vertex is easily computed because we have chosen to limit ourselves to the case $\mathbf{q} = 0$; non-zero values of $\mathbf{q}$ lead to a much more complicated algebra. We obtain from (7.16), by doing the angular integration explicitly

$$^*\Gamma_4(P-Q,P) = \left[1 - \frac{m_f^2}{i\varpi p}\delta Q_0\right]\gamma_4 + \frac{m_f^2}{\varpi p}\delta Q_1 \gamma \cdot \hat{\mathbf{p}} \qquad (7.103)$$

$$^*\Gamma_i(P-Q,P) = \left[1 + \frac{m_f^2}{3i\varpi p}(\delta Q_0 - \delta Q_2)\right]\gamma_i$$

$$+ \frac{m_f^2}{\varpi p}\delta Q_1 \hat{p}_i\gamma_4 + \frac{m_f^2}{i\varpi p}\delta Q_2\, \hat{p}_i\,\gamma\cdot\hat{\mathbf{p}} \qquad (7.104)$$

with $\varpi = -q_4$, $\omega = -p_4$ and $\omega' = \omega - \varpi$ and we have defined

$$\delta Q_n = Q_n\!\left(\frac{i\omega}{p}\right) - Q_n\!\left(\frac{i\omega'}{p}\right) \qquad (7.105a)$$

Q_n being a Legendre function of the second kind (see (6.49) for $n = 1, 2$) while

$$Q_2(x) = \frac{1}{2}\Big(3xQ_1(x) - Q_0(x)\Big) \qquad (7.105b)$$

Inserting the explicit expression (7.90) of $^*S(P)$ and the results (7.103) and (7.104) for $^*\Gamma_\mu$, $\Pi_{\mu\mu}$ reduces to a sum of terms of the form $g_1(\omega)g_2(\omega - \varpi)$. In order to compute the imaginary part we use (2.84) transposed to fermions (see also exercise 5.7 and (A.44)):

$$\text{Im } T \sum_n g_1(i\omega_n) g_2(i(\omega_n - \varpi)) = \pi(1 - e^{\beta q_0}) \int_{-\infty}^{\infty} \frac{dp_0}{2\pi}$$

$$\times \int_{-\infty}^{\infty} \frac{dp_0'}{2\pi} \tilde{f}(p_0) \tilde{f}(p_0') \delta(q_0 - p_0 - p_0') \rho_1(p_0) \rho_2(-p_0') \qquad (7.106)$$

where ρ_1 and ρ_2 are the spectral functions of g_1 and g_2. In the computation of (7.101), we encounter terms which can be written schematically, with $x = p_0/p$, $P_i(x)=$ polynomials in x:

$$\frac{m_f^2 P_1(x) Q_0(x)}{p^2 P_2(x) + m_f^2 P_3(x) Q_0(x)} = F(p_0, p) \qquad (7.107)$$

where the numerator comes from the vertices and propagators (see (7.104) and (7.105)), and the denominators from the propagators (see (6.141)). It is straightforward to check that the spectral function of F is expressed in terms of the spectral function ρ of the denominator:

$$\text{Im } F = -\frac{p^2}{2} P_1(x) \rho(x) \frac{P_2(x)}{P_3(x)} \qquad (7.108)$$

Thus the discontinuity may be expressed in terms of the spectral functions ρ_+ and ρ_- of the quark propagator. Other combinations of propagators and vertices involve $(Q_0)^2$ in the numerator of (7.107), and their discontinuities lead to terms linear in $\rho_\pm$ in the final result which reads

$$\frac{d\Gamma}{dq_0 d^3 q}\bigg|_{q=0} = \frac{10\alpha^2}{9\pi^4 q_0^2} \int_0^\infty p^2\, dp \int_{-\infty}^\infty \frac{dp_0}{2\pi} \int_{-\infty}^\infty \frac{dp_0'}{2\pi} \tilde{f}(p_0)\tilde{f}(p_0')$$

$$\times \delta(q_0 - p_0 - p_0')\left(4\left[1 - \frac{p_0^2 - p_0'^2}{2pq_0}\right]^2 \rho_+(p_0, p)\rho_-(p_0', p)\right.$$

$$+ \left[1 + \frac{p_0^2 + p_0'^2 - 2p^2 - 2m_f^2}{2pq_0}\right]^2 \rho_+(p_0, p)\rho_+(p_0', p)$$

$$+ \left[1 - \frac{p_0^2 + p_0'^2 - 2p^2 - 2m_f^2}{2pq_0}\right]^2 \rho_-(p_0, p)\rho_-(p_0', p)$$

$$+ \theta(p^2 - p_0^2)\frac{m_f^2}{4pq_0^2}\left(1 - \frac{p_0^2}{p^2}\right)\left[\left(1 + \frac{p_0}{p}\right)\rho_+(p_0', p)\right.$$

$$\left.\left. + \left(1 - \frac{p_0}{p}\right)\rho_-(p_0', p)\right]\right) \qquad (7.109)$$

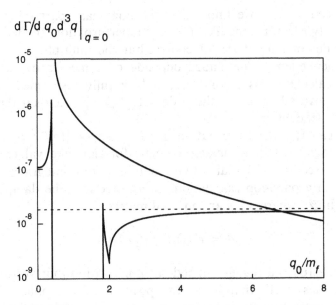

Fig. 7.9. Dilepton production (adapted from Braaten, Pisarski and Yuan (1990)). Dashed line: one-loop result (7.100).

The range of integration in (7.109) is actually finite, since the spectral functions vanish for $|p_0| \leq \omega_{\pm}(p)$. Note that ρ_+ and ρ_- are both positive functions, and (7.109) defines a positive production rate.

Each of the terms in (7.109) has a simple physical interpretation. The spectral functions $\rho_{\pm}$ contain contributions from the $\chi = +1$ and $\chi = -1$ quasi-particle poles and from cuts. The terms with two powers of ρ give three types of contribution: pole–pole, pole–cut and cut–cut. The pole–pole contribution exhibits a remarkable structure (fig. 7.9), which is discussed in detail in Braaten, Pisarski and Yuan (1990).

7.3.3 The gluon damping rate

Historically, the calculation of the gluon damping rate was very important in our understanding of thermal QCD. The calculation in bare perturbation theory was known to be gauge-dependent, and the damping rate was even found to be negative in some gauges. This was sometimes interpreted as a signal of plasma instability. Yet one knows from general field theoretical arguments that the position of the pole of the gluon propagator, whose imaginary part gives the damping rate, is gauge-independent. Thus a consistent calculation should give a gauge-independent result. Note that the damping rate is always computed for a collective excitation at rest: otherwise one encounters infrared divergences linked to transverse gluons, which will be examined in section 10.4.

From section 7.3.1 we know that the imaginary part of Π is of the order of $g(gT)^2$ for time-like Q. Bare perturbation theory gives the correct order of magnitude, of course, but the multiplicative coefficient is gauge-dependent. This gauge dependence comes from the fact that the bare calculation is incomplete; only a fully resummed calculation is able to give all terms of the order of $g(gT)^2$, and then the result is gauge-independent.

Let us call $\Pi'_{\mu\nu}$ the contribution of the order of $g(gT)^2$ to $\Pi_{\mu\nu}$; if we are only interested in its imaginary part for time-like values of Q, we know from section 7.3.1 that we may obtain it from integration over soft momenta in a one-loop calculation. More precisely, the damping rate is given by $\mathrm{Im}\mathscr{T}$, where $\mathscr{T}$ is an 'on-shell matrix element':

$$\mathscr{T} = e^{\mu}(Q)\Pi_{\mu\nu}(Q)e^{\nu}(Q) \qquad (7.110)$$

In (7.110), $e_{\mu}(Q)$ is a (gauge-dependent) polarization four-vector, Q being taken on mass-shell, that is, at the position of the pole. There are, of course, three linearly independent $e_{\mu}(Q)$s: one longitudinal and two transverse. In the HTL approximation, $e_{\mu}(Q)$ obeys

$$(\Delta_{\mu\nu}^{-1}(Q) - \delta\Pi_{\mu\nu}(Q))e^{\mu}(Q) = 0 \qquad (7.111)$$

where

$$\Delta_{\mu\nu}^{-1}(Q) = Q^2 g_{\mu\nu} - Q_{\mu}Q_{\nu} \qquad (7.112)$$

One may prove the following results:

(i) $Q^{\mu}\Pi'_{\mu\nu}(Q)Q^{\nu} = 0$ for arbitrary values of Q;
(ii) $Q^{\mu}\Pi'_{\mu\nu}e^{\mu}(Q) = 0$, for Q on mass-shell;
(iii) $\mathscr{T}$ is the same in Coulomb and covariant gauges. This confirms the gauge independence of $\mathscr{T}$.

Result (i) is not sufficient to ensure the decomposition (6.20) of $\Pi'_{\mu\nu}$; in addition to the transverse and longitudinal components Π_T and $\tilde{\Pi}_L$, there is a third function $\tilde{\Pi}_L$. It turns out that this third function does not contribute to the propagator in the strict Coulomb gauge, where the propagator reads

$$D'_{\mu\nu}(Q) = \frac{P_{\mu\nu}^T}{Q^2 + \delta\Pi_T + \Pi'_T} + \frac{\delta_{\mu4}\delta_{\nu4}}{q^2 + \delta\Pi_L + \Pi'_L} \qquad (7.113)$$

After continuation to Minkowski space, the mass-shell conditions which determine the dispersion relations $\omega_T(q)$ and $\omega_L(q)$ for the transverse and longitudinal gluonic excitations are

$$\omega_T^2(q) - q^2 + \delta\Pi_T(\omega_T + i\eta, q)$$
$$+ \Pi_T'(\omega_T + i\eta, q) = 0 \tag{7.114a}$$

$$q^2 + \delta\Pi_L(\omega_L + i\eta, q)$$
$$+ \Pi_L'(\omega_L + i\eta, q) = 0 \tag{7.114b}$$

The damping rates $\gamma_T(q)$ and $\gamma_L(q)$ are given by the imaginary part of the poles of the propagators:

$$\gamma_T(q) = -\mathrm{Im}\,\omega_T(q) \quad \gamma_L(q) = -\mathrm{Im}\,\omega_L(q) \tag{7.115}$$

The imaginary part of the hard thermal loop vanishes for time-like Q so that only Π_T' and Π_L' contribute to the damping rate. To lowest order in g, the damping rates for collective excitations at zero momentum are

$$\gamma_T(0) = -\frac{1}{2\omega_P}\mathrm{Im}\,\Pi_T'(\omega_P + i\eta, 0) \tag{7.116}$$

$$\gamma_L(0) = \frac{1}{2\omega_P}\lim_{q\to 0}\frac{\omega_P^2}{q^2}\,\mathrm{Im}\,\Pi_L'(\omega_P + i\eta, q) \tag{7.117}$$

At $\mathbf{q} = 0$, there is no distinction between transverse and longitudinal excitations, and we expect that $\gamma_T(0) = \gamma_L(0) = \gamma$. This result can also be derived from the requirement that $\Pi_{ij}'(Q)$ has a smooth limit when $q \to 0$.

For the explicit evaluation of $\Pi_{\mu\nu}'$ (see fig. 7.6) we need the three- and four-gluon effective vertices (7.92) and (7.94). The one-loop resummed expression has already been given in (7.96). In computing the imaginary part, we rely on techniques already described in section 7.3.2. The algebra is rather tedious and we limit ourselves to giving the final result, written conventionally in the form

$$\gamma = a\frac{g^2 N T}{24\pi} \tag{7.118a}$$

where a is obtained with the approximation $f(k_0) \simeq T/k_0$:

$$a = 9\int_0^\infty dk \int_{-\infty}^{+\infty}\frac{dk_0}{2\pi k_0}\int_{-\infty}^{+\infty}\frac{dk_0'}{2\pi k_0'}\,\delta(\omega_P - k_0 - k_0')$$
$$\times \Bigg(2(k^2 + k_0 k_0')^2\,\rho_T(k_0, k)\rho_T(k_0', k)$$
$$+ k^4\rho_L(k_0, k)\rho_L(k_0', k) + \frac{k^2}{\omega_P^2}(k^2 - k_0^2)^2\rho_T(k_0, k)\rho_L(k_0', k)$$
$$+ \frac{k_0}{6k^3}(k^2 - k_0^2)^2\theta(k^2 - k_0^2)\rho_T(k_0', k)\Bigg) \tag{7.118b}$$

Since ρ_T/k_0 and ρ_L/k_0 are positive, γ is clearly a positive quantity. Furthermore, the δ-function in (7.118), together with the behaviour of the spectral functions, effectively restricts the integration region to soft momenta. The numerical value of a is: $a \simeq 6.64$. A similar calculation for the damping rate of the fermionic excitation at rest yields

$$\gamma_f(0) = a(N, N_f) \frac{g^2 T C_F}{4\pi} \tag{7.119}$$

with $a(N, N_f) \simeq 1.40$ for $N = 3$ and $N_f = 2$.

Problems with the dependence of the result on the gauge parameter ξ in covariant gauges have been raised. This gauge dependence is not present if one uses an infrared regulator, for example, by working in dimension $D = 4 + 2\varepsilon$, and if one takes the mass-shell limit before the infrared cut off is removed: the limits of going to the mass-shell and removing the infrared cut-off do not commute.

7.3.4 Higher-order calculations

The resummation program has been tested with two higher-order calculations. The first one attempts to compute corrections of order g to the plasma frequency. Since we are now interested in the real part of Π, we need to take into account all corrections of order g which have been enumerated in section 7.3.1:

  (i) one-loop with hard loop momentum;
 (ii) two-loop with hard momenta;
(iii) one-loop with soft loop momentum.

An explicit calculation shows that, at least in the evaluation of the plasma frequency, cases (i) and (ii) do not result in a contribution of the order of g. At the time of writing, it is not known whether this property is true in general, in which case one would have to understand why power counting overestimates the contribution of the first two cases.

There is a non-zero contribution of the order of g from case (iii). The algebra which leads to the final result is lengthy; one finds for $N_f = 0$

$$\omega_P^2 \simeq \frac{1}{9} g^2 C_A T^2 (1 - 0.18 g \sqrt{C_A}) \tag{7.120}$$

The negative sign in (7.120) could have been expected intuitively; collective excitations should disappear at the critical temperature, and if we take (7.120) (too) seriously, we deduce that the critical value of g should be $g_{cr} \simeq 3.2$ for $C_A = 3$. Of course, we do not believe that perturbation theory remains valid close to the critical temperature, and for such large values of g; one can only conclude that perturbation theory points towards the right direction. A similar conclusion could be reached from the perturbative expression of the pressure.

The second higher-order calculation addresses the problem of the Debye mass. Let us assume that we may write a formula which is the QCD analogue of the potential energy of two static charges (see (6.63)):

$$V(r) = q_1 q_2 \int \frac{d^3 q}{(2\pi)^3} \frac{e^{i\mathbf{q}\cdot\mathbf{r}}}{q^2 - \Pi_{00}(q_0 = 0, q)} \qquad (7.121)$$

If screening is controlled by the pole of the integrand in (7.121), which is defined by

$$q^2 = \Pi_{00}(q_0 = 0, q^2 = -m_D^2) \qquad (7.122)$$

then the asymptotic behaviour of the potential energy is

$$V(r) \simeq \frac{q_1 q_2}{4\pi r} e^{-m_D r} \qquad (7.123)$$

The position of the pole is a gauge-invariant quantity, and so is the Debye mass m_D. On the other hand, $\Pi_{00}(q_0 = 0, q)$ is a gauge-dependent quantity, and, except to leading order, it is incorrect to define m_D^2 by $m_D^2 = -\Pi_{00}(q_0 = 0, q \to 0)$. Even in QED, where $\Pi_{\mu\nu}$ is gauge-independent, there is a difference of order $e^4 T^2$ between the location of the pole and $-\Pi_{00}(q_0 = 0, q \to 0)$. Actually, $\Pi_{00}(q_0 = 0, q \to 0)$ can be deduced from the pressure (see exercise 6.7):

$$-\Pi_{00}(q_0 = 0, q \to 0) = e^2 \frac{\partial^2 P}{\partial \mu^2}$$

$$= \frac{e^2 T^2}{3} \left(1 - \frac{3e^2}{8\pi^2} + \frac{\sqrt{3}\,e^3}{4\pi^3} + O(e^4) \right) \qquad (7.124)$$

while m_D^2 is given by

$$m_D^2 = \frac{e^2 T^2}{3} \left(1 - \frac{3e^2}{8\pi^2} - \frac{e^2}{12\pi^2} \left[\ln \frac{4\tilde{\mu}^2}{\pi T^2} + \gamma - 1 \right] + \frac{\sqrt{3}\,e^3}{4\pi^3} + O(e^4) \right) \qquad (7.125)$$

where $\tilde{\mu}$ is the subtraction point in the $\overline{MS}$-scheme and γ is the Euler constant. Since $de/d\ln\tilde{\mu} = e^3/(12\pi^2) + O(e^5)$, (7.125) is renormalization-group-invariant, and is thus a candidate for being a physical quantity, in contradistinction to (7.124), which is not renormalization-group-invariant.

After this digression on QED, let us revert to QCD. Since we are interested in a static quantity, the calculation is greatly simplified; it is sufficient to sum over the static modes (in general it is not possible to separate the static modes, even for soft external momenta, because such a separation precludes the possibility of a straightforward analytic continuation of the external frequencies). Furthermore, since all gluon lines are static, HTL corrections to the vertices vanish. We write the static, resummed gluon propagator, in a covariant gauge, as follows:

$$^*D_{\mu\nu}(\mathbf{q}) = \frac{1}{q^2 + 2m^2}\delta_{\mu 4}\delta_{\nu 4} + \frac{1}{q^2}P_{\mu\nu}^T + \xi \frac{Q_\mu Q_\nu}{Q^4}\Big|_{q_4 = 0} \tag{7.126}$$

The computation of $\Pi_{00}(q_0 = 0, q)$ shows that it is indeed gauge-dependent; however, the gauge dependence vanishes at the pole. The correction δm_D^2 to the Debye mass squared is infrared-divergent. In order to regularize this divergence, it is customary to conjecture that a magnetic mass $m_{\mathrm{mag}} \sim g^2 T$, which will be discussed further in chapter 10, plays the role of an infrared cut off. One then finds

$$\delta m_D^2 = 2gm^2 \frac{C_A}{2\pi}\left(\frac{3}{C_A + N_f/2}\right)^{1/2} \ln\frac{1}{g} + O(g^2) \tag{7.127}$$

The correction of order g to the leading approximation of the Debye mass is large and positive. This result is in qualitative agreement with lattice simulations.

7.4 Kinetic theory of hard thermal loops

Up to now our starting point in the study of hard thermal loops has been conventional perturbation theory, although we have had to reorganize it somewhat in order to obtain meaningful results. There is another powerful approach, which starts from the equation of motion for Green's functions and uses a consistent set of approximations relying on the existence of two length scales, $1/T$, the average distance between particles, and $1/gT \gg 1/T$, the characteristic wavelength of collective excitations. These collective excitations correspond to non-zero gauge and fermion *mean fields* A_μ and ψ; at equilibrium these mean fields vanish of course.

In the limit of long wavelengths, one finds a system of equations which couples mean fields and particles. A first group of equations couples the fields to induced sources, a second group expresses sources in terms of off-equilibrium two-point functions, and the system closes thanks to equations which relate induced sources to mean fields. These last equations are derived by truncating in a consistent way the Schwinger–Dyson equations of motion for N-point functions, retaining only two-point functions. These equations are then written in the form of kinetic equations for Wigner transforms of two-point functions; in the non-Abelian case, these Wigner transforms are colour density matrices. As an example, let us consider the induced colour current $J_a^\mu(X)$, which may be written as an integral over K of a Wigner transform $J_a^\mu(X, K)$:

$$J_a^\mu(X) = \int dK \, J_a^\mu(X, K) \tag{7.128}$$

with

$$dK = \frac{d^4K}{(2\pi)^3} 2\theta(k^0)\delta(K^2) \tag{7.129}$$

Then one can show that $J_a^\mu(X,K)$ obeys a kinetic equation which generalizes the results of section (6.4):

$$\left[K \cdot D_X, J^\mu(X,K)\right] = 2g^2 K^\mu K^\nu F_{\nu 0} \frac{d}{dk_0}\left[Nn(k_0) + N_f \tilde{n}(k_0)\right] \tag{7.130}$$

We have used the matrix notations $A^\mu = A_a^\mu T_a$, $J^\mu = J_a^\mu T_a$, $D^\mu = \partial^\mu + igA^\mu$ for the covariant derivative and $F^{\mu\nu} = [D^\mu, D^\nu]/(ig)$ for the field strength tensor, where T_a represents the Lie algebra generators in the adjoint representation of the colour group. In the Abelian case $D_X = \partial_X$, $N = 0$ and $N_f \rightarrow 2$: (6.110) is then a direct consequence of (7.130).

It is possible to give an intuitive derivation of (7.130) which closely parallels that of section 6.4 by using a classical transport theory of the QCD plasma. However, one must understand that, although the derivation is formulated in terms of 'particles', its purpose is to uncover general features of the kinetic equations, and not to describe the trajectories of real (classical) particles: what is classical in the problem is not the motion of individual particles, but rather their average long-range behaviour.

With these restrictions in mind, let us consider a particle with mass M bearing a non-Abelian colour charge Q_a which follows a world line $X^\mu(\tau)$, where τ is the proper time. We ignore all effects linked to spin, as they can be shown to be negligible in this problem. The evolution of the dynamical variables X^μ, K^μ and Q_a, where K^μ is the kinetic momentum, is then described by the following equations:

$$M\frac{dX^\mu}{d\tau} = K^\mu \tag{7.131a}$$

$$M\frac{dK^\mu}{d\tau} = -gQ_a F_a^{\mu\nu} K_\nu \tag{7.131b}$$

$$M\frac{dQ_a}{d\tau} = gf_{abc}K_\mu A_b^\mu Q_c \tag{7.131c}$$

The usual phase-space (X,K) is enlarged to (X,K,Q), leading to a distribution function $f(X,K,Q)$. The integration measure over K is given by (7.129) with $\delta(K^2) \rightarrow \delta(K^2 - M^2)$. It ensures the positivity of the energy and mass-shell evolution; the integration measure over the colour charge must enforce the conservation of the Casimir invariants of the group. Taking, for simplicity, the $SU(2)$ case, where there is only one Casimir invariant $Q_aQ_a = q_2$, we make a change of variables $Q_a \rightarrow (\varphi, J, \pi)$:

$$Q_1 = \cos\varphi\sqrt{J^2 - \pi^2} \quad Q_2 = \sin\varphi\sqrt{J^2 - \pi^2} \quad Q_3 = \pi \tag{7.132}$$

and we find

$$dQ = d\varphi \, d\pi \, J dJ \, d(J^2 - q_2) \tag{7.133}$$

Integrating over the constrained variable J gives the proper canonical volume element $d\varphi \, d\pi$. An analogous, but lengthier, construction may also be given in the $SU(3)$ case.

The Boltzmann equation which generalizes (6.104) reads

$$K_\mu \left(\frac{\partial}{\partial X_\mu} + g Q_a F_a^{\mu\nu} \frac{\partial}{\partial K^\nu} \right.$$

$$\left. + g f_{abc} A_b^\mu Q_c \frac{\partial}{\partial Q_a} \right) f(X, K, Q) = C[f] \tag{7.134}$$

As in section 6.4 we set the collision integral $C[f]$ to zero and obtain a set of self-consistent Vlasov equations by adding the field equations:

$$[D_\nu, F^{\nu\mu}](X) = J^\mu(X) \tag{7.135}$$

The total colour current $J_a^\mu(X)$ is obtained by summing over all particle species:

$$J_a^\mu(X) = \sum_{\text{part., spins}} j_a^\mu(X) \tag{7.136}$$

and each component $j_a^\mu(X)$ is computed from the corresponding distribution function as

$$j_a^\mu(X) = g \int dK \, dQ \, K^\mu Q_a f(X, K, Q)$$

$$= g \int dK \, j_a^\mu(X, K) \tag{7.137}$$

This equation is the non-Abelian generalization of (6.106); comparing (6.106) and (7.130) allows one to check the normalization of the integration measure dK. Using the collisionless Boltzmann equation, one can check that the current j_a^μ is covariantly conserved:

$$[D_\mu, j^\mu](X) = 0 \tag{7.138}$$

We note that the Boltzmann equation can be shown to be invariant under gauge transformations: the distribution function $f(X, K, Q)$ transforms as a scalar and the current $j_a^\mu(X)$ as a gauge vector.

The approximate solution of the system of equations (7.134)–(7.137) relies on the following physical picture, which was alluded to at the beginning of the present section. There are in the plasma long wavelength excitations, corresponding to non-zero mean fields $A_a^\mu(X)$, $\psi(X)$ and $\overline{\psi}(X)$, with frequencies and wave vectors of order gT, $g \ll 1$. The space-time derivative ∂_X acting on these mean fields or on distribution functions is thus of order gT; remember that this derivative gives zero when applied to equilibrium distributions. Since we wish a gauge-covariant description, the term gA of the covariant derivative must, for consistency, be at most of order gT, which means that A is at most of order T and F of order gT^2. Another way of arriving at this conclusion is to remark that when a coloured particle is submitted during a time $1/(gT)$ to a force $\sim gF$, its momentum change is of order F/T. If we want this change to be much smaller than the typical momentum $\sim T$, we need $F \sim gT^2$. It can also be shown that $\bar{\psi}\psi \sim gT^3$: see the discussion following (7.162b).

As in section 6.4, we expand $f(X, K, Q)$ in powers of g:

$$f = f^{(0)} + gf^{(1)} + g^2 f^{(2)} + \cdots \qquad (7.139)$$

where $f^{(0)}$ is the equilibrium distribution

$$f^{(0)} = C\, n(k_0) \quad \text{or} \quad f^{(0)} = \tilde{C}\, \tilde{n}(k_0) \qquad (7.140)$$

C and $\tilde{C}$ being normalization constants. At leading order in g the induced colour current is, from (7.137),

$$j_a^\mu(X) = g^2 \int dK\, dQ\, K^\mu Q_a f^{(1)}(X, K, Q)$$

$$= \int dK\, j_a^\mu(X, K) \qquad (7.141)$$

while the Boltzmann equation reduces to a form which takes consistently into account all contributions of order g, while preserving the original non-Abelian symmetry

$$K_\mu \left(\frac{\partial}{\partial X_\mu} + g f_{abc} A_b^\mu Q_c \frac{\partial}{\partial Q_a} \right) f^{(1)}(X, K, Q)$$

$$= K_\mu Q_a F_a^{\mu\nu} \frac{\partial}{\partial K^\nu} f^{(0)}(k_0) \qquad (7.142)$$

From (7.141) and (7.142) the colour current $j_a^\mu(X, K)$ obeys

$$K_\nu D_{ab}^\nu j_b^\mu(X, K) = g^2 K^\mu K_\nu F_b^{\nu\rho} \frac{\partial}{\partial K_\rho} \left[\int dQ\, Q_a Q_b f^{(0)}(k_0) \right] \qquad (7.143)$$

The colour integral on the RHS of (7.143) is given by

$$\int dQ\, Q_a Q_b f^{(0)}(k_0) = N n(k_0) \delta_{ab} \quad \text{(gluons)}$$

$$= \frac{1}{2} \tilde{n}(k_0) \delta_{ab} \quad \text{(quarks)} \qquad (7.144)$$

which, upon summation over all particle species, yields (7.130).

Our task now is to solve (7.130) with retarded boundary conditions. As in preceding sections we introduce the light-like vector $\hat{K} = (1, \hat{\mathbf{k}})$ and solve as a preliminary step the equation

$$\left[\hat{K} \cdot D_X, W^\mu(X, \hat{K})\right] = F^{\mu\nu}(X)\hat{K}_\nu \tag{7.145}$$

The solution of (7.145) is expressed in terms of the parallel transporter:

$$U(X, Y) = P \exp\left(-ig \int_Y^X dz^\mu A_\mu(z)\right) \tag{7.146}$$

where P is the symbol for path ordering; $U(X, Y)$ obeys the partial differential equations

$$\hat{K} \cdot \partial_X U(X, Y) = -ig\hat{K} \cdot A(X)U(X, Y) \tag{7.147a}$$
$$\hat{K} \cdot \partial_Y U(X, Y) = igU(X, Y)\hat{K} \cdot A(Y) \tag{7.147b}$$

It is checked by inspection that the retarded solution of (7.145) is

$$W^\mu(X, \hat{K}) = \int_0^\infty d\tau\, U(X, X - \hat{K}\tau)$$
$$\times F^{\mu\nu}(X - \hat{K}\tau)\hat{K}_\nu U(X - \hat{K}\tau, X) \tag{7.148}$$

Comparing (7.130) and (7.145) we see that the induced current $j_\mu^{\text{ind}}(X)$

$$j_\mu^{\text{ind}}(X) = \int dK\, J^\mu(X, K) \tag{7.149}$$

is given by

$$j_\mu^{\text{ind}}(X) = 2m^2 \int \frac{d\Omega}{4\pi} \hat{K}_\mu W^0(X, \hat{K}) \tag{7.150}$$

Note that we may restrict ourselves to the component W^0 of W^μ because we are using an isotropic distribution. The integral over the modulus of $\mathbf{k}$ has been performed immediately, giving the gluon thermal mass m of (6.161) and leaving us with an angular integral. The explicit form of (7.150) is then

$$j_\mu^{\text{ind}}(X) = 2m^2 \int \frac{d\Omega}{4\pi} \hat{K}_\mu \int_0^\infty d\tau\, U(X, X - \hat{K}\tau)$$
$$\times \hat{\mathbf{k}} \cdot \mathbf{E}(X - \hat{K}\tau)U(X - \hat{K}\tau, X) \tag{7.151}$$

where m is the gauge boson thermal mass (6.161) and $\mathbf{E}$ is the chromo-electric field. Contrary to the Abelian result (6.110), j_μ^{ind} is not linear in A; it is remarkable that the solution (7.151) respects the non-Abelian gauge symmetry. The proper vertices in the HTL approximation follow from functional differentiation of the induced current (see (6.61) in the

Abelian case):

$$\delta\Pi_{ab}^{\mu\nu}(X,Y) = \frac{\delta j_a^{\mu,\text{ind}}(X)}{\delta A_{vb}(Y)}\bigg|_{A=0} \tag{7.152}$$

$$g\delta\Gamma_{abc}^{\mu\nu\rho}(X,Y,Z) = \frac{\delta^2 j_a^{\mu,\text{ind}}(X)}{\delta A_{vb}(Y)\delta A_{\rho c}(Z)}\bigg|_{A=0} \tag{7.153}$$

In the Abelian case, U and $\mathbf{E}$ are c-numbers rather than matrices, and (7.151) reduces to (6.110), which may be written as

$$j_\mu^{\text{ind}}(X) = 2m^2 \int \frac{d\Omega}{4\pi}\hat{K}_\mu \int_0^\infty d\tau\, \hat{\mathbf{k}}\cdot\mathbf{E}(X - \hat{K}\tau) \tag{7.154}$$

This equation shows immediately that in QED hard thermal loops with more than two external photon lines and no external electron lines vanish identically. One can also see at once that the self-energy $\delta\Pi_{\mu\nu}$ in QED and QCD differs only by the value of m^2. As a simple exercise, let us compute $\delta\Pi_{\mu\nu}$ from (7.152) and (7.154):

$$\delta\Pi^{\mu\nu}(X,Y) = 2m^2 \int \frac{d\Omega}{4\pi}\hat{K}_\mu \int_0^\infty d\tau$$
$$\times \hat{K}^\rho(g_{\rho\nu}\partial_0 - g_{0\rho}\partial_\nu)\delta^{(4)}(X - Y - \hat{K}\tau) \tag{7.155}$$

The Fourier transform is

$$\delta\Pi^{\mu\nu}(Q) = \int d^4(X - Y)e^{iQ\cdot(X-Y)}\Pi^{\mu\nu}(X,Y) \tag{7.156}$$

Using

$$\int_0^\infty d\tau e^{i(Q\cdot\hat{K})\tau} = \frac{i}{Q\cdot\hat{K} + i\eta} \tag{7.157}$$

yields the retarded self-energy

$$\delta\Pi_R^{\mu\nu}(Q) = 2m^2\left(-g^{\mu 0}g^{\nu 0} + q_0 \int \frac{d\Omega}{4\pi}\frac{\hat{K}^\mu\hat{K}^\nu}{Q\cdot\hat{K} + i\eta}\right) \tag{7.158}$$

which is the continuation to real time of (6.43). Note that (7.155) implies $x^0 > y^0$, and we do obtain the retarded self-energy, hence the $+i\eta$ in (7.158). Higher-order proper vertices with external gluons only follow from functional differentiation of $j_\mu^{\text{ind}}(X)$; the Ward identities of section 7.2.3 are easily derived from the fact that $j_\mu^{\text{ind}}(X)$ is covariantly conserved:

$$\left[D^\mu, j_\mu^{\text{ind}}\right](X) = 0 \tag{7.159}$$

Equation (7.151) also yields the effective action δS of hard thermal loops:

$$j_{\mu a}^{\text{ind}}(X) = \frac{\delta S}{\delta A_a^\mu(X)} \tag{7.160}$$

which can be written in the form

$$\delta S = m^2 \int \frac{\mathrm{d}\Omega}{4\pi} \int \mathrm{d}^4 X \int \mathrm{d}^4 Y$$

$$\times \mathrm{Tr}\left[F_{\mu\lambda}(X) < X\left|\frac{\hat{K}^\mu \hat{K}_\nu}{(\hat{K}\cdot D)^2}\right|Y > F^{\nu\lambda}(Y)\right] \qquad (7.161)$$

where D is the covariant derivative in the adjoint representation. This effective action is clearly identical to (7.85), and it is defined only on the space of fields A such that $\hat{K}\cdot D$ does not vanish: as in the Abelian case, this condition corresponds to propagation without dissipation.

Let us now turn to the fermion case, following schematically the field theoretical approach alluded to at the beginning of the present section; we take, for simplicity, the QED case. There is, first, an equation which couples the mean field $\psi(X)$ to the external source $\eta^{\mathrm{cl}}(X)$ and to an induced source $\eta^{\mathrm{ind}}(X)$:

$$i\slashed{D}\psi(X) = \eta^{\mathrm{cl}}(X) + \eta^{\mathrm{ind}}(X) \qquad (7.162a)$$

and an equation coupling the electromagnetic mean field to its sources

$$\partial^\nu_X F_{\nu\mu}(X) - e\overline{\psi}(X)\gamma_\mu\psi(X) = j^{\mathrm{ind}}_\mu(X) + j^{\mathrm{cl}}_\mu(X) \qquad (7.162b)$$

As usual, the external sources $\eta^{\mathrm{cl}}(X)$ and $j^{\mathrm{cl}}_\mu(X)$ allow one to obtain Green's functions from functional differentiation, and they are set to zero at the end of the calculation; D is the covariant derivative. The induced electromagnetic current is $e\langle\overline{\psi}(X)\gamma_\mu\psi(X)\rangle_c$ and the induced fermionic source $\eta^{\mathrm{ind}}(X)$ is the mean value $e\langle\slashed{A}\psi(X)\rangle_c$, where c stands for 'connected'; it is non-vanishing only in the presence of a non-zero fermionic mean field (remember that in the present section F and ψ denote mean fields). Since on the LHS of (7.162b) $\partial^\nu_X F_{\nu\mu}$ is of order $e^2 T^3$, we must have for consistency $\langle\overline{\psi}\psi\rangle$ at most of order eT^3.

The induced source $\eta^{\mathrm{ind}}(X)$ is expressed as an integral of a Wigner transform $\mathscr{K}(X,K)$ of the two-point function $e\langle\slashed{A}\psi(X)\rangle_c$:

$$\eta^{\mathrm{ind}}(X) = \int \frac{\mathrm{d}^4 K}{(2\pi)^4}\, \mathscr{K}(X,K) \qquad (7.163)$$

and the Schwinger–Dyson equations, truncated at the level of the two-point function, are transformed into a kinetic equation for $\mathscr{K}(X,K)$, which links it to the mean field $\psi(X)$

$$i(\hat{K}\cdot D_X)\mathscr{K}(X,K) = e\, 2\pi\, \delta(K^2)[n(k_0) + \tilde{n}(k_0)]\slashed{K}\psi(X) \qquad (7.164)$$

In the QCD case, $e \to gC_F$ in (7.164). The solution of (7.164) is written in terms of the parallel transporter U and one finds for the induced source

$$\eta^{\mathrm{ind}}(X) = -im^2_f \int \frac{\mathrm{d}\Omega}{4\pi}\hat{\slashed{K}} \int_0^\infty \mathrm{d}\tau\, U(X - \hat{K}\tau)\psi(X - \hat{K}\tau) \qquad (7.165)$$

where m_f is the fermionic thermal mass (6.163); the induced source transforms as $\psi(X)$ in a gauge transformation. Hard thermal loops with two external fermion lines are given by functional differentiation of $\eta^{\text{ind}}(X)$; the self-energy $\delta\Sigma(X, Y)$ is

$$\delta\Sigma(X, Y) = \frac{\delta\eta^{\text{ind}}(X)}{\delta\psi(Y)}\bigg|_{\psi=A=0} \qquad (7.166)$$

while for the quark–gluon vertex we have

$$g\delta\Gamma_a^{\mu}(X, Y, Z) = \frac{\delta^2\eta^{\text{ind}}(X)}{\delta A_{\mu a}(Y)\delta\psi(Z)}\bigg|_{\psi=A=0} \qquad (7.167)$$

These equations allow us to recover (6.128) for $\delta\Sigma(P)$ and the analytical continuation of (7.93) for the quark–gluon vertex ($Q = P_2 - P_1$):

$$\delta\Gamma_a^{\mu}(Q; P_1, P_2) = t_a m_f^2 \int \frac{d\Omega}{4\pi} \frac{\hat{K}\hat{K}^{\mu}}{(\hat{K} \cdot P_1 + i\eta)(\hat{K} \cdot P_2 + i\eta)} \qquad (7.168)$$

The imaginary parts in the denominator reflect the time ordering of the external lines in the vertex; there are several possible analytical continuations of the imaginary-time vertex function (7.93), but here the solution of the kinetic equations with well-specified boundary conditions singles out (7.168) in a unique way.

As a final comment, let us show how one can derive a different form of the effective action. Define the field (in matrix form)

$$a^{\mu}(X, \hat{K}) = A^{\mu}(X) + W^{\mu}(X, \hat{K}) \qquad (7.169)$$

which obeys the partial differential equation

$$[\hat{K} \cdot D, a^{\mu}] = \partial^{\mu}(\hat{K} \cdot A) \qquad (7.170)$$

which shows that a^{μ} is a functional of $\hat{K} \cdot A$ only. From (7.148), the solution of (7.170) is

$$a^{\mu}(X, \hat{K}) = \int_0^{\infty} d\tau\, U(X, X - \hat{K}\tau)$$
$$\times \partial^{\mu}\!\left(\hat{K} \cdot A(X - \hat{K}\tau)\right) U(X - \hat{K}\tau, X) \qquad (7.171)$$

The field strength tensor $f^{\mu\nu}$ built from a^{μ} obeys $f^{\mu\nu} = 0$, which means that the field a^{μ} has zero curvature. This remark allows one to prove the gauge invariance of the effective action first obtained by Taylor and Wong (1990); this form of the effective action may be derived from the expression (7.150) of the induced current.

Finally, by projecting $f^{\mu\nu} = 0$ on the hyperplane defined by the four-vectors $\hat{K}^{\mu}$ and $\hat{K}'^{\mu} = (1, -\hat{\mathbf{k}})$ and defining

$$a_+ = \hat{K} \cdot a \qquad a_- = \hat{K}' \cdot a \qquad (7.172)$$

one obtains the equation

$$\partial_+ a_- - \partial_- a_+ + ig[a_+, a_-] = 0 \qquad (7.173)$$

which is the starting point for a formal analogy between the dynamics of colour fields in the QCD plasma and a Chern–Simons theory in 2+1 dimension.

References and further reading

The fundamental papers on hard thermal loops and resummed perturbation theory were written by Braaten and Pisarski (1990a,b) and by Frenkel and Taylor (1990). A first form of the effective action for hard thermal loops was discovered by Taylor and Wong (1990); the form of the effective action quoted in section 7.1.3 is due to Braaten and Pisarski (1992a) and to Frenkel and Taylor (1992). The production rate of soft dileptons at rest in a quark–gluon plasma was computed by Braaten, Pisarski and Yuan (1990); their result was generalized to non-zero three-momentum of the dilepton by Wong (1992). The gluon damping rate was shown to be a gauge-independent quantity by Kobes, Kunstatter and Rebhan (1990) and the explicit computation to order $g^2 T$ was performed by Braaten and Pisarski (1990c,d). The difficulty with gauge invariance was pointed out by Baier, Kunstatter and Schiff (1992) and the solution was given by Rebhan (1992). The fermion damping rate for an excitation at rest has been computed by Kobes, Kunstatter and Mak (1992) and by Braaten and Pisarski (1992b). The correction (7.120) to the plasma frequency was derived by Schulz (1994) and the correction (7.127) to the Debye mass was derived by Rebhan (1994); see also Braaten and Nieto (1994). Section 7.4 is a (too) short account of an approach developed in a series of papers by Blaizot and Iancu (1993a,b, 1994a,b,c), and to which we refer for further details. From (7.131)–(7.144) we have relied on work by Wong (1970), Elze and Heinz (1989) and Kelly, Liu, Lucchesi and Manuel (1994); see also Jackiw and Nair (1993). The analogy with Chern–Simons theory was pointed out by Efraty and Nair (1992, 1993). Scalar QED has been studied by Kraemmer, Rebhan and Schulz (1995) and by Blaizot, Iancu and Parwani (1995).

Exercises

7.1 Consider a neutral scalar field in $D = 6$ space-time dimension, with a self-interaction $g\varphi^3/3!$; ignore tadpole diagrams. Show that the hard thermal loop of the self-energy is

$$\Pi(Q) = \frac{g^2 T^2}{192\pi} \int_{-1}^{+1} \mathrm{d}\cos\theta \, (1 - \cos^2\theta) \left(\frac{i\omega}{Q \cdot \hat{K}} - \frac{3}{2} \right)$$

Are there other hard thermal loops?

7.2 By performing the angular integrations explicitly, demonstrate the equivalence between (7.34) and (7.37).

7.3 Check equations (7.103) and (7.104) for the various components of the quark–photon vertex and the argument leading to (7.108).

7.4 Show that in bare perturbation theory the photon damping rate for an excitation at rest is

$$\gamma = \frac{e^2}{24\pi}\omega_P \sim e^3 T$$

This result is gauge-independent but is incorrect: the numerical coefficient of $e^3 T$ is modified by resummation.

7.5 Derive the Ward identities of hard thermal loops from the covariant conservation law for the induced current (7.159).

7.6 (a) Show that the retarded solution of

$$i(\hat{K} \cdot D_X)G_R(X, Y ; \hat{K}) = \delta^{(4)}(X - Y)$$

is given by

$$G_R(X, Y ; \hat{K}) = -i \int_0^\infty d\tau\, \delta^{(4)}(X - Y - \hat{K}\tau)U(X, Y)$$

where the parallel transporter $U(X, Y)$ is defined in (7.146).
(b) Show that

$$\frac{\delta G_R(X, Y ; \hat{K})}{\delta A_a^\mu(Z)} = g\hat{K}_\mu G_R(X, Z ; \hat{K})t_a G_R(Z, Y ; \hat{K})$$

(c) Express $\eta^{\text{ind}}(X)$ in terms of G_R and use (b) to prove (7.168).

7.7 Scalar QED is a useful laboratory with which to test various topics in perturbation theory. The Lagrangian density is

$$\mathcal{L} = (D_\mu\varphi)^*(D^\mu\varphi) - \frac{1}{4}F_{\mu\nu}F^{\mu\nu} - \frac{1}{2\xi}(\partial_\mu A^\mu)^2$$

where φ is a massless charged scalar field and $D_\mu = \partial_\mu - ieA_\mu$ is the covariant derivative. One could add a $\lambda(\varphi^*\varphi)^2$ interaction without any essential modification.
(a) Show that the photon self-energy $\Pi_{\mu\nu}$ is given in imaginary time by

$$\Pi_{\mu\nu}(Q) = -m^2\delta_{\mu\nu} + e^2 \int \frac{d^4K}{(2\pi)^4}(K_\mu K_\nu - Q_\mu Q_\nu)$$

with $m^2 = e^2 T^2/6$. Compute the HTL of $\Pi_{\mu\nu}$ and obtain the longitudinal and transverse components

$$F = 2m^2 \left(1 - \frac{q_0^2}{q^2}\right)\left(1 - \frac{q_0}{2q}\ln\frac{q_0+q}{q_0-q}\right)$$

$$G = \frac{1}{2}(2m^2 - F)$$

(b) Using kinetic theory, give a one-line proof that the thermal mass is the same as in standard QED.

(c) Show that the hard thermal loop of the scalar self-energy is momentum-independent and adds only a thermal mass μ:

$$\mu^2 = \frac{1}{4}e^2 T^2$$

(d) Show that the effective Lagrangian density for the scalar particle reduces to

$$\delta\mathcal{L}_s = \frac{1}{2}\mu^2\varphi^*\varphi$$

so that the only non-trivial hard thermal loops occur in two-point amplitudes.

(e) Compute $\Pi_{ii}(0, q \to 0)$ in bare perturbation theory:

$$\Pi_{ii}^{\text{bare}}(0, q \to 0) = \frac{1}{8}e^2 qT + O(q^2)$$

and its resummed counterpart

$$\Pi_{ii}^{\text{res.}}(0, q \to 0) = -\frac{e\mu}{2\pi}\left[2\mu - \frac{q^2 + 4\mu^2}{q}\tan^{-1}\frac{q}{2\mu}\right] + O(e^2\mu^2)$$

Check the absence of magnetic mass in the two cases and show that the resummed result reduces to the bare one for $q \gg \mu$. The bare result is gauge-fixing independent, but it is nevertheless wrong!

8

Dynamical screening

This chapter deals with three important applications of Braaten and Pisarski's resummation program. These applications depend on the so-called 'dynamical screening' phenomenon, first discovered in the case of the photon. Since the photon is massless in the vacuum, many processes exhibit infrared singularities when the momentum Q of an exchanged photon becomes soft, at least in naïve (or bare) perturbation theory. However, we know from chapter 6 that, for soft values of Q, we should use a fully dressed, or resummed, photon propagator. This resummed propagator is expressed in terms of transverse and longitudinal propagators, $\Delta_T(Q)$ and $\Delta_L(Q)$, which are given in (6.74) and (6.75). Let us consider the zero frequency limit $q_0 \to 0$ of Δ_T and Δ_L; the scaling variable $x = q_0/q$ goes to zero in (6.74) and (6.75). We obtain

$$\Delta_L(0,q) = \frac{-1}{q^2 + 2m^2} \tag{8.1}$$

This leads to the familiar Debye screening: for static interactions, longitudinal photons acquire an effective mass $m_D^2 = 2m^2 = e^2 T^2/3$, which screens infrared singularities.

The situation for $\Delta_T(Q)$ is more involved. At first sight

$$\Delta_T(0,q) = \frac{1}{q^2} \tag{8.2}$$

which implies that static transverse (or magnetic) photons are not screened. However, if we retain the leading term in x, we obtain for $x \to 0$

$$\Delta_T(q_0,q) \simeq \frac{1}{q^2 - i\pi m^2 x/2} \tag{8.3}$$

This equation shows that there is a frequency-dependent screening, with a frequency-dependent cut-off $q_c = (\pi m^2 x/2)^{1/2}$ or $q_c = (\pi m^2 q_0/2)^{1/3}$. In some, but not all, situations this cut-off is able to screen infrared singularities of bare perturbation theory so that finite results are obtained: this is dynamical screening. We shall work out two examples in detail: the energy loss of a heavy fermion propagating in a QED plasma (section 8.1), and transport coefficients in a QCD plasma (section 8.3).

Dynamical screening may also occur in the case of infrared singularities due to the exchange of massless fermions; this will be explained in section 8.2, where we examine the production of hard real photons from a QCD plasma.

Other examples where dynamical screening yields finite results are:

• the Primakoff production of axions from a QED plasma;
• the production of massive photon pairs from a quark–gluon plasma.

8.1 Energy loss of a heavy fermion in a QED plasma

Our first illustration of dynamical screening will be a study of the energy loss of a heavy fermion ('muon') propagating in a QED plasma. We assume that this muon has a mass $M \gg T$ and retains a momentum $p \gg T$ when travelling through the plasma: thus the muon is never thermalized. The electrons of the plasma are assumed to be massless, or, more precisely, to have a mass $\ll eT$. Energy loss occurs through Compton scattering on the photons and Coulomb scattering on the electrons and positrons of the plasma; we assume for simplicity that the chemical potential vanishes. It can easily be checked that the two graphs of Compton scattering interfere destructively, and that their contribution is suppressed by a factor $(T/M)^2$ with respect to Coulomb scattering (exercise 8.1). Then the only relevant graph is that of Coulomb scattering (fig. 8.1), from which one can read the kinematics. If the initial muon has four-momentum $P = (E, \mathbf{p})$, the interaction rate $\Gamma(E)$ is given by

$$\Gamma(E) = \frac{2}{E} \int \frac{\mathrm{d}^3 p'}{(2\pi)^3} \frac{1}{2E'} \int \frac{\mathrm{d}^3 k}{(2\pi)^3} \frac{1}{2k} \tilde{n}(k)$$
$$\times \int \frac{\mathrm{d}^3 k'}{(2\pi)^3} \frac{1}{2k'} (1 - \tilde{n}(k'))$$
$$\times (2\pi)^4 \delta^{(4)} (P + K - P' - K') \overline{|\mathcal{M}|^2} \qquad (8.4)$$

where $\mathcal{M}$ is the transition matrix element corresponding to the Feynman graph of fig. 8.1; $\overline{|\mathcal{M}|^2}$ is summed over final spins and averaged over initial spins; equation (8.4) takes into account the scattering on both electrons and positrons. The average time between two muon interactions is $1/\Gamma$, so the average distance (or mean free path) travelled by the muon between

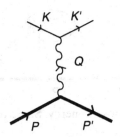

Fig. 8.1.   Scattering graph with photon exchange for the energy loss of a heavy fermion. Heavy lines: 'muons'. Wavy line: photon. Solid lines: electrons.

two successive interactions is $\Delta z = v/\Gamma$, where $v = p/E$ is the muon velocity. The average energy lost by the muon per interaction is

$$\Delta E = \frac{1}{\Gamma} \int_M^\infty dE' (E - E') \frac{d\Gamma}{dE'} \tag{8.5}$$

The integral extends over all final values of $E' \geq M$, and not only $E' < E$, because there is a small probability that the muon will gain energy in the collision. The rate of energy loss dE/dz per distance travelled is the ratio $\Delta E/\Delta z$:

$$\frac{dE}{dz} = \frac{1}{v} \int_M^\infty dE' \, (E - E') \frac{d\Gamma}{dE'} \tag{8.6}$$

Thus, in order to obtain dE/dz, one has only to insert $(E - E')/v = q_0/v$ in the integrand of (8.4).

Before moving on to the explicit evaluation of (8.4), we would like to make the connection between $\Gamma(E)$ and the self-energy $\Sigma(P)$ of the muon to lowest order in (bare) perturbation theory (fig. 8.2). We rewrite (8.4) by using K and $Q = P - P' = K' - K$ as integration variables:

$$\Gamma(E) = \frac{2}{E} \int \frac{d^4Q}{(2\pi)^4} \, 2\pi\delta_+((P - Q)^2 - M^2) \int \frac{d^4K}{(2\pi)^4}$$
$$\times 2\pi\delta_+(K^2) 2\pi\delta_+((Q + K)^2)\tilde{n}(k)(1 - \tilde{n}(k'))|\overline{\mathscr{M}}|^2 \tag{8.7}$$

with

$$\delta_+(K^2) = \theta(k_0)\delta(K^2) \tag{8.8}$$

and

$$\overline{|\mathscr{M}|^2} = \frac{e^4}{4(Q^2)^2} \mathrm{Tr}\Big[(\slashed{P} + M)\gamma_\mu(\slashed{P} - \slashed{Q} + M)\gamma_\nu\Big]$$
$$\times \mathrm{Tr}\Big[\slashed{K}\gamma^\mu(\slashed{Q} + \slashed{K})\gamma^\nu\Big] \tag{8.9}$$

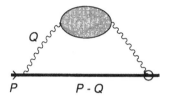

Fig. 8.2 Heavy fermion self-energy. Same definitions as in fig. 8.1.

We compare (8.7) with $\Sigma^>(P)$, which is computed from (4.62), suitably generalized to photons and fermions. Furthermore, we must take into account the fact that the outgoing muon is not thermalized and has positive energy; thus the muon propagator to be used in the calculation of $\Sigma^>(P)$ is not the cut thermal propagator $S_F^>(P-Q)$, but rather the cut $T = 0$ propagator $S_{F,T=0}^>(P-Q)$. We get for $\Sigma^>(P)$ (fig. 8.2)

$$\Sigma^>(P) = -2e^2 \int \frac{\mathrm{d}^4 Q}{(2\pi)^4} 2\pi \delta_+((P-Q)^2 - M^2)\varepsilon(q_0)(1 + f(q_0))$$

$$\times \frac{1}{(Q^2)^2} \gamma^\mu (\not P - \not Q + M)\gamma^\nu \operatorname{Im} \Pi_{\mu\nu}(Q) \tag{8.10}$$

where we have taken into account the fact that the kinematics at the muon–photon vertex imposes $Q^2 < 0$, so that $\operatorname{Re} \Pi_{\mu\nu}$ does not contribute. The expression for the imaginary part of the photon self-energy, $\operatorname{Im} \Pi_{\mu\nu}$, is

$$\operatorname{Im} \Pi_{\mu\nu} = -\frac{e^2}{2}\varepsilon(q_0) \int \frac{\mathrm{d}^4 K}{(2\pi)^4} \operatorname{Tr} \left(\not K \gamma_\mu (\not Q + \not K)\gamma_\nu\right)$$

$$\times \varepsilon(k_0)\varepsilon(q_0 + k_0)2\pi\delta(K^2)$$

$$\times 2\pi\delta((Q+K)^2)(\tilde f(k_0) - \tilde f(q_0 + k_0)) \tag{8.11}$$

Let us now look at some simple kinematics: as $K^2 = K'^2 = 0$, we have

$$Q^2 = -2k_0 k_0'(1 - \varepsilon(k_0 k_0')\hat{\mathbf{k}} \cdot \hat{\mathbf{k}}') \tag{8.12}$$

so that k_0 and $k_0' = q_0 + k_0$ have the same sign. The case $k_0 > 0$ corresponds to incident electrons and the case $k_0 < 0$ to incident positrons; since the chemical potential is assumed to vanish, both particles give identical contributions and we may concentrate on the case $k_0 > 0$. Using the identity

$$(1 + f(q_0))(\tilde f(k_0) - \tilde f(k_0 + q_0)) = \tilde n(k)(1 - \tilde n(k')) \tag{8.13}$$

we may readily compare (8.7) with (8.10) and obtain

$$\Gamma(E) = \frac{1}{4E} \operatorname{Tr}\left[(\not P + M)\Sigma^>(P)\right] \tag{8.14}$$

We have derived equation (8.14) in the particular case of Coulomb scattering; let us try to generalize it. Assume, as in section 4.4.3, that the muon interacts with the photon with a small coupling constant $\lambda \ll e$. Then we know that (8.14) is exact to lowest order in λ, but to all orders in e (note, incidentally, that Coulomb scattering would be of the order of $\lambda^2 e^2$, while Compton scattering would be of the order of λ^4, and thus negligible!). In our 'realistic' case, we only know that, to lowest order in e, $\Gamma(E)$ is given by (8.14). However, by appealing to the power-counting arguments of chapter 7, we may argue that the leading corrections to our bare perturbative calculation of order e^4 will be given by using a photon propagator fully dressed by hard thermal loops or, in other words, a resummed photon propagator. All neglected terms are down by further powers of e. We may thus conclude that the leading contribution to $\Gamma(E)$ is calculable from $\Sigma^>(P)$, provided that one uses a resummed photon propagator.

After these preliminary considerations, let us now turn to the actual evaluation of $\Gamma(E)$ and dE/dz. The interesting feature of the calculation is the occurrence of an infrared singularity in the region $q \to 0$. In order to exhibit this singularity, let us start from the expression of $\Sigma^>(P)$, which we compute by using the full photon cut propagator $D_{\mu\nu}^>(Q)$ written in the Coulomb gauge:

$$D_{\mu\nu}^>(Q) = (1 + f(q_0))\left[\rho_L(Q)P_{\mu\nu}^C + \rho_T(Q)P_{\mu\nu}^T\right] \qquad (8.15a)$$

with

$$P_{\mu\nu}^C = g_{\mu 0}g_{\nu 0} \qquad P_{ij}^T = \delta_{ij} - \hat{q}_i\hat{q}_j \qquad (8.15b)$$

$\Sigma^>(P)$ is now obtained from a suitable generalization of (4.64) to photons and fermions, remembering, however, that the muon cut propagator $S^>(P - Q)$ is the $T = 0$ cut propagator:

$$\Sigma^>(P) = \int \frac{d^4Q}{(2\pi)^4} 2\pi\delta_+\left((P-Q)^2 - M^2\right)$$
$$\times \gamma^\mu(\not{P} - \not{Q} + M)\gamma^\nu D_{\mu\nu}^>(Q) \qquad (8.16)$$

A simple trace calculation yields

$$\text{Tr}\left[(\not{P} + M)\Sigma^>(P)\right] = 4e^2 \int \frac{d^3q}{(2\pi)^3} \, dq_0 \, \frac{1}{2E'}\delta(E - E' - q_0)$$
$$\times (1 + f(q_0))\left[\left(2E^2 - Eq_0 - \mathbf{p}\cdot\mathbf{q}\right)\rho_L(q_0, q)\right.$$
$$\left. + 2\left(p^2 - Eq_0 + \mathbf{p}\cdot\mathbf{q} - (\mathbf{p}\cdot\hat{\mathbf{q}})^2\right)\rho_T(q_0, q)\right] \qquad (8.17)$$

This expression is exact, provided that we use the exact values of ρ_T and ρ_L, needed in the region $Q^2 < 0$, and not their HTL approximations.

However, as we are interested in the low q region, we perform the kinematical approximation

$$\delta(E - E' - q_0) \simeq \delta(q_0 - \mathbf{v} \cdot \mathbf{q}) \tag{8.18}$$

The angular integral over the directions of $\mathbf{q}$ is then readily performed; using the scaling variable $x = q_0/q$, the result for $\Gamma(E)$ is

$$\begin{aligned}
\Gamma(E) &= \frac{1}{2\pi} \frac{e^2}{v} \int q^2 dq \int_{-v}^{v} \frac{dx}{2\pi} \left(1 + f(q\,x) \right) \\
&\times \left(\rho_L(q_0, q) + (v^2 - x^2) \rho_T(q_0, q) \right)
\end{aligned} \tag{8.19}$$

where it is understood that the q-integration is limited to the region of q where our kinematical approximations are valid, namely, for q small enough; the region where q is large will be examined later on; see exercise 8.4 for an alternative derivation of (8.19), which can also be obtained from the classical current associated with fermion propagation (exercise 6.8). Let us introduce a scale q^*, intermediate between the soft and hard scales:

$$eT \ll q^* \ll T \tag{8.20}$$

(for example, $q^* = \sqrt{e}\,T$); this scale allows us to separate the integration over q into two regions: the soft one $(0 \le q \le q^*)$ and the hard one $(q^* \le q < \infty)$. In the region $q \sim q^*$, we may use the approximations (6.93)–(6.96) for ρ_T and ρ_L; bare perturbation theory to order e^4 is recovered by simply neglecting m^2 in the denominators, an approximation which is valid in the region $q \sim q^*$:

$$(2\pi)^{-1} \rho_L(q_0, q) \simeq \frac{m^2 x}{q^4} \qquad (2\pi)^{-1} \rho_T(q_0, q) \simeq \frac{m^2 x}{2q^4(1 - x^2)} \tag{8.21}$$

As we are interested in the strongest infrared singularities in q^*, we use the approximation

$$1 + f(qx) = \frac{T}{qx} + \frac{1}{2} + O\left(\frac{qx}{T}\right) \tag{8.22}$$

The first term in (8.22) leads to the most singular contribution to $\Gamma(E)$. If q varies in the region $q \sim q^*$, elementary integrations give

$$\Gamma_{\mathrm{sing}}(E) = \frac{e^2 m^2 T}{2\pi} \int \frac{dq}{q^3} \left[2 + \left(1 - \frac{1 - v^2}{2v} \ln \frac{1+v}{1-v} \right) \right] \tag{8.23}$$

The factor of 2 in the square bracket of (8.23) comes from longitudinal photons; the remaining term comes from transverse photons. Equation (8.23) shows that bare perturbation theory suffers from a dq/q^3 divergence, which is of course typical of Coulomb scattering.

Because the integrand must be even in x, it is the second term in (8.22) which gives the most singular contribution to dE/dz (remember that one goes from $\Gamma(E)$ to dE/dz by introducing a factor $q_0/v = qx/v$ in the integrand). The result, when q varies in the region $q \sim q^*$, is

$$\frac{dE}{dz} = \frac{e^2 m^2}{4\pi v} \int \frac{dq}{q} \left[1 - \frac{1-v^2}{2v} \ln \frac{1+v}{1-v} \right] \tag{8.24}$$

where longitudinal photons contribute $2v^2/3$ to the square bracket of (8.24). One sees that the Coulomb divergence dq/q^3 of $\Gamma(E)$ has been softened into a logarithmic divergence.

Let us now examine how resummation affects the divergences in the soft region $0 \le q \le q^*$. In order to keep the discussion simple (see exercise 8.2 for an alternative argument, which relies on (8.3)), we assume that $v \to 1$, which allows us to use the sum rules of section 6.3.3. By separating out the pole contribution in $\rho_{L,T}(q_0, q)$, we can rewrite, for example, the sum rule (6.79) as

$$\int_{-q}^{q} \frac{dq_0}{2\pi} \frac{\rho_L(q_0, q)}{q_0} + \frac{2Z_L(q)}{\omega_L(q)} = \frac{1}{q^2} - \frac{1}{q^2 + 2m^2} \tag{8.25}$$

Using the approximations (6.73) and (6.88), we obtain for $q \to 0$

$$\int_{-1}^{1} \frac{dx}{2\pi} \frac{\rho_L(q_0, q)}{x} \simeq \frac{2}{5m^2} \tag{8.26}$$

The other equations we need read

$$\int_{-1}^{1} \frac{dx}{2\pi} \frac{\rho_T(q_0, q)}{x} \simeq \frac{1}{q^2} - \frac{3}{2m^2} \tag{8.27a}$$

$$\int_{-1}^{1} \frac{dx}{2\pi} x\rho_T(q_0, q) \simeq \frac{3}{10m^2} \tag{8.27b}$$

The main difference between (8.26) and (8.27a) comes from the fact that, in the longitudinal case, the sum rule is dominated by the pole in the $q \to 0$-region, while the corresponding sum rule for ρ_T is dominated by the continuum. With these results in hand, the relevant integrals for $\Gamma(E)$ in (8.19) become, in the $q \to 0$ region (i.e. $q^2 \ll m^2$),

$$\int_0^q q^2 dq \int_{-1}^{1} \frac{dx}{2\pi} \frac{T}{qx} \rho_L(q_0, q) \simeq \frac{2T}{5m^2} \int_0^q q\, dq \tag{8.28a}$$

$$\int_0^q q^2 dq \int_{-1}^{1} \frac{dx}{2\pi} \frac{T}{qx} \rho_T(q_0, q) \simeq T \int_0^q \frac{dq}{q} \tag{8.28b}$$

Thus longitudinal photons give a finite contribution (it can easily be checked that this follows directly from Debye screening), while transverse photons give a logarithmic singularity. Although the convergence is improved, since we go from a dq/q^3 singularity to a dq/q one, the result is still divergent: this divergence will be examined in section 10.4. However, the result for the energy loss dE/dz is now convergent:

$$\frac{1}{2}\int_0^{q^*} q^3 dq \int_{-1}^{1}\frac{dx}{2\pi} x\rho_T(q_0,q) \simeq \frac{3}{20m^2}\int_0^{q^*} q^3 dq \tag{8.29}$$

The complete result for the soft region $0 \le q \le q^*$ is

$$\frac{dE}{dz}\bigg|_{\text{soft}} = \frac{e^4 T^2}{24\pi v}\left[\left(1 - \frac{1-v^2}{2v}\ln\frac{1+v}{1-v}\right)\ln\frac{q^*}{eT} + A_{\text{soft}}(v)\right] \tag{8.30}$$

where the coefficient which multiplies $\ln(q^*/eT)$ is determined from (8.24); the argument of the logarithm follows from the fact that $m \sim eT$ is the only scale at our disposal in the soft region. In the evaluation of the hard contribution to the energy loss, one uses (8.4) without any kinematical approximation and one obtains

$$\frac{dE}{dz}\bigg|_{\text{hard}} = \frac{e^4 T^2}{24\pi v}\left[\left(1 - \frac{1-v^2}{2v}\ln\frac{1+v}{1-v}\right)\ln\frac{ET}{Mq^*} + A_{\text{hard}}(v)\right] \tag{8.31}$$

The functions $A_{\text{soft}}(v)$ and $A_{\text{hard}}(v)$ are obtained after a tedious computation, and part of it has to be performed numerically. It should be clear from our study of the region $q \sim q^*$ that q^* disappears from the final result, as it acts as a lower (upper) cut-off in the hard (soft) region, with identical integrands; the final result reads

$$\frac{dE}{dz} = \frac{e^4 T^2}{24\pi v}\left[\left(1 - \frac{1-v^2}{2v}\ln\frac{1+v}{1-v}\right)\ln\frac{E}{eM} + A_{\text{hard}}(v) + A_{\text{soft}}(v)\right] \tag{8.32}$$

In the weak coupling limit $\alpha = e^2/(4\pi) \to 0$, the leading term in (8.32) is proportional to $\alpha^2 T^2 \ln(1/\alpha)$ and the dominant region of integration contributing of the logarithm is $eT \lesssim q \lesssim T$.

There is another interesting quantity related to $\Gamma(E)$ and to dE/dz, namely, the relaxation time τ for the heavy fermions (see section 4.4.4). In order to define τ, assume that the heavy fermions are in equilibrium with the electrons and photons of the thermal bath with an equilibrium distribution $\tilde{n}_{\text{eq}}(E) \ll 1$. Then, if one brings this distribution slightly out of equilibrium: $\tilde{n}(E) \to \tilde{n}_{\text{eq}}(E) + \delta\tilde{n}(E)$, $\delta\tilde{n}(E)$ obeys the equation (see (4.86))

$$\frac{d\delta\tilde{n}}{dt} = -\tilde{n}\Gamma^{>} + (1-\tilde{n})\Gamma^{<} = -\frac{1}{\tau}\delta\tilde{n} \tag{8.33}$$

The relaxation time is given by a convergent integral (exercise 8.4), and one finds in the weak coupling limit $1/\tau \sim \alpha^2 T \ln(1/\alpha)$.

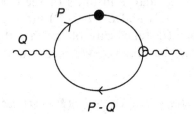

Fig. 8.3.   Imaginary part of the photon self-energy. The heavy dot indicates a resummed propagator.

8.2   Production of hard real photons from a quark–gluon plasma

Our second example deals with the screening of a quark mass singularity in the production of real photons from a quark–gluon plasma. To lowest order of perturbation theory, photons can be produced either from quark–antiquark annihilation or from quark–gluon scattering. Both reactions lead to production rates which are singular when the quark mass vanishes, and, as in the preceding section, it is worth checking the finiteness of the production rate when one uses resummed perturbation theory. We shall assume that the photons have hard momenta: this assumption leads to important simplifications, since, as we shall see shortly, we need not use dressed vertices. The problem of soft photon production seems to be a difficult one; it has been addressed recently by several authors, but it seems that one cannot avoid a collinear singularity, as we shall see in section 10.3.

We compute the production rate from the imaginary part of the photon self-energy. The graph of order e^2 does not contribute to the rate because no phase-space is available when the photon is real. Since the photon momentum Q is hard, at least one quark (or antiquark) line in the graph of fig. 8.3 is hard, but the other line may be soft. Let us call P the four-momentum of the soft line. We know that a resummed propagator must be used for this line. However, as only one line of the vertices is soft, we may use bare vertices. Obviously, this need not be the case for soft photon production, where all three lines may be soft.

We use (5.152) to relate the production rate to the photon self-energy $\Pi_{\mu\nu}(Q)$, $q_0 > 0$:

$$q_0 \frac{d\Gamma}{d^3q} = -\frac{1}{2(2\pi)^3} e^{-\beta q_0} g^{\mu\nu} \Pi^{>}_{\mu\nu}(Q) \tag{8.34}$$

where

$$\Pi^{>}_{\mu\nu}(Q) = -2e^2 e_q^2 N \int \frac{d^4P}{(2\pi)^4} \mathrm{Tr}\left(\gamma_\mu \,{}^{*}S^{>}(P)\, \gamma_\nu\, S^{<}_F(P-Q)\right) \tag{8.35}$$

In (8.35) e_q is the quark charge in units of the electron charge, N is the number of colours, $^*S^>(P)$ and $S^<(P - Q)$ are the dressed and bare cut propagators; a factor of 2 takes care of the fact that either the quark or the antiquark line may be dressed. We recall the expression (cf. (5.45b)) of $S_F^<(P - Q)$

$$S_F^<(P - Q) = -\varepsilon(p_0 - q_0)\tilde{f}(p_0 - q_0)(\slashed{P} - \slashed{Q})2\pi\delta\left((P - Q)^2\right) \qquad (8.36)$$

while $^*S^>(P)$ is deduced from (5.38) and (7.90) continued to real values of the energy:

$$^*S^>(P) = \frac{1}{2}(1 - \tilde{f}(p_0))\left[(\gamma_0 - \gamma \cdot \hat{\mathbf{p}})\rho_+(p_0, p)\right.$$
$$\left. + (\gamma_0 + \gamma \cdot \hat{\mathbf{p}})\rho_-(p_0, p)\right] \qquad (8.37)$$

where the spectral functions ρ_+ and ρ_- are given in (6.141). In order to eliminate trivial multiplicative factors, we define $\Pi^>$ by

$$\Pi^>(Q) = -\int \frac{d^4P}{(2\pi)^4}\text{Tr}\left(\gamma^\mu \, ^*S^>(P) \, \gamma_\mu \, S^<(P - Q)\right) \qquad (8.38)$$

As in the preceding section we introduce an intermediate scale p^*:

$$gT \ll p^* \ll T \qquad (8.39)$$

where g is the QCD coupling constant, and we examine the region in phase space where $p \le p^*$. The factor $\varepsilon(p_0 - q_0)\delta((P - Q)^2)$ which occurs in $S^<(P - Q)$ may be written

$$\varepsilon(p_0 - q_0)\delta\left((P - Q)^2\right) = \frac{1}{2E'}\left[\delta(p_0 - q_0 - E') - \delta(p_0 - q_0 + E')\right] \quad (8.40)$$

with $E' = |\mathbf{p} - \mathbf{q}|$. If $p_0, p \ll q$, we have $E' \simeq (q - py)$, where $y = \hat{\mathbf{p}} \cdot \hat{\mathbf{q}}$, which entails either $p_0 \simeq 2q$ or $p_0 \simeq py$. The first solution does not obey the initial assumption, so we concentrate on the second case and write

$$\varepsilon(p_0 - q_0)\delta\left((P - Q)^2\right) \simeq -\frac{1}{2q}\delta(p_0 - py) \qquad (8.41)$$

Furthermore, the Fermi–Dirac factor $(1 - \tilde{f}(p_0))$ in (8.37) may be approximated by $1/2$. We perform the trace calculation and integrate over the direction of $\mathbf{p}$ thanks to (8.41) with the result ($x = p_0/p$)

$$\Pi_{\text{soft}}^>(Q) = -\frac{1 - \tilde{f}(q_0)}{4\pi}\int_0^{p^*} p\,dp \int_{-p}^p \frac{dp_0}{2\pi}\left[(1 - x)\rho_+(p_0, p)\right.$$
$$\left. + (1 + x)\rho_-(p_0, p)\right] \qquad (8.42)$$

In order to extract the infrared singularities, we revert to the bare calculation by neglecting the thermal quark mass in the denominator of $\rho_\pm$:

$$(1 - x)\rho_+(p_0, p) \simeq (1 + x)\rho_-(p_0, p) \simeq \frac{\pi m_f^2}{p^3} \tag{8.43}$$

with the logarithmically divergent result

$$\Pi_{\text{bare}}^>(Q) = -(1 - \tilde{f}(q_0))\frac{m_f^2}{2\pi}\int_0^{p^*}\frac{dp}{p} \tag{8.44}$$

To leading order in perturbation theory, the integrand in (8.44) is correct for $p \sim p^*$, but it is of course modified by resummation in the region $p \sim gT \sim m_f$. The effect of resummation is most easily obtained by appealing to the sum rules (6.146) and (6.147) written in the form

$$\int_{-p}^{p}\frac{dp_0}{2\pi}(1 \mp x)\rho_\pm(p_0, p)$$
$$= -\left[(1 \mp \frac{\omega_\pm}{p})Z_\pm(p) + (1 \pm \frac{\omega_\mp}{p})Z_\mp(p)\right] \tag{8.45}$$

where $Z_\pm(p)$ is the residue at the quasi-particle pole $\omega_\pm(p)$. We now use the explicit expression of $Z_\pm(p)$ and obtain the contribution from the soft region $0 \leq p \leq p^*$ to $\Pi_{\text{soft}}^>$:

$$\Pi_{\text{soft}}^>(Q) = -\frac{1 - \tilde{f}(q_0)}{4\pi m_f^2}\int_0^{p^*} dp\left[(\omega_+ - p)\right.$$
$$\left. \times (\omega_+^2 - p^2) - (\omega_- + p)(\omega_-^2 - p^2)\right] \tag{8.46}$$

From the behaviour (6.140) of ω_+ and ω_- when $p \gg m_f$, one recovers the result (8.44) of the bare theory. However, the p-integral, which was divergent for $p \to 0$ in the bare theory, is now convergent as ω_+ and ω_- tend to finite values when $p \to 0$. Thanks to the identity

$$(\omega_\pm \mp p)(\omega_\pm^2 - p^2)/m_f^2 = \omega_\pm - p\frac{d\omega_\pm}{dp} \tag{8.47}$$

the production rate in the soft region can be cast in the form ($\alpha = e^2/4\pi$, $\alpha_s = g^2/4\pi$)

$$q_0\frac{d\Gamma}{d^3q}\bigg|_{\text{soft}} = \frac{e_q^2\alpha\alpha_s}{2\pi^2}\frac{T^2}{e^{\beta q_0} + 1}\left[\ln\left(\frac{p^*}{m_f}\right)^2 - 1\right.$$
$$\left. + \int_0^{p^*} dp\left(\frac{2(\omega_+ - \omega_-)}{m_f^2} - \frac{2}{p + m_f}\right)\right] \tag{8.48}$$

where terms of order p^*/T have been neglected. The integral has to be worked out numerically, its value being -0.31.

The hard contribution can be computed from the reactions

$$q + \bar{q} \rightarrow g + \gamma$$

and

$$q\,(\bar{q}) + g \rightarrow q\,(\bar{q}) + \gamma$$

working out the exact kinematics with the constraint $p > p^*$. As in the preceding section, the factor of $\ln p^*$ cancels between the soft and hard contributions, and the final result is finite; in the region $q_0 \gg T$

$$q_0 \frac{d\Gamma}{d^3 q} = \frac{e_q^2 \alpha \alpha_s}{2\pi^2}\, T^2 e^{-q_0/T} \ln \frac{c q_0}{\alpha_s T} \qquad (8.49)$$

where the constant c is given by

$$c \simeq \frac{2}{3\pi} \exp\left(\frac{1}{2} + \frac{1}{3}\ln 2 - \gamma + \frac{\zeta'(2)}{\zeta(2)} - 0.31\right) \qquad (8.50)$$

In (8.50), γ is the Euler constant and ζ the Riemann function.

8.3 Screening and transport phenomena

Let us now turn to the problem of transport coefficients: we shall see that dynamical screening is able to yield finite transport cross-sections in almost all cases, while those cross-sections would diverge in the bare theory. For simplicity the following discussion will be held in the case of a pure $SU(3)$ gauge theory, and the extension to the case where quarks are included will be mentioned only briefly. When computing the gluon–gluon differential cross-section, one encounters the standard Rutherford-like behaviour at small angles due to the exchange of bare gluons:

$$\frac{d\sigma}{d\Omega^*} \sim \frac{1}{\sin^4 \theta^*} \qquad (8.51)$$

where θ^* is the scattering angle in the centre-of-mass system. This singular behaviour at $\theta^* = 0$ does not allow one to compute the total cross-section σ which is needed, for example, to evaluate the mean-free path $\lambda = 1/(n\sigma)$. One often takes the 90° cross-section $\sigma \sim (d\sigma/d\Omega^*)_{\theta^*=\pi/2}$ in order to obtain a rough estimate of λ, but this is not really satisfactory. Fortunately, when one wants to compute transport cross-sections, $d\sigma/d\Omega^*$ is weighted in almost all cases by a factor $(1 - \cos\theta^*)$: when computing the viscosity coefficients, for example, this factor means physically that small-angle cross-sections are very inefficient for exchanging momentum. Then the transport cross-section is only logarithmically divergent:

$$\sigma_{\text{tr}} \sim \int d\Omega^* \frac{d\sigma}{d\Omega^*}(1 - \cos\theta^*) \sim \int \frac{d\theta^*}{\theta^*} \tag{8.52}$$

If one now exchanges dressed gluons, longitudinal gluons lead to a convergent transport cross-section thanks to Debye screening, as in non-relativistic plasmas. The exchange of transverse gauge particles is usually neglected in non-relativistic plasmas, where magnetic effects are small. On the contrary, magnetic effects are *a priori* as important as electric effects in ultrarelativistic plasmas, and we have to cope with the absence of static screening for the exchange of transverse gauge particles.

Let us assume a Boltzmann equation for the one-particle distribution $f_1 = f(\mathbf{x}_1, \mathbf{p}_1, t)$, which we write

$$Df_1 \equiv \frac{\partial f_1}{\partial t} + \mathbf{v}_1 \cdot \frac{\partial f_1}{\partial \mathbf{x}_1} + \mathbf{F} \cdot \frac{\partial f_1}{\partial \mathbf{p}_1} = C[f_1] \tag{8.53}$$

The collision term $C[f_1]$ is given by

$$C[f_1] = \frac{v_g}{2\varepsilon_1} \int d\tilde{p}_2 \, d\tilde{p}_3 \, d\tilde{p}_4 \, (2\pi)^4 \delta^{(4)}(P_1 + P_2 - P_3 - P_4)$$
$$\times \left[f_3 f_4 (1 + f_1)(1 + f_2) - f_1 f_2 (1 + f_3)(1 + f_4) \right] \overline{|\mathcal{M}|^2} \tag{8.54}$$

where $v_g = 16$ represents the sum over the target colour and spin degrees of freedom, $\varepsilon = |\mathbf{p}|$,

$$d\tilde{p} = \frac{d^3 p}{(2\pi)^3 2\varepsilon} \tag{8.55}$$

and $\overline{|\mathcal{M}|^2}$ is the gluon–gluon scattering matrix element squared, summed over spins and colours in the final state and averaged over initial spins and colours. In terms of the usual Mandelstam invariants:

$$s = (P_1 + P_2)^2 \quad t = (P_1 - P_3)^2 \quad u = (P_1 - P_4)^2 \tag{8.56}$$

$\overline{|\mathcal{M}|^2}$ reads to lowest order of bare (B) perturbation theory

$$\overline{|\mathcal{M}|^2_B} = \frac{9}{2} g^4 \left(3 - \frac{ut}{s^2} - \frac{us}{t^2} - \frac{ts}{u^2} \right) \tag{8.57}$$

We shall be interested in near forward scattering, and it will prove useful to introduce the four-vectors P, P' and Q by

$$\begin{aligned} P_1 &= P + Q/2 & P_2 &= P' - Q/2 \\ P_3 &= P - Q/2 & P_4 &= P' + Q/2 \end{aligned} \tag{8.58}$$

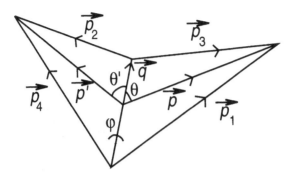

Fig. 8.4   Kinematics of the two-body collision.

The corresponding three vectors are shown in fig. 8.4: θ (θ') denotes the angle between $\mathbf{q}$ and $\mathbf{p}$ ($\mathbf{p}'$) and φ the angle between the $(\mathbf{p}_1, \mathbf{p}_3)$ and $(\mathbf{p}_2, \mathbf{p}_4)$ planes. In the case of near-forward scattering we have $q \ll (p, p')$ and we can make some kinematical approximations. We first note that

$$q_0 \simeq \hat{\mathbf{p}} \cdot \mathbf{q} \simeq \hat{\mathbf{p}}' \cdot \mathbf{q} \tag{8.59}$$

so that $\theta \simeq \theta'$, while

$$s \simeq 2pp'(1 - x^2)(1 - \cos \varphi) \tag{8.60}$$

where we have defined $x = \cos \theta = q_0/q$. The expression for t is

$$t = Q^2 = -q^2(1 - x^2) \tag{8.61}$$

so that $\overline{|\mathcal{M}|_B^2}$ in the near-forward direction reads

$$\overline{|\mathcal{M}|_B^2} \simeq \frac{9}{2} g^4 \frac{s^2}{t^2} \simeq 18 g^4 \frac{p^2 p'^2}{q^4} (1 - \cos \varphi)^2 \tag{8.62}$$

Let us correct the above expression at once in order to take the re-summation of the gluon propagator into account. The matrix element for resummed gluon exchange is given in terms of the external currents J_μ^1 and J_ν^2 by

$$\mathcal{M} \propto J_\mu^1 D^{\mu\nu} J_\nu^2 = J_L^1 \Delta_L(q_0, q) J_L^2 + \mathbf{J}_T^1 \cdot \mathbf{J}_T^2 \Delta_T(q_0, q) \tag{8.63}$$

where $\mathbf{J}_T$ and J_L are the transverse and longitudinal components of the external currents with respect to $\mathbf{q}$, while Δ_T and Δ_L are the transverse and longitudinal components of the gluon propagator defined in (6.74) and (6.75). In the case of near-forward scattering (8.63) reduces to

$$\mathcal{M} \propto \Delta_L(q_0, q) + \hat{\mathbf{p}}_{1T} \cdot \hat{\mathbf{p}}_{2T} \Delta_T(q_0, q)$$
$$= \Delta_L(q_0, q) + (1 - x^2) \cos \varphi \, \Delta_T(q_0, q) \tag{8.64}$$

The normalization of $|\mathcal{M}|^2$ is obtained by taking the limit of bare gluon exchange ($m = 0$ in (6.74) and (6.75)) and comparing with (8.62); one finds

$$\overline{|\mathcal{M}|^2} = 18g^4 p^2 p'^2 |\Delta_L(q_0, q) + (1 - x^2)\cos\varphi\,\Delta_T(q_0, q)|^2 \qquad (8.65)$$

After these preliminaries, let us come back to the Boltzmann equation (8.54). As an example of a transport coefficient, we shall compute the shear viscosity; once this calculation has been understood, it is straightforward to generalize it to other transport coefficients. We linearize the Boltzmann equation around a local equilibrium solution, corresponding to a plasma in motion with a time-independent flow velocity $\mathbf{u}, u \ll 1$:

$$f^{(0)} = \left(\exp\left[\beta(\varepsilon - \mathbf{p}\cdot\mathbf{u})\right] - 1\right)^{-1} \qquad (8.66)$$

We assume that the x-component of $\mathbf{u}$ is a function of the coordinate y and we write

$$f = f^{(0)} + \frac{\partial f^{(0)}}{\partial\varepsilon}\Phi\frac{\partial u_x}{\partial y} = f^{(0)}(1 + \Phi') \qquad (8.67)$$

with $\Phi' \ll 1$. Energy–momentum conservation gives the relation

$$\frac{f_1^{(0)}}{1 + f_1^{(0)}}\frac{f_2^{(0)}}{1 + f_2^{(0)}} = \frac{f_3^{(0)}}{1 + f_3^{(0)}}\frac{f_4^{(0)}}{1 + f_4^{(0)}} \qquad (8.68)$$

so that $f^{(0)}$, with uniform $\mathbf{u}$, is a solution of the Boltzmann equation: $Df_1^{(0)} = C[f_1^{(0)}] = 0$. With non-uniform $\mathbf{u}$, the LHS of the Boltzmann equation is to leading order

$$Df_1 = -p_{1x}v_{1y}\frac{\partial f_1^{(0)}}{\partial\varepsilon_1}\frac{\partial u_x}{\partial y} \simeq \beta p_{1x}v_{1y}n_1(1 + n_1)\frac{\partial u_x}{\partial y} \qquad (8.69)$$

Using (8.67) and (8.68), equation (8.54) is cast in the form

$$p_{1x}v_{1y} = \frac{v_g}{2\varepsilon_1}\int d\tilde{p}_2\,d\tilde{p}_3\,d\tilde{p}_4\,(2\pi)^4\delta^{(4)}(P_1 + P_2 - P_3 - P_4)$$
$$\times \overline{|\mathcal{M}|^2}\,\frac{n_2(1 + n_3)(1 + n_4)}{1 + n_1}\left[\Phi_1 + \Phi_2 - \Phi_3 - \Phi_4\right] \qquad (8.70)$$

The xy-component P_{xy} of the pressure tensor is, by definition,

$$P_{xy} = v_g\int\frac{d^3p_1}{(2\pi)^3}p_{1x}v_{1y}\frac{\partial f^{(0)}}{\partial\varepsilon}\Phi_1\frac{\partial u_x}{\partial y} = -\eta\frac{\partial u_x}{\partial y} \qquad (8.71)$$

so that the shear viscosity η reads

$$\eta \simeq -v_g\int\frac{d^3p_1}{(2\pi)^3}p_{1x}v_{1y}\frac{dn_1}{d\varepsilon_1}\Phi_1 \qquad (8.72)$$

On the other hand, we may also use the expression (8.70) of $p_{1x}v_{1y}$, so that an alternative expression for η is

$$
\eta = \frac{\beta}{4} v_g^2 \int d\tilde{p}_1 \, d\tilde{p}_2 \, d\tilde{p}_3 \, d\tilde{p}_4 \, n_1 \, n_2(1 + n_3)(1 + n_4)
$$

$$
\times (2\pi)^4 \delta^{(4)}(P_1 + P_2 - P_3 - P_4)\overline{|\mathcal{M}|^2}\left[\Phi_1 + \Phi_2 - \Phi_3 - \Phi_4\right]^2
\tag{8.73}
$$

In writing (8.73) we have used symmetry properties of the collision term. A convenient formula for η is now

$$
\frac{1}{\eta} = \left(\int \frac{d^3 p_1}{(2\pi)^3} \, p_{1x}v_{1y}\frac{dn_1}{d\varepsilon_1}\Phi_1\right)^{-2}\left(\frac{\beta}{4}\int d\tilde{p}_1 \, d\tilde{p}_2 \, d\tilde{p}_3 \, d\tilde{p}_4\right.
$$

$$
\times (2\pi)^4 \delta^{(4)}(P_1 + P_2 - P_3 - P_4)n_1 n_2(1 + n_3)(1 + n_4)
$$

$$
\left.\times \overline{|\mathcal{M}|^2}\left[\Phi_1 + \Phi_2 - \Phi_3 - \Phi_4\right]^2\right)
\tag{8.74}
$$

Equation (8.74) is the starting point for a variational treatment of the viscosity calculation: if instead of the solution Φ we take an arbitrary function Ψ in (8.74), it can be shown that one obtains a value $\eta[\Psi]$ which is lower than the exact value (exercise 8.5). From rotational invariance we must choose

$$
\Psi = A(p)p_x p_y = pA(p)p_x v_y
\tag{8.75}
$$

and the variational treatment can be applied to $A(p)$. In order to obtain a simple estimate, we choose $A(p) = $ constant; this constant is then irrelevant as it cancels between the numerator and denominator in (8.74), and it may be taken equal to one. The denominator in (8.74) is easily evaluated:

$$
-\int \frac{d^3 p}{(2\pi)^3} \, p_x^2 \, p_y^2 \frac{1}{p}\frac{dn}{dp} = \frac{4T^5\zeta(5)}{\pi^2}
\tag{8.76}
$$

where $\zeta(s)$ is the Riemann ζ-function.

The evaluation of the numerator in (8.74) is more involved. We choose as integration variables $\mathbf{p}$, $\mathbf{p}'$ and $\mathbf{q}$, and we remark that $\overline{|\mathcal{M}|^2}$ depends only on the magnitude and on the relative orientation of $\mathbf{p}$, $\mathbf{p}'$ and $\mathbf{q}$. We may then average $\left[\Psi_1 + \Psi_2 - \Psi_3 - \Psi_4\right]^2$ over the angles by keeping the magnitude and the relative orientation of $\mathbf{p}$, $\mathbf{p}'$ and $\mathbf{q}$ fixed and by rotating this system over the three Euler angles in a fixed reference frame. In order to perform this angular average, we note that for two arbitrary vectors $\mathbf{A}$ and $\mathbf{B}$:

$$
\langle A_i A_j B_k B_l\rangle = \alpha\delta_{ij}\delta_{kl} + \beta[\delta_{ik}\delta_{jl} + \delta_{il}\delta_{jk}]
\tag{8.77}
$$

where $\langle\ldots\rangle$ denotes an angular average; the coefficients α and β are given by

$$
\alpha = \frac{1}{15}(2\mathbf{A}^2\mathbf{B}^2 - (\mathbf{A}\cdot\mathbf{B})^2) \qquad \beta = \frac{1}{30}(-\mathbf{A}^2\mathbf{B}^2 + 3(\mathbf{A}\cdot\mathbf{B})^2)
\tag{8.78}
$$

so that

$$\langle[\Psi_1 + \Psi_2 - \Psi_3 - \Psi_4]^2\rangle = q^2 Q(p, p'; x, \varphi)$$

$$= \frac{q^2}{15}\left[3(p^2 + p'^2) + x^2(p - p')^2 - 6pp'(\hat{\mathbf{p}} \cdot \hat{\mathbf{p}}')\right] \qquad (8.79)$$

We transform the Bose–Einstein factors by making use of (4.35)

$$n_1(1 + n_3) = n\left(p + \frac{q_0}{2}\right)\left(1 + n\left(p - \frac{q_0}{2}\right)\right)$$

$$= f(q_0)\left(f\left(p - \frac{q_0}{2}\right) - f\left(p + \frac{q_0}{2}\right)\right)$$

$$\simeq -q_0 f(q_0)\frac{dn}{dp} \qquad (8.80)$$

and similarly

$$n_2(1 + n_4) \simeq q_0 f(-q_0)\frac{dn}{dp'} \qquad (8.81)$$

Let us now examine the phase space integration. We have, first,

$$\delta(\varepsilon_1 + \varepsilon_2 - \varepsilon_3 - \varepsilon_4) \simeq \delta(\mathbf{q} \cdot (\hat{\mathbf{p}} - \hat{\mathbf{p}}')) = \frac{1}{q}\delta(\cos\theta - \cos\theta') \qquad (8.82)$$

We may then integrate over $d\Omega'_p$ since we have performed the angular average (8.79):

$$\int d^3p_1 \ldots d^3p_4 (2\pi)^4 \delta^{(4)}(P_1 + P_2 - P_3 - P_4)$$

$$= 2(2\pi)^7 \int q \, dq \, p^2 dp \, p'^2 dp' dx \frac{d\varphi}{2\pi} \qquad (8.83)$$

With $\alpha_s = g^2/4\pi$, the numerator in (8.74) can be written as

$$N = \frac{9\beta\alpha_s^2}{32\pi^3}\int_0^\infty p^2\frac{dn}{dp}dp \int_0^\infty p'^2\frac{dn}{dp'}dp' \int_{-1}^{+1} dx \int_0^{2\pi}\frac{d\varphi}{2\pi}$$

$$\times \int_0^\infty dq \, q^3 (qx)^2 f(qx)(1 + f(qx))Q(p, p'; x, \varphi)$$

$$\times |\Delta_L + (1 - x^2)\cos\varphi \, \Delta_T|^2 \qquad (8.84)$$

In bare perturbation theory:

$$|\Delta_L + (1 - x^2)\cos\varphi \, \Delta_T|^2 \to \frac{1}{q^4}(1 - \cos\varphi)^2 \qquad (8.85)$$

so that (8.84) exhibits a dq/q logarithmic divergence. One notes that (8.84) does involve the transport cross-section: the angular average in (8.79) contributes an extra factor of q^2 ($q^2(1 - x^2) = -t \propto (1 - \cos\theta^*)$).

In the region $q \ll T$, we also note the approximation

$$(qx)^2 f(qx)(1 + f(qx)) \simeq T^2 \tag{8.86}$$

but of course the Bose–Einstein factors cut-off the q-integration at $q_{max} \sim T$. In the case of a resummed gluon propagator, we may thus write the (x, q)-integral schematically as

$$\int_{-1}^{+1} h(x) dx \int dq\, q^3 |\Delta_L + (1 - x^2) \cos \varphi\, \Delta_T|^2 \tag{8.87}$$

where the function $h(x)$ is regular. The contribution of $|\Delta_L|^2$ is finite due to Debye screening. Let us examine the contribution of $|\Delta_T|^2$: since the function $h(x)$ is regular, and since no divergence can arise from Δ_T for non-zero values of x, it suffices to examine the region $x \to 0$, where we can use the approximate expression (8.3) for Δ_T:

$$\int_0^T \frac{q^3 dq}{|q^2 - i\pi m^2 x/2|^2} = \ln \frac{T}{m} - \frac{1}{2} \ln \frac{\pi x}{2} \tag{8.88}$$

Since the function $h(x)$ in (8.56) is regular at $x = 0$, the factor of $\ln x$ does not lead to any singular behaviour when $x \to 0$, and one recovers, as in the case of Debye screening, a $\ln(T/m)$ screening factor. This factor is not affected by the possible existence of a magnetic mass. Thus, despite the absence of screening of static magnetic fields, transverse gluon exchange is effectively cut off in the infrared by the thermal mass m. The contribution from the transverse–longitudinal interference term can be treated in a similar way, and in the leading logarithmic approximation the q-integral is $(1 - \cos \varphi)^2 \ln(T/m)$. It is now easy to perform the x- and φ-integrals in (8.84):

$$\int_{-1}^{+1} dx \int_0^{2\pi} \frac{d\varphi}{2\pi} Q(p, p'; x, \varphi)(1 - \cos \varphi)^2 = \frac{2}{3}(p^2 + p'^2) \tag{8.89}$$

The p- and p'-integrals in (8.84) are readily performed, yielding for N the result:

$$N = \frac{\alpha_s^2 T^7 \pi^3}{30} \ln \left(\frac{T}{m} \right) \tag{8.90}$$

From (8.76) and (8.90) we obtain for $1/\eta$, to leading logarithmic accuracy,

$$\frac{1}{\eta} = \frac{\pi^7}{960(\zeta(5))^2 T^3} \alpha_s^2 \ln \frac{1}{\alpha_s} \simeq \frac{2.93}{T^3} \alpha_s^2 \ln \frac{1}{\alpha_s} \tag{8.91}$$

A variational estimate of η, with the choice $A(p) = p^\nu$ in (8.75) leads to the optimal value $\nu \simeq 0.104$ (exercise 8.6), and the result (8.91) differs from the optimal value by less than 1%.

We may associate with the viscosity a viscous relaxation time τ_η, which is defined by the relaxation time approximation of the Boltzmann equation:

$$Df_1 = -\frac{f_1 - f_1^{(0)}}{\tau_\eta} \qquad (8.92)$$

or

$$p_{1x}v_{1y} = \frac{1}{\tau_\eta}\Phi_1 \qquad (8.93)$$

These equations lead to the estimate

$$\frac{1}{\tau_\eta} \simeq -\frac{v_g}{\eta}\int \frac{d^3 p}{(2\pi)^3} p_x^2 p_y^2 \frac{dn}{dp} \simeq \frac{1.402\, T^4}{\eta} \qquad (8.94)$$

or

$$\frac{1}{\tau_\eta} \simeq 4.11\, T\alpha_s^2 \ln \frac{1}{\alpha_s} \qquad (8.95)$$

Taking $T = 200$ Mev and $\alpha_s \simeq 0.3$ gives the numerical estimate $\tau_\eta \simeq 2.5\,\mathrm{fm/c}$.

The preceding results are easily generalized when quarks are included, because the cross-sections for the three relevant processes: $gg \to gg$, $qg \to qg$, $qq \to qq$ are proportional in the forward direction. One finds that the total viscosity η_{tot} is simply the sum of the gluon and quark viscosities: $\eta_{\mathrm{tot}} = \eta_g + \eta_q$, with η_g and η_q being given by

$$\eta_g = \frac{\eta}{1 + N_f/6} \qquad \eta_q = \frac{3^5 \times 5^2}{2^9 \times 7} N_f \eta_g \simeq 1.70 N_f \eta_g \qquad (8.96)$$

with η from (8.91). With $N_f = 2$ the quark viscosity is about 3.4 times the gluon viscosity. This rather large factor stems mainly from the fact that the $gg \to gg$, $qg \to qg$, $qq \to qq$ cross-sections are in proportion to $\frac{9}{2} : 2 : \frac{8}{9}$. From (8.94) we also see that the gluon viscous relaxation time is smaller than the estimate (8.94) by a factor $1 + N_f/6 = 4/3$, and that it is smaller than the quark relaxation time by a factor 2.6: in the weak coupling approximation, the viscous relaxation time is about 2 fm/c for gluons and 5 fm/c for quarks. However, it must be emphasized that realistic relaxation times could be much shorter, because $\alpha_s = 0.3$ $(g = 2!)$ is not really a weak coupling.

Finally, it must be mentioned that in the case of colour transportation, dynamical screening is not sufficient to give a finite result. The basic reason is that quarks and gluons can easily change colour directions in forward scattering by colour exchange, and one does not obtain the extra q^2-factor of other transport cross-sections. In order to obtain a finite result, one has to appeal to a magnetic mass of the order of $g^2 T$, and one finds for the coefficient of colour diffusion

$$D_{\text{colour}}^{-1} \simeq 4.9 \frac{1 + 7 N_f / 33}{1 + N_f / 6} \, T \alpha_s \ln \frac{1}{\alpha_s} \tag{8.97}$$

One should note the dependence of D_{colour}^{-1} on the coupling constant: $\alpha_s \ln(1/\alpha_s)$ instead of $\alpha_s^2 \ln(1/\alpha_s)$: indeed transverse gluons give a contribution of the form

$$\alpha_s^2 \int_{g^2 T}^{T} \int_{-1}^{+1} dx \, \frac{q \, dq}{|q^2 - i\pi m^2 x/2|^2} \sim \frac{\alpha_s}{T^2} \ln \frac{1}{\alpha_s} \tag{8.98}$$

where the factor α_s^2 comes from the matrix element squared. The dominant range of integration is $g^2 T \lesssim q \lesssim g T$, while in the preceding case it was $g T \lesssim q \lesssim T$.

References and further reading

The phenomenon of dynamical screening in an ultrarelativistic plasma seems to have been first noticed by Weldon (1982a). The study of the energy loss of a heavy fermion in a quark–gluon plasma follows the work of Braaten and Thoma (1991). The production rate of hard photons from a quark–gluon plasma has been computed by Baier, Nakkagawa, Niégawa and Redlich (1992a) and by Kapusta, Lichard and Seibert (1991). The Primakoff production of axions is discussed by Altherr (1991) and by Braaten and Yuan (1991). The production of massive photon pairs was studied by Baier, Nakkagawa, Niégawa and Redlich (1992b). The importance of dynamical screening for transport cross-sections was first discussed by Baym, Monien, Pethick and Ravenhall (1990); see also Heiselberg and Pethick (1993). Colour transportation is discussed by Selikhov and Gyulassy (1993) and by Heiselberg (1994).

Exercises

8.1 Show that in the kinematical conditions of section 8.1, the contribution of Compton scattering is down by a factor $(T/M)^2$ with respect to Coulomb scattering.

8.2 Computing ρ_T from (8.3), show that dynamical screening makes the energy loss finite, but is not sufficient to cancel the infrared divergence in the reaction rate. Hint: start from (8.19) and interchange the x- and q-integrations. Also examine the longitudinal photons.

8.3 Derive the results of section 8.2 by using the imaginary-time formalism. Hint: compute the discontinuities from (7.106).

8.4 In the soft approximation $q \ll k, p$, one may write $\bar{u}(\mathbf{p}')\gamma_\mu u(\mathbf{p}) \simeq 2P_\mu$ by using the Gordon identity. From this result obtain $\Gamma_{\text{soft}}(E)$ as

$$\Gamma_{\text{soft}}(E) = \frac{e^4 T^2}{2\pi v} \int_0^{q^*} q \, dq \int_{-v}^{v} dx \left(1 + f(q\,x)\right)$$
$$\times \left(|\Delta_L|^2 + \frac{1}{2}(v^2 - x^2)(1 - x^2)|\Delta_T|^2\right)$$

and recover (8.19). Next estimate the relaxation time τ in (8.33); the leading term

$$\frac{1}{\tau} = \frac{e^2}{2\pi v} \int_0^{q^*} q^2 dq \int_{-v}^{v} \frac{dx}{2\pi} \left(\rho_L + (v^2 - x^2)\rho_T\right)$$

vanishes because the integrand is odd in x. Show that in the bare theory τ is logarithmically divergent and that in the resummed theory $1/\tau \sim \alpha^2 T \ln(1/\alpha)$.

8.5 One writes equation (8.70) schematically as $p_x v_y = I[\Phi]$, so that $\eta = (\Phi, I[\Phi])$, where the (positive definite) scalar product (Ψ', Ψ) is defined by

$$(\Psi', \Psi) = -\int \frac{d^3 p}{(2\pi)^3} \Psi'(p) \frac{dn}{dp} \Psi(p)$$

Use Schwarz's inequality to show that $\eta[\Psi] \leq \eta$.

8.6 Starting from (8.76) with an arbitrary function $A(p)$, show that (8.89) becomes

$$\frac{2}{3}p^2 A^2(p) + \frac{4}{15}p^3 A(p)A'(p) + \frac{1}{15}p^4(A'(p))^2 + [p \to p']$$

Use this result to compute $\eta[\Psi]$ for $A(p) = 1/p$ and show that the numerical coefficient in (8.91) is then equal to 5.95. Make a variational estimate of $\eta[\Psi]$ for $A(p) = p^v$.

8.7 Compute the thermal conductivity in the case of a quarkless plasma, starting from a local equilibrium distribution

$$f^{(0)} = \frac{1}{e^{\beta(z)\varepsilon} - 1}$$

with a z-dependent temperature. Write

$$f = f^{(0)} + \frac{df^{(0)}}{d\varepsilon} \Phi \frac{1}{T} \frac{\partial T}{\partial z}$$

and choose as a trial function $\Psi = A(p)\varepsilon p_z$.

9

Neutrino emission from stars

There are some interesting applications of thermal field theory to astrophysical processes; to be fair, all necessary computations may also be performed in the framework of standard kinetic theory, but experience with thermal field theory has recently allowed us to improve earlier calculations, resulting in large corrections in some circumstances. We shall be interested in the energy losses of stars due to the emission of weakly interacting particles. The relevant astrophysical systems are the following.

(i) The core of type II supernovae, which is a plasma with temperature $T \sim 30\text{--}60$ MeV and density $\rho \sim 10^{15}$ g cm^{-3}. The electron chemical potential is $\mu \sim 350$ MeV and the plasma frequency is $\omega_P \sim 20$ MeV.

(ii) The core of red giants before the 'helium flash': in this case $T \sim 10^8$ K (10 keV), $\rho \simeq 10^6$ g cm^{-3}, corresponding to an almost degenerate electron gas with Fermi momentum $p_F \simeq 400$ keV/c and a plasma frequency $\omega_P \simeq 20$ keV.

(iii) The core of young white dwarves, in which typical conditions are $T \sim 10^6 - 10^7$ K (0.1–1 keV) and $\rho \simeq 2 \times 10^6$ g cm^{-3}, which again corresponds to an almost degenerate electron gas with Fermi momentum $p_F \simeq 500$ keV.

In cases (ii) and (iii), one is interested in the energy losses due to neutrino emission, while axion emission is relevant for cases (i) and (ii). As the axion is still a hypothetical particle, we shall limit ourselves to neutrino emission. Many processes are involved in neutrino emission from stars, but the dominant one is often the so-called 'plasmon' decay into a neutrino pair; remember, however, that our terminology is rather 'transverse photon' and 'longitudinal photon'. In this chapter we shall thus compute the decay rate of these quasi-particles into neutrino pairs, but we first need to improve our knowledge of the dispersion laws.

9.1 The photon dispersion relations revisited

The expressions which were derived in chapter 6 for the photon dispersion relations when the electron mass m_e is much smaller than the temperature and/or the chemical potential: $m_e \ll T$ and/or $m_e \ll \mu$, may be generalized to simple approximate analytic formulae which become exact in the ultrarelativistic, degenerate and classical limits, and which interpolate smoothly between these limits. These expressions allow efficient numerical computation of the processes we study in the present chapter. Our starting point is the exact expression to order one-loop of the T-dependent part of the photon polarization tensor $\Pi_{\mu\nu}(Q)$. Starting, for example, from the expression derived in exercise 5.7 and performing an analytical continuation to real time yields the following expression in Minkowski space:

$$
\begin{aligned}
\Pi_{\mu\nu}(Q) = 16\pi\alpha \int & \frac{d^3 p}{(2\pi)^3} \frac{1}{2E} (\tilde{n}_+(E) + \tilde{n}_-(E)) \\
& \times \frac{P \cdot Q(P_\mu Q_\nu + P_\nu Q_\mu) - Q^2 P_\mu P_\nu - (P \cdot Q)^2 g_{\mu\nu}}{(P \cdot Q)^2 - (Q^2/2)^2}
\end{aligned}
\tag{9.1}
$$

where $Q_\mu = (q_0, \mathbf{q})$ is the external photon momentum, $P_\mu = (E, \mathbf{p})$, $E = (p^2 + m_e^2)^{1/2}$ is the loop momentum, $\alpha = e^2/4\pi$, while

$$
\tilde{n}_\pm(E) = \frac{1}{e^{\beta(E \mp \mu)} + 1}
\tag{9.2}
$$

represents, as in preceding chapters, the Fermi–Dirac equilibrium distribution for electrons and positrons at temperature T and chemical potential μ. The electrical neutrality of the plasma is ensured by protons or light ions whose effects may be taken into account if necessary.

If one works consistently to lowest order in α, a detailed analysis shows that one must drop the term $(Q^2/2)^2$ in the denominator of (9.1). Indeed, in all kinematical configurations, one can show that the effect of this term is at most comparable to corrections of the order of α^2 to the polarization tensor. The physical interpretation is that this approximation corresponds to calculating plasma corrections by using forward scattering amplitudes for electrons and positrons in the vacuum. As a result of dropping this term, the dispersion relations always give real values for the transverse and longitudinal frequencies $\omega_T(q)$ and $\omega_L(q)$, while, superficially, the reaction $\gamma \to e^+ - e^-$ would give complex values when the plasma frequency ω_P became larger than the threshold $2m$ for this reaction. However, this threshold is unphysical because in a plasma electrons and positrons acquire a mass which is large enough to prevent the decay of a photon into an $e^+ - e^-$ pair.

As in chapter 6 we use the decomposition (6.20) of $\Pi_{\mu\nu}$ into transverse and longitudinal components; instead of the functions F and G of (6.20), we sometimes use Π_T and Π_L:

$$\Pi_T = \frac{1}{2}(\delta_{ij} - \hat{\mathbf{q}}_i\hat{\mathbf{q}}_j)\Pi_{ij} = G$$

$$\Pi_L = \Pi_{00} = \frac{q^2}{Q^2}F \tag{9.3}$$

Once the $(Q^2/2)^2$ term has been dropped, the denominator of (9.1) never vanishes for $Q^2 > 0$; the angular integration is readily performed and one finds for Π_T and Π_L

$$\Pi_T(q_0, q) = \frac{4\alpha}{\pi} \int_0^\infty \frac{p^2 dp}{E}(\tilde{n}_+(E) + \tilde{n}_-(E))$$
$$\times \left(\frac{q_0^2}{q^2} - \frac{q_0^2 - q^2}{q^2}\frac{q_0}{2vq}\ln\frac{q_0 + vq}{q_0 - vq} \right) \tag{9.4}$$

$$\Pi_L(q_0, q) = \frac{4\alpha}{\pi} \int_0^\infty \frac{p^2 dp}{E}(\tilde{n}_+(E) + \tilde{n}_-(E))$$
$$\times \left(\frac{q_0}{vq}\ln\frac{q_0 + vq}{q_0 - vq} - 1 - \frac{q_0^2 - q^2}{q_0^2 - v^2q^2} \right) \tag{9.5}$$

with $v = p/E$. The transverse and longitudinal dispersion relations are obtained, as in chapter 6, by solving the equations

$$\omega_T^2(q) - q^2 - \Pi_T(\omega_T(q), q) = 0 \tag{9.6}$$

$$q^2 - \Pi_L(\omega_L(q), q) = 0 \tag{9.7}$$

When $q \to 0$, one can no longer distinguish transverse from longitudinal excitations, so that the plasma frequency ω_P is derived either from Π_T:

$$\omega_P^2 = \lim_{q \to 0} \Pi_T(\omega_T(q), q) \tag{9.8}$$

or from Π_L:

$$\omega_P^2 = \lim_{q \to 0} \frac{\omega_L^2(q)}{q^2}\Pi_L(\omega_L(q), q) \tag{9.9}$$

The result for ω_P is

$$\omega_P^2 = \frac{4\alpha}{\pi} \int_0^\infty \frac{p^2 dp}{E}(\tilde{n}_+(E) + \tilde{n}_-(E))\left(1 - \frac{1}{3}v^2\right) \tag{9.10}$$

Let us recall that in the ultrarelativistic case $\omega_P \sim eT$ or $e\mu$. It is also interesting to note that for $q \to \infty$ the transverse photon dispersion relation is $\omega_T^2(q) \simeq q^2 + m_T^2$, where the effective mass m_T reads

$$m_T^2 = \frac{4\alpha}{\pi} \int_0^\infty dp \frac{p^2}{E}(\tilde{n}_+(E) + \tilde{n}_-(E)) \tag{9.11}$$

and that $\omega_P \leq m_T \leq \sqrt{3/2}\,\omega_P$. As q increases, the dispersion relation $\omega_L(q)$ crosses the light cone at a point $q_{\text{crit.}}$ given by

$$q_{\text{crit.}}^2 = \frac{4\alpha}{\pi} \int_0^\infty \frac{p^2 \mathrm{d}p}{E} (\tilde{n}_+(E) + \tilde{n}_-(E)) \left(\frac{1}{v} \ln \frac{1+v}{1-v} - 1 \right) \qquad (9.12)$$

What happens around this value of q is the subject of exercise 9.2.

The integrals in (9.4) and (9.5) have been evaluated analytically in the ultrarelativistic case in section 6.3.4, for non-zero T and μ. An analytical integration is also possible in the degenerate limit ($\tilde{n}(E) \simeq \theta(\mu - E)$, $T \ll (\mu - m_e)$): an integration by parts allows one to rewrite Π_T and Π_L as

$$\Pi_T(q_0, q) = -\frac{4\alpha}{\pi} \int_0^\infty \frac{p^3 \mathrm{d}p}{E} \frac{\mathrm{d}}{\mathrm{d}p} (\tilde{n}_+(E) + \tilde{n}_-(E))$$
$$\times \left(\frac{1}{2v^2} \left(\frac{q_0^2}{q^2} - \frac{q_0^2 - v^2 q^2}{q^2} \frac{q_0}{2vq} \ln \frac{q_0 + vq}{q_0 - vq} \right) \right) \qquad (9.13)$$

$$\Pi_L(q_0, q) = -\frac{4\alpha}{\pi} \int_0^\infty \frac{p^3 \mathrm{d}p}{E} \frac{\mathrm{d}}{\mathrm{d}p} (\tilde{n}_+(E) + \tilde{n}_-(E))$$
$$\times \left(\frac{1}{v^2} \left(\frac{q_0}{2vq} \ln \frac{q_0 + vq}{q_0 - vq} - 1 \right) \right) \qquad (9.14)$$

Then, if $T \ll (\mu - m_e)$, the p-derivative in (9.13) and (9.14) peaks at $p = p_F$, where p_F is the Fermi momentum, which allows a straightforward calculation (exercise 9.2). Finally, it is also easy to evaluate (9.4) and (9.5) in the classical, non-degenerate limit $T \ll m_e$, $T \ll (m_e - \mu)$, where the Fermi–Dirac distribution is approximated by a Boltzmann distribution.

Let us derive an approximate analytical expression for (9.13) and (9.14). If one pulls out the v-dependent factor from the integrand, the remaining integral is proportional to the plasma frequency, as can be seen by integrating (9.10) by parts:

$$\omega_P^2 = -\frac{4\alpha}{3\pi} \int_0^\infty \frac{p^3 \mathrm{d}p}{E} \frac{\mathrm{d}}{\mathrm{d}p} (\tilde{n}_+(E) + \tilde{n}_-(E)) \qquad (9.15)$$

One may then conclude that if one approximates v by a suitable constant v_*, which is interpreted as some average velocity of the particles in the plasma, one obtains a reasonable approximation to (9.13) and (9.14). The value of v_* is fixed by requiring that the first two terms of the Taylor expansion of $\omega_T(q)$ and $\omega_L(q)$ in powers of q coincide with the exact result. This leads to the choice $v_* = \omega_1/\omega_P$, where ω_1 is given by (exercise 9.3)

$$\omega_1^2 = \frac{4\alpha}{\pi} \int_0^\infty \frac{p^2 \mathrm{d}p}{E} (\tilde{n}_+(E) + \tilde{n}_-(E)) \left(\frac{5}{3} v^2 - v^4 \right) \qquad (9.16)$$

and lies in the range $0 \leq \omega_1 \leq \omega_P$. Note that $v_* = 1$ in the ultrarelativistic case and $v_* = v_F$ in the degenerate case. The results for Π_T and Π_L are

$$
\begin{aligned}
\Pi_T(q_0, q) = \omega_P^2 \frac{3}{2v_*^2} \Bigg(& \frac{q_0^2}{q^2} - \frac{q_0^2 - v_*^2 q^2}{q_0^2} \\
& \times \frac{q_0}{2v_* q} \ln \frac{q_0 + v_* q}{q_0 - v_* q} \Bigg)
\end{aligned}
\tag{9.17}
$$

$$
\Pi_L(q_0, q) = \omega_P^2 \frac{3}{v_*^2} \left(\frac{q_0}{2v_* q} \ln \frac{q_0 + v_* q}{q_0 - v_* q} - 1 \right)
\tag{9.18}
$$

This yields, for the quasi-particle frequencies,

$$
\begin{aligned}
\omega_T^2 = q^2 + \omega_P^2 \frac{3\omega_T^2}{2v_*^2 q^2} \Bigg(& 1 - \frac{\omega_T^2 - v_*^2 q^2}{\omega_T^2} \frac{\omega_T}{2v_* q} \\
& \times \ln \frac{\omega_T + v_* q}{\omega_T - v_* q} \Bigg)
\end{aligned}
\tag{9.19}
$$

$$
\omega_L^2 = \omega_P^2 \frac{3\omega_L^2}{v_*^2 q^2} \left(\frac{\omega_L}{2v_* q} \ln \frac{\omega_L + v_* q}{\omega_L - v_* q} - 1 \right)
\tag{9.20}
$$

We shall also need the generalizations of (6.87) and (6.90) for the residues at the quasi-particle poles:

$$
\begin{aligned}
Z_T(q) &= \left[\frac{\partial}{\partial q_0} \left(Q^2 - \Pi_T(q_0, q) \right) \right]^{-1}_{q_0 = \omega_T(q)} \\
&= \frac{\omega_T(\omega_T^2 - v_*^2 q^2)}{3\omega_P^2 \omega_T^2 + (\omega_T^2 + q^2)(\omega_T^2 - v_*^2 q^2) - 2\omega_T^2(\omega_T^2 - q^2)}
\end{aligned}
\tag{9.21}
$$

and

$$
\begin{aligned}
Z_L(q) &= - \left[\frac{\partial \Pi_L(q_0, q)}{\partial q_0} \right]^{-1}_{q_0 = \omega_L(q)} \\
&= \frac{\omega_L(\omega_L^2 - v_*^2 q^2)}{q^2 [3\omega_P^2 - (\omega_L^2 - v_*^2 q^2)]}
\end{aligned}
\tag{9.22}
$$

Finally, neutrino emission also involves an axial component Π_A (see (9.31)):

$$
\begin{aligned}
\Pi_A(q_0, q) = 8\pi\alpha \frac{Q^2}{q^2} \int & \frac{d^3 p}{(2\pi)^3 2E} (\tilde{n}_+(E) - \tilde{n}_-(E)) \\
& \times \frac{P \cdot Q q_0 - Q^2 E}{(P \cdot Q)^2 - (Q^2/2)^2}
\end{aligned}
\tag{9.23}
$$

which has to be evaluated for $q_0 = \omega_T(q)$:

$$
\Pi_A(\omega_T(q), q) = \omega_A q \frac{\omega_T^2 - q^2}{\omega_T^2 - v_*^2 q^2} \frac{3\omega_P^2 - 2(\omega_T^2 - q^2)}{\omega_P^2}
\tag{9.24}
$$

where ω_A is given by

$$\omega_A = \frac{2\alpha}{\pi} \int_0^\infty \frac{p^2 dp}{E} (\tilde{n}_+(E) - \tilde{n}_-(E)) \left(1 - \frac{2}{3}v^2\right) \tag{9.25}$$

9.2 Neutrino emission from photon decay

When neutrinos propagate in dense matter, they scatter coherently on the charged particles, such as electrons and positrons, of the plasma: this process leads to an additional inertia of neutrinos. Another possiblity is that they diffuse coherently on electrons and positrons, which in turn couple to the electromagnetic field A_β. This leads to an effective coupling of the neutrino current to the electromagnetic waves propagating in the plasma through its interactions with the spectator particles, so that longitudinal and transverse massive photons can decay into neutrino pairs. Our aim now is to compute this effective photon–neutrino coupling. We start from the Fermi approximation of the Glashow–Weinberg–Salam electron–neutrino Lagrangian:

$$\mathscr{L} = \frac{G_F}{\sqrt{2}} \{\overline{\psi}_\nu \gamma_\alpha (1 - \gamma_5)\psi_\nu\} \{\overline{\psi}_e \gamma^\alpha (g_V - g_A \gamma_5)\psi_e\} \tag{9.26}$$

where G_F is the weak coupling constant

$$g_V = \frac{1}{2} + 2\sin^2\theta_W \qquad g_A = \frac{1}{2} \tag{9.27}$$

for electron neutrinos, while

$$g_V = -\frac{1}{2} + 2\sin^2\theta_W \qquad g_A = -\frac{1}{2} \tag{9.28}$$

for muon and tau neutrinos; θ_W is the Weinberg angle. The effective electromagnetic coupling of neutrinos can be expressed as a sum of the amplitudes of coherent diffusion on electrons and positrons (see fig. 9.1): $\mathscr{L}_{\text{eff}} = \mathscr{L}_{\text{eff}}^{(\text{el})} + \mathscr{L}_{\text{eff}}^{(\text{pos})}$. We have, for the electrons,

$$\mathscr{L}_{\text{eff}}^{(\text{el})} = \frac{eG_F}{\sqrt{2}} \{\overline{\psi}_\nu \gamma_\alpha (1 - \gamma_5)\psi_\nu\} A_\beta \sum_s \int \frac{d^3p}{(2\pi)^3 2E} \tilde{n}_+(E)$$

$$\times \left(\left[\overline{u}_s(\mathbf{p})\gamma^\alpha (g_V - g_A \gamma_5) \frac{1}{\not{P} + \not{Q} - m} \gamma^\beta u_s(\mathbf{p}) \right] \right.$$

$$\left. + \left[\overline{u}_s(\mathbf{p})\gamma^\beta \frac{1}{\not{P} - \not{Q} - m} \gamma^\alpha (g_V - g_A \gamma_5)u_s(\mathbf{p}) \right] \right) \tag{9.29}$$

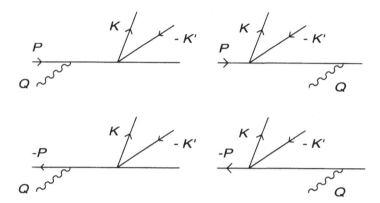

Fig. 9.1. Coherent diffusion of the neutrino current on electrons and positrons and coupling to the electromagnetic field.

$\mathscr{L}_{\text{eff}}^{(\text{pos})}$ is obtained from (9.29) by the substitutions $P \to -P$, $\tilde{n}_+(E) \to -\tilde{n}_-(E)$ and $u_s(\mathbf{p}) \to v_s(\mathbf{p})$, where u and v are positive and negative energy Dirac spinors. This yields an effective photon–neutrino coupling (fig. 9.2):

$$\mathscr{L}_{\text{eff}} = \frac{G_F}{\sqrt{2}} \{\overline{\psi}_\nu \gamma_\alpha (1 - \gamma_5) \psi_\nu\} \Gamma^{\alpha\beta} A_\beta \tag{9.30}$$

where the tensor $\Gamma^{\alpha\beta}$ is given by

$$
\begin{aligned}
\Gamma^{\alpha\beta} = {}& 4eg_V \int \frac{\mathrm{d}^3 p}{(2\pi)^3 2E} (\tilde{n}_+(E) + \tilde{n}_-(E)) \\
& \times \frac{P \cdot Q(P^\alpha Q^\beta + P^\alpha Q^\beta) - Q^2 P^\alpha P^\beta - (P \cdot Q)^2 g^{\alpha\beta}}{(P \cdot Q)^2 - (Q^2/2)^2} \\
& - 2ieg_A \varepsilon^{\alpha\beta\rho\sigma} \int \frac{\mathrm{d}^3 p}{(2\pi)^3 2E} \\
& \times (\tilde{n}_+(E) - \tilde{n}_-(E)) \left[\frac{P_\rho Q_\sigma Q^2}{(P \cdot Q)^2 - (Q^2/2)^2} \right]
\end{aligned}
\tag{9.31}
$$

Let us concentrate on the first term in (9.31), which we denote by $\Gamma_V^{\alpha\beta}$; we recognize, up to a multiplicative factor, the photon polarization tensor

$$\Gamma_V^{\alpha\beta} = \frac{g_V}{e} \Pi^{\alpha\beta} \tag{9.32}$$

The matrix element for the decay of a photon with polarization λ into a neutrino pair reads

$$\mathscr{M}_\lambda = \frac{G_F}{\sqrt{2}} \left[\Gamma_V^{\alpha\beta} e_\beta(q, \lambda) \right] \left[\overline{u}(K) \gamma_\alpha (1 - \gamma_5) v(K') \right] \tag{9.33}$$

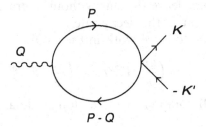

Fig. 9.2   The effective photon–neutrino vertex.

where $e_\beta(q, \lambda)$ is the photon polarization four-vector and k and k' are the final neutrino momenta. In the vacuum, the decay of a massive elementary photon with polarization λ would be given by

$$R_\lambda = \frac{1}{2q_0} \int \frac{\mathrm{d}^3 k}{(2\pi)^3 2k_0} \int \frac{\mathrm{d}^3 k'}{(2\pi)^3 2k_0'} (2\pi)^4 \delta^{(4)}(Q - K - K') |\mathcal{M}_\lambda|^2 \quad (9.34)$$

An elementary calculation yields, with massless neutrinos,

$$\int \frac{\mathrm{d}^3 k}{(2\pi)^3 2k_0} \int \frac{\mathrm{d}^3 k'}{(2\pi)^3 2k_0'} (2\pi)^4 \delta^{(4)}(Q - K - K')$$
$$\times |\bar{u}(k)\gamma^\alpha(1 - \gamma_5)v(k)|^2$$
$$= \frac{1}{3\pi}(Q^\alpha Q^\delta - Q^2 g^{\alpha\delta}) \quad (9.35)$$

Since $\Gamma^{\alpha\beta}$ is transverse ($Q_\alpha \Gamma^{\alpha\beta} = 0$), only the term that is proportional to $g^{\alpha\delta}$ contributes to the decay rate.

Let us first consider a transverse polarization e_T^β : $e_T^0 = 0$, $\mathbf{e}_T \cdot \mathbf{q} = 0$. The decay rate R_T is, from (9.3), (9.32), (9.34) and (9.35),

$$R_T = \frac{1}{2q_0} \frac{1}{3\pi} \left(\frac{g_V G_F}{e\sqrt{2}} \right)^2 Q^2 (-g_{\alpha\beta} e_T^\alpha e_T^\beta) \, \Pi_T^2 \quad (9.36)$$

For a transverse photon in the plasma, $q_0 = \omega_T$ and from (9.6) we may replace Π_T^2 by $Q^4 = (\omega_T^2 - q^2)^2$; furthermore, the phase space factor $\mathrm{d}^4 Q \, \delta_+(Q^2 - m^2)$ for an elementary photon in the vacuum must be replaced by the appropriate factor for a quasi-particle:

$$\mathrm{d}^4 Q \, \delta_+(Q^2 - m^2) \to \mathrm{d}^4 Q \, \theta(q_0)\delta(Q^2 - \Pi_T) \quad (9.37)$$

or, from (9.21),

$$\frac{\mathrm{d}^3 q}{2q_0} \to Z_T(q)\mathrm{d}^3 q \quad (9.38)$$

This yields, for R_T,

$$R_T(q) = \frac{g_V^2 G_F^2}{24\pi^2\alpha} Z_T(q)(\omega_T^2 - q^2)^3 \quad (9.39)$$

Note that the neutrinos leave the star without interacting, so that there is no Pauli blocking factor in the decay rate.

For longitudinal photons we work in the Coulomb gauge: $e_\beta^L = (1,0)$:

$$\Gamma_L^\alpha = \left(\frac{g_V}{e}\right)\Pi^{\alpha\beta}e_\beta^L = \left(\frac{g_V}{e}\right)FP_L^{\alpha 0} \qquad (9.40)$$

where $P_L^{\alpha 0}$ is the $(\alpha 0)$ component of the longitudinal projector P_L (6.52). Using

$$P_L^{\alpha 0}P_{\alpha 0}^L = -P_L^{00} = -\frac{q^2}{Q^2} \qquad (9.41)$$

and following the same steps as for the transverse photons we obtain for R_L

$$R_L(q) = \frac{g_V^2 G_F^2}{24\pi^2\alpha}Z_L(q)q^2(\omega_L^2 - q^2)^2 \qquad (9.42)$$

In order to complete the calculation we have to take the second term in (9.31) into account. Because of the $\varepsilon^{\alpha\beta\rho\sigma}$ factor, only transverse photons contribute. Let us call $\tilde{\Pi}_{\rho\sigma}$ the last integral in (9.31); only the combination

$$\tilde{\Pi}_i = \tilde{\Pi}_{0i} - \tilde{\Pi}_{i0} \qquad (9.43)$$

is relevant for the total rate. The expression for $\tilde{\Pi}_i$ is

$$\tilde{\Pi}_i = Q^2 q_i \int \frac{d^3p}{(2\pi)^3 2E}\left(\tilde{n}_+(E) - \tilde{n}_-(E)\right)\frac{E - q_0(\hat{\mathbf{p}}\cdot\hat{\mathbf{q}})}{(P\cdot Q)^2 - (Q^2/2)^2} \qquad (9.44)$$

so that

$$q_i\tilde{\Pi}_i = \frac{q}{8\pi\alpha}\Pi_A \qquad (9.45)$$

The correction to the decay rate due to Π_A is

$$R_A(q) = \frac{g_V^2 G_F^2}{24\pi^2\alpha}Z_T(q)\frac{\omega_T^2 - q^2}{\omega_T(q)}\Pi_A^2(\omega_T(q), q) \qquad (9.46)$$

Integrating over q yields for the total rate

$$R = \sum_\nu \int_0^\infty \frac{q^2 dq}{2\pi^2}\left(2n(\omega_T(q))[R_T(q) + R_A(q)] + n(\omega_L(q))R_L(q)\right) \qquad (9.47)$$

where the factor of 2 takes care of the two polarization states of transverse photons and $\sum_\nu$ sums over the three neutrino families. The integral in (9.46) may be evaluated analytically in various limits.

It is interesting to give an intuitive interpretation of the results. In the soft limit (i.e. $q \ll \mu$), one can write an effective photon–neutrino Lagrangian:

$$\mathscr{L}_{\text{eff}} = -e\frac{\langle r^2\rangle}{6}Q^2\left\{\overline{\psi}_\nu\gamma_\alpha(1 - \gamma_5)\psi_\nu\right\}A^\alpha \qquad (9.48)$$

where the neutrino couples to the photon with a charge radius $\langle r^2 \rangle$:

$$\langle r^2 \rangle = \frac{3G_F}{\sqrt{2}\,\pi\alpha} g_V \tag{9.49}$$

In the hard limit ($q \sim \mu$ but $Q^2 \simeq \omega_P^2$!) (9.48) translates into a coupling with an effective neutrino charge Q_v:

$$Q_v = \frac{G_F g_V}{2\sqrt{2}\,\pi\alpha} m_T^2 \tag{9.50}$$

where m_T is the transverse mass (9.11). In the case of red giants for example, with $p_F = 410$ keV, this effective charge is $\simeq 6.4 \times 10^{-14}$ in units of the proton charge.

References and further reading

The importance of plasmon decay in neutrino emission from stars was first pointed out by Adams, Ruderman and Woo (1963). Braaten (1991) showed that the use of exact dispersion laws was essential in the case of neutron stars; detailed calculations can be found in Braaten and Segel (1993). The interpretation in terms of effective charge and charge radius is due to Altherr and Kainulainen (1991) and Altherr and Salati (1994). Axion emission has been studied by Braaten and Yuan (1991) and by Altherr, Petitgirard and del Rio Gaztelurrutia (1993).

Exercises

9.1 (a) Derive equations (9.10)–(9.12) from the expressions (9.4) and (9.5) of Π_T and Π_L.
(b) Show that the Debye mass is:

$$m_D^2 = \frac{4\alpha}{\pi} \int_0^\infty dp\, \frac{2p^2 + m_e^2}{E} (\tilde{n}_+(E) + \tilde{n}_-(E))$$

9.2 Derive the expressions of Π_T and Π_L in the degenerate case, without any assumption on kinematics. Discuss the value of $q_{crit.}$ and show that there is a region where there is no Landau damping and where the phase velocity of the longitudinal photons is larger than the velocity of light. What is the group velocity?

9.3 Expand Π_T and Π_L for small values of k and show that the first two terms of this expansion agree with (9.17) and (9.18).

9.4 Derive equation (9.35).

10
Infrared problems at finite temperature

This chapter deals with some infrared problems which arise in gauge theories at finite temperature. Most of these problems are still open, and represent interesting – and difficult – challenges for future investigations.

At zero temperature, the existence of singularities in field theories with massless particles has been known for a long time. It is convenient to classify these singularities into infrared and mass singularities. In order to be definite, let us take QCD as an example.

(i) A quark can emit a soft gluon, whose momentum $k \to 0$: because the gluon is massless, this leads to infrared singularities, even when the quark is massive.
(ii) A massless quark can emit a gluon whose momentum makes a small angle θ with the initial quark momentum. The region $\theta \to 0$ leads to mass (or collinear) singularities, even if the gluon momentum k remains finite.

In both cases, the singularities arise from the fact that the final state quark + gluon is degenerate with the initial one when $k \to 0$ in case (i) and when $\theta \to 0$ in case (ii).

Much is known about these infrared and mass singularities at $T = 0$; the most important information is contained in the Bloch–Nordsieck and Kinoshita–Lee–Nauenberg (KLN) theorems, which allow, in well-defined situations, control of these singularities.

At $T \neq 0$, the status of these singularities is far from settled, although some results have been obtained in recent years. In the real-time formalism, because of the presence of the Bose–Einstein factor

$$n(k_0) = \frac{1}{e^{\beta |k_0|} - 1} \underset{k \to 0}{\sim} \frac{1}{\beta k} \tag{10.1}$$

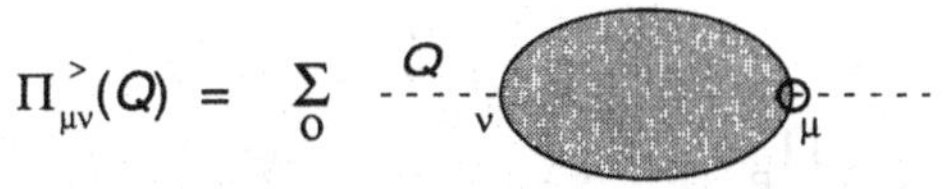

$$\Pi_{\mu\nu}^{>}(Q) = \sum_{0} \quad \underset{\nu}{\overset{Q}{\text{------}}} \bigodot_{\mu} \text{------}$$

Fig. 10.1. Kobes–Semenoff rules for the calculation of $\Pi_{\mu\nu}^{>}(Q)$. Dashed lines: photons.

for massless particles, one expects to find stronger singularities than at zero temperature. The same conclusion is also easily obtained in the imaginary-time formalism: the leading infrared singularities are controlled by the static mode $\omega_n = 0$, since $2\pi n T$, $n \neq 0$ acts as an infrared cut-off for the other Matsubara frequencies. Then one works effectively in three-dimensional field theory, which is known to be much more infrared singular than the original four-dimensional one.

In spite of this enhanced infrared singular behaviour, there exists a plausible generalization of the Kinoshita–Lee–Nauenberg theorem at non-zero temperature; at least explicit calculations have shown that it holds in specific cases, and an example will be worked out in section 10.1. Section 10.2 will be devoted to the problem of low-momentum transverse gluons and its relation with a possible magnetic mass. Finally in sections 10.3 and 10.4 we shall revert to the resummation program, with two open problems: the production of soft photons from a quark–gluon plasma and the moving fermion damping rate.

10.1 Kinoshita–Lee–Nauenberg theorem in lepton pair production

In recent years there have been many attempts to check the validity of the KLN theorem at non-zero temperature. Most calculations take the case of non-thermalized lepton pair, or, equivalently, heavy photon production, from a quark–gluon plasma. From the KLN theorem, this process is infrared safe at $T = 0$; of course the theorem deals with heavy photon decay in this case. The total production rate Γ_{tot} was given in (5.158)

$$\Gamma_{\text{tot}} = -\frac{\alpha e^{-\beta q_0}}{24\pi^4 Q^2} g^{\mu\nu} \Pi_{\mu\nu}^{>}(Q) \tag{10.2}$$

$\Pi_{\mu\nu}^{>}(Q)$ can be computed thanks to the Kobes–Semenoff cutting rules (fig. 10.1). We assume for simplicity that the photon has zero three-momentum in the plasma rest frame: $Q_\mu = (q, \mathbf{0})$.

We choose to perform the calculation in the real time formalism and in the Coulomb gauge. We recall the expression for the real-time gluon

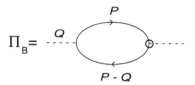

Fig. 10.2   The one-loop approximation.

propagator:

$$D_{11}^{\mu\nu}(K) = (D_{22}^{\mu\nu}(K))^*$$

$$= \left[i\mathbf{P}\left(\frac{1}{K^2}\right) + \pi F(k_0)\,\delta(K^2) \right] P_T^{\mu\nu} + \frac{i}{k^2} P_c^{\mu\nu} \qquad (10.3)$$

where $\mathbf{P}$ denotes a principal value, while the cut propagator $D^>(K)$ reads:

$$D_{\mu\nu}^>(K) = 2\pi\,\varepsilon(k_0)\,(1 + f(k_0))\,\delta(K^2) P_{\mu\nu}^T \qquad (10.4)$$

In these equations, $F(k_0)$ and $f(k_0)$ are Bose–Einstein factors :

$$F(k_0) = 1 + 2n(k_0) = \coth\frac{\beta|k_0|}{2} \qquad (10.5)$$

$$f(k_0) = \frac{1}{e^{\beta k_0} - 1} \qquad (10.6)$$

$\varepsilon(k_0)$ is the sign function, $P_{\mu\nu}^T$ is the transverse projector and $P_c^{\mu\nu} = g^{\mu 0}g^{\nu 0}$. In the fermionic case, we have for massless quarks:

$$S_{11}(P) = \slashed{P}\left[i\mathbf{P}\left(\frac{1}{P^2}\right) + \pi \tilde{F}(p_0)\,\delta(P^2) \right] \qquad (10.7)$$

$$S^>(P) = 2\pi\,\varepsilon(p_0)\,(1 - \tilde{f}(p_0))\slashed{P}\,\delta(P^2) \qquad (10.8a)$$

$$S^<(P) = -2\pi\,\varepsilon(p_0)\,\tilde{f}(p_0)\slashed{P}\,\delta(P^2) \qquad (10.8b)$$

where the statistical factors are now

$$\tilde{F}(p_0) = 1 - 2\tilde{n}(p_0) = \tanh(\beta|p_0|/2) \qquad (10.9)$$

$$\tilde{f}(p_0) = \frac{1}{e^{\beta p_0} + 1} \qquad (10.10)$$

From these equations, it is easy to compute the one-loop (or Born) approximation drawn in Fig. 10.2:

$$\Pi_B = -g_{\mu\nu}\Pi_B^{\mu\nu>} = -2e^2 \int \frac{d^4P}{(2\pi)^4} \,\text{Tr}\,[S^>(P)\,S^<(P - Q)]$$

$$= \frac{2e^2\,\kappa^2}{\pi}\,[1 - \tilde{n}(\kappa)]^2 \qquad (10.11)$$

with $\kappa = q/2$, a formula already quoted in chapter 7.

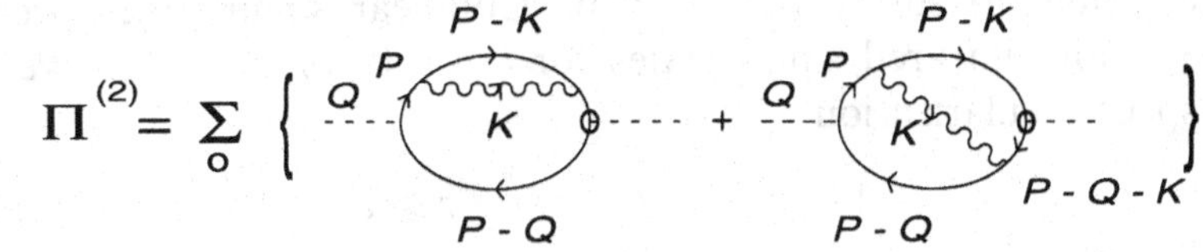

$$\Pi^{(2)} = \sum_{0} \left\{ \quad + \quad \right\}$$

Fig. 10.3   The two-loop approximation. Wavy lines: gluons. Solid lines: quarks.

10.1.1 The two-loop approximation

In contrast with section 7.3.2, we are now interested in a situation where q_0 is not soft, for example $q_0 \sim T$, and where naïve perturbation theory is *a priori* valid. Let us examine the two-loop approximation $\Pi^{(2)>}$. It is convenient to distinguish between 'self-energy' and 'vertex' graphs (fig. 10.3). Moreover it is possible to interpret the $T \neq 0$ cutting rules as at $T = 0$ in terms of unitarity cuts and to distinguish between 'real' and 'virtual' graphs, thanks to equation (4.62): see fig. 10.4. In order to eliminate irrelevant multiplicative factors, we define Π by

$$\Pi = \frac{\pi}{\alpha_S \, C_F \, \Pi_B} \left(-g^{\mu\nu} \Pi^{(2)>}_{\mu\nu} \right) \tag{10.12}$$

with $\alpha_S = g^2/(4\pi)$ and $C_F = \frac{4}{3}$.

As we work in the Coulomb gauge, mass singularities appear in self-energy graphs only. Below we discuss only that part of the result which depends on the Bose–Einstein factor $F(k_0)$; the other part is not infrared singular. It is convenient to define the scaling variable $z = k/\kappa$, while y denotes the cosine of the angle between **p** and **k**: $y = \hat{\mathbf{p}} \cdot \hat{\mathbf{k}}$. There are four distinct kinematical regions, according to the signs of k_0 ($\eta = \varepsilon(k_0)$), $p_0 - k_0$ and $q - p_0$:

(a)	$\eta = 1$	$z \leq 1$	$\varepsilon(q - p_0) \, \varepsilon(p_0 - k_0) = 1$
(b)	$\eta = 1$	$z > 1$	$\varepsilon(q - p_0) \, \varepsilon(p_0 - k_0) = -1$
(c)	$\eta = -1$	$z \geq 0$	$\varepsilon(q - p_0) \, \varepsilon(p_0 - k_0) = 1$
(d)	$\eta = 1$	$z > 1$	$\varepsilon(q - p_0) \, \varepsilon(p_0 - k_0) = -1$

$$\Pi_R = \qquad \qquad \Pi_V =$$

(a)   (b)

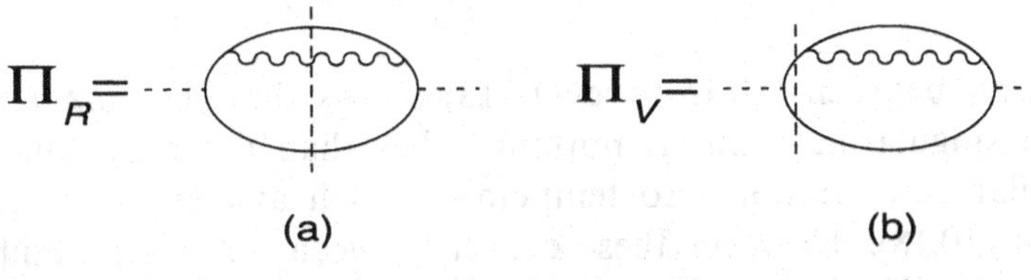

Fig. 10.4   Cutting rules for the self-energy graph.

Only region (a) survives at $T = 0$. Collinear singularities occur for $y \to \pm 1$, and infrared singularities for $k \to 0$ (or $z \to 0$). We use a phase-space regularization

$$1 - y \geq \varepsilon \qquad\qquad z \geq \varepsilon' \qquad\qquad (10.13)$$

where ε and ε' are arbitrarily small numbers. The following change of variables:

$$u = \frac{\eta - y}{2\,(1 - z\,(\eta + y)/2)} \qquad\qquad (10.14)$$

allows us to write the real graphs in a very convenient form, in the different regions previously listed; recall that we write only that part of the graphs which involves $F(k_0)$:

$$\Pi_R^{(a)} = \int_{\varepsilon'}^1 \frac{dz}{z}\, \mathscr{A}_+(z)\, F(\kappa z) \int_{\varepsilon(z)}^1 \frac{du}{u}\, g(u, z) \qquad\qquad (10.15a)$$

$$\Pi_R^{(b)} = \int_1^\infty \frac{dz}{z}\, \mathscr{A}_+(z)\, F(\kappa z) \int_{\varepsilon(z)}^\infty \frac{du}{u}\, g(u, z) \qquad\qquad (10.15b)$$

$$\Pi_R^{(c)} = \int_{\varepsilon'}^\infty \frac{dz}{z}\, \mathscr{A}_-(z)\, F(\kappa z) \int_{\varepsilon(z)}^1 \frac{du}{u}\, g(u, z) \qquad\qquad (10.15c)$$

$$\Pi_R^{(d)} = -\int_1^\infty \frac{dz}{z}\, \mathscr{A}_+(z)\, F(\kappa z) \int_1^\infty \frac{du}{u}\, g(u, z) \qquad\qquad (10.15d)$$

In (10.15), $\mathscr{A}_\pm(z)/z$ is nothing other than the Altarelli–Parisi splitting function:

$$\mathscr{A}_\eta(z) = 1 - \eta\, z + z^2/2 \qquad\qquad (10.16)$$

$\varepsilon(z) = \varepsilon/(2\,|1 - \eta\, z|)$ and

$$g(u, z) = \frac{1 + \cosh(\beta\, \kappa)}{\cosh(\beta\, \kappa\, u\, z) + \cosh(\beta\, \kappa)} \qquad\qquad (10.17)$$

From (10.15), one obviously recognizes collinear singularities when $u \to 0$, and infrared singularities when $z \to 0$. The evaluation of virtual graphs gives

$$\Pi_V = -\int_{\varepsilon'}^\infty \frac{dz}{z}\, F(\kappa z) \int_{\varepsilon/2}^1 du \left[\frac{1}{u}\,(2 + z^2) + z^2 \right] \qquad\qquad (10.18)$$

Comparison between (10.15) and (10.18) shows that the sum $\Pi_R + \Pi_V$ is free of singularities: this is nothing other than the KLN theorem in a particular case, at non-zero temperature. Ultraviolet divergences are hidden in (10.18). However these cancel between self-energy and vertex graphs due to Ward identities, so they are easily dealt with.

(a) (b)

Fig. 10.5. Resummed calculation. The heavy dot indicates a resummed gluon propagator.

The calculation has been pushed to three-loop order but, due to the complexity of QCD, a renormalizable scalar theory, $g\,\varphi_6^3$, was used in the actual computation. Moreover, the computation was performed only in a particular (self-energy) configuration, this configuration being, however, the most singular. The most interesting feature is the appearance of a new type of infrared singularity, arising from the behaviour of the one-loop self-energy $\Sigma(K)$ when $k \to 0$:

$$\Sigma(K) \sim \left(\frac{T}{k}\right)^2 \tag{10.19}$$

while $g\,\varphi_6^3$ is infrared regular at $T = 0$!

Using a dimensional regularization $D = 6 + 2\varepsilon$, one finds the cancellation of the strongest (power-law) infrared singularities between real and virtual graphs, while the next-to-leading singularities, which behave as ε^{-3}, cancel between the regions $k_0 > 0$ and $k_0 < 0$, within real or virtual graphs. The collinear singularities (in ε^{-2} and ε^{-1}) also cancel out, giving a finite result. Thus the KLN theorem seems again to be satisfied at non-zero temperature.

10.1.2 Resummation and infrared safe processes

The results just described were obtained in the framework of bare (or naïve) perturbation theory. However, we have emphasized that, because of the infrared behaviour of propagators and vertices, one should, rather, use an effective (or resummed) perturbative expansion.

In this section, we shall check that bare perturbation theory is valid at order g^2 in the presence of resummation, and we shall evaluate the next correction for a hard external photon which turns out to be of the order of g^3. As the external photon line is hard it is easy to convince oneself that resummation can only affect the propagators; since we are going to be interested in the infrared behaviour of our results, it is enough to consider the case where the gluon propagator is soft (fig. 10.5). The gluon propagator is thus dressed by hard thermal loops. The expressions for the dressed gluon propagator $^*D_{11}^{\mu\nu}(K)$ and its cut version $^*D_{\mu\nu}^>(K)$ read:

$$
\begin{aligned}
^*D_{11}^{\mu\nu}(K) &= \left[i\,b_T(K) + \frac{1}{2}F(k_0)\,\varepsilon(k_0)\rho_T(K) \right] P_T^{\mu\nu} \\
&\quad + \left[i\,b_L(K) + \frac{1}{2}F(k_0)\,\varepsilon(k_0)\rho_L(K) \right] P_c^{\mu\nu}
\end{aligned}
\tag{10.20}
$$

$$
^*D_{\mu\nu}^>(K) = (1 + f(k_0))\,[\,\rho_T(K)\,P_{\mu\nu}^T + \rho_L(K)\,P_{\mu\nu}^L\,]
\tag{10.21}
$$

In the preceding expressions $\rho_{T,L}(K) = 2\,\mathrm{Im}\,\Delta_{T,L}(K)$ and $b_{T,L}(K) = -\mathrm{Re}\,\Delta_{T,L}(K)$, where the (retarded) functions $\Delta_{T,L}(K)$ have been defined in (6.74) and (6.75):

$$
\Delta_T(k_0, k) = \frac{-1}{k^2\,(x^2 - 1) - m^2 \left[x^2 + \frac{x}{2}\,(1 - x^2)\log\frac{x+1}{x-1} \right]}
\tag{10.22}
$$

$$
\Delta_L(k_0, k) = \frac{-1}{k^2 - m^2 \left[x\log\frac{x+1}{x-1} - 2 \right]}
\tag{10.23}
$$

with $x = k_0/k$, and the thermal mass m is given in (6.161):

$$
m^2 = \frac{1}{6}\left(C_A + \frac{1}{2}N_f \right) (g\,T)^2
\tag{10.24}
$$

The real graphs are computed by generalization of the change of variables (10.14):

$$
u = \frac{x - y + z\,(1 - x^2)/2}{2\,(1 - z\,(x + y)/2)}
\tag{10.25}
$$

The result is given by a complicated integral over x, y and z. Here we shall only give the result for longitudinal gluons in the infrared region $z\,|x| \ll 1$. We find for self-energy graphs:

$$
\begin{aligned}
\tilde{\Pi}_s^L &= \frac{\kappa^2}{4\pi}\left(1 - \frac{(\beta\,\kappa)^2}{2\,(1 + \cosh(\beta\,\kappa))} \right) \\
&\quad \times \int_0^\infty dz\, z^3 \int_{-\infty}^\infty dx\, F(\kappa\,z\,x)\,\rho_L(|k_0|, k)
\end{aligned}
\tag{10.26}
$$

and for vertex graphs* we obtain:

$$\tilde{\Pi}_v^L = -\frac{\kappa^2}{4\pi} \int_0^\infty dz\, z^3 \int_{-\infty}^\infty dx\, F(\kappa\, z\, x)\, \rho_L(|k_0|, k)$$

$$\times \left[1 + \frac{(\beta\,\kappa)^2}{2\,(1 + \cosh(\beta\,\kappa))} (1 - 2\,x\, Q_0(x)) \right] \qquad (10.27)$$

Adding (10.26) and (10.27) yields

$$\tilde{\Pi}^L = \tilde{\Pi}_s^L + \tilde{\Pi}_v^L$$

$$= \frac{\beta\,\kappa^3}{1 + \cosh(\beta\,\kappa)} \int_0^\infty dz\, z^2$$

$$\times \int_{-\infty}^\infty \frac{dx}{2\pi} \frac{\rho_L(k_0, k)}{x} Q_1(x) \qquad (10.28)$$

One must use the new sum rule

$$\int_0^\infty dk\, k^2 \int_{-\infty}^\infty \frac{dx}{2\pi} \frac{\rho_L(k_0, k)}{x} (2\, x\, Q_0(x)) = \frac{\pi^2}{\sqrt{6}} m \qquad (10.29)$$

to obtain

$$\tilde{\Pi}^L \simeq -\frac{\pi}{\sqrt{2}} \left(1 - \frac{\pi}{2\sqrt{3}} \right) \frac{\beta\, m}{(1 + \cosh(\beta\,\kappa))} \equiv g\, f_1(\beta\,\kappa) \sim g \qquad (10.30)$$

The main feature of this result is that it is proportional to $m \sim g\, T$: we find terms of the order of g^3 in the total rate. This is clearly reminiscent of the perturbative expansion of the free energy where the g^3 term arises from ring diagrams. In the case of transverse gluons, one can show that the bare result is recovered at order g^2, and that higher-order terms are of the order of g^4 at least. Thus we have shown that the lowest order of bare perturbation theory is valid, but that it breaks down beyond this lowest non-trivial order.

10.2 Transverse gluons in the static limit

10.2.1 The magnetic mass

This section deals with the zero momentum limit of the transverse part of the gluon polarization tensor and its relation with a difficulty first pointed out by Linde and which is not yet understood. Let us discuss briefly the behaviour of non-leading terms in the gluon polarization tensor, in bare

---

* In fact, for reasons of computational simplicity, a heavy scalar photon has been used in this calculation instead of a spin one photon. This does not make any difference for self-energy graphs, but it makes the vertex graphs much simpler, without any essential modification to the physics.

perturbation theory, and in the limit $k_0 = 0$, $k \to 0$. The longitudinal part $F = -\Pi_{00}$ behaves in a pure $SU(N)$ gauge theory as

$$F(k_0 = 0, \, k \to 0) = \frac{1}{3}g^2NT^2 - \frac{1}{4}g^2NTk + \ldots \tag{10.31}$$

The first term is of course the leading term which was examined in great detail in chapter 6, and which gives rise to the Debye screening length $r_D = m_{el}^{-1}$. When one calculates the contribution of ring diagrams to the partition function Z, this term is at the origin of the g^3-behaviour. It is thus certainly gauge-independent, as is already known from other arguments. Similarly, the second term in (10.31) gives rise to terms of order $g^4 \ln g^2$ in Z and must also be gauge-independent. Although higher-order contributions to Z have not been computed yet, it should be noted that one does not expect these higher-order contributions to be independent of the gauge parameter and, more generally, of the gauge adopted in the calculation. Indeed, the dependence with respect to the gauge parameter can be absorbed in the gauge dependence of the coupling constant. Calculations in various gauges confirm the gauge independence of the first two terms of the expansion (10.31).

Difficulties appear when one looks at the transverse part $G = \frac{1}{2}\Pi_{ii}$ of the gluon polarization tensor. In TAG one finds

$$G(k_0 = 0, k \to 0) = -\frac{5}{16}g^2NTk + \ldots \tag{10.32}$$

while in a general covariant gauge with gauge parameter ξ the result is

$$G(k_0 = 0, k \to 0) = -\frac{1}{64}(8 + (1 + \xi)^2)g^2NTk + \ldots \tag{10.33}$$

Clearly, even the first term in the expansion of G is not gauge-independent! The result of bare perturbation theory is wrong, as can be seen in a simple model, scalar QED: resummation changes the $k \to 0$ behaviour of G completely (exercise 7.7). Nevertheless one finds in scalar QED that $G(k_0 = 0, k)$ vanishes in the limit $k = 0$, even after resummation. If one extrapolates (boldly) this result to QCD, this means that *static* (chromo)-magnetic fields are not screened: there is no magnetic mass, at least to lowest order of perturbation theory. It has been argued that the magnetic mass, if it is non-zero, cannot be computed in perturbation theory; from dimensional arguments one may infer that m_{mag} should be related to the mass gap in the $D = 3$ theory, and should thus be of the order of g^2T. Although the arguments sound plausible, this result cannot be considered as being firmly established. Note that in QED, no magnetic mass can appear, as is shown in exercise 6.7.

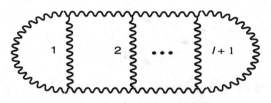

Fig. 10.6 Linde's diagram.

10.2.2 Linde's problem

Consider the $(l+1)$-loop graph of fig. 10.6 contributing Z_l to the partition function, and let us try to evaluate its leading infrared behaviour. We write each propagator, including the summation over Matsubara frequencies, as

$$\sum_n \frac{1}{\omega_n^2 + k^2 + m^2} \qquad \omega_n = \frac{2\pi n}{\beta} \qquad (10.34)$$

where m is some screening mass. The leading IR divergences occur for $n = 0$, since ω_n for $n \neq 0$ acts as an infrared cut-off. The leading IR behaviour can then be estimated from power counting

$$g^{2l}(T \int d^3k)^{l+1} k^{2l}(k^2 + m^2)^{-3l} \qquad (10.35)$$

The various factors in (10.35) are obvious:

- vertices lead to $g^{2l}k^{2l}$;
- propagators lead to $(k^2 + m^2)^{-3l}$;
- loop integrations lead to $(T \int d^3k)^{l+1}$

For $l < 3$, the behaviour is IR regular, while divergences occur for $l \geq 3$:

$$l = 3 \qquad g^6 T^4 \ln \frac{T}{m} \qquad (10.36)$$

$$l > 3 \qquad g^6 T^4 \left(\frac{g^2 T}{m} \right)^{l-3} \qquad (10.37)$$

For longitudinal gluons, the screening mass is the electric mass $m_{el} \sim gT$, and thus for $l > 3$:

$$Z_l \sim g^{l+3} T^4 \qquad (10.38)$$

Thus, instead of the expected g^{2l}-behaviour of ordinary perturbation theory, one finds a g^{l+3}-behaviour. This is reminiscent of what occurs in ring diagrams, and corresponds to what could be called a mild breakdown of perturbation theory. However, if we assume a magnetic mass $m_{mag} \sim g^2 T$, the situation is much worse for transverse gluons:

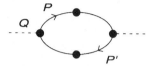

Fig. 10.7. Diagram for soft photon production. Heavy dots indicate resummed quark propagators and resummed quark–photon vertices.

$$Z_l \sim g^6 T^4 \qquad (10.39)$$

independently of l. For $l > 3$, all graphs are of order g^6, independently of the order of perturbation theory. This is of course a complete failure of perturbation theory!

One does not know exactly how to interpret this difficulty. It could be that power counting is misleading and that some cancellation occurs which invalidates the argument. It could be that the result is true, but that it does not affect physics in an essential way since the phenomenon occurs at a rather high order of perturbation theory (remember that even the term of order g^4 has yet to be computed!). The most pessimistic view would be of course that perturbation theory is completely unreliable and should be abandoned. It is worth noting that no problem arises in QED, despite the absence of magnetic screening: in QED a photon can only give an $e^+ - e^-$ pair, and for fermions the Matsubara frequencies $\omega_n = \pi(2n + 1)$ can never vanish, so that no infrared problem arises.

10.3 Soft photon production

We now re-examine a problem which has already been studied in section 8.2, that of real photon production from a quark–gluon plasma. We recall that the production rate is related to the photon polarization tensor by

$$q_0 \frac{\mathrm{d}\Gamma}{\mathrm{d}^3 q} = \frac{g^{\mu\nu} \operatorname{Im} \Pi_{\mu\nu}(Q)}{(2\pi)^3 (e^{\beta q_0} - 1)} \qquad (10.40)$$

However, in contrast with section 8.2, we now assume that the photon momentum Q is soft: $q_0 = q \sim gT$. Then the two quark lines arriving at the vertices in fig. 10.7 may both be soft, so that one needs to dress the photon–quark vertices. This will turn out to be an essential complication.

Leaving out the unimportant multiplicative factor $e^2 e_q^2 N$, where e_q is the quark charge in units of the electron charge, we compute in Euclidean space:

$$\Pi_{\mu\mu}(q_0, q) = \int \frac{\mathrm{d}^4 P}{(2\pi)^4} \operatorname{Tr} \left[{}^*S(P) {}^*\Gamma_\mu(P, P') {}^*S(P') {}^*\Gamma_\mu(P', P) \right] \qquad (10.41)$$

where $^*S(P)$ is the effective quark propagator (7.90):

$$^*S(P) = \frac{1}{2}\Delta_+(P)(i\gamma_4 + \gamma \cdot \hat{\mathbf{p}}) + \frac{1}{2}\Delta_-(P)(i\gamma_4 - \gamma \cdot \hat{\mathbf{p}})$$

$$= \frac{1}{2}\Delta_+(P)\not{P}_+ + \frac{1}{2}\Delta_-(P)\not{P}_- \tag{10.42}$$

with $P_\pm = (i, \pm\hat{\mathbf{p}})$, and $^*\Gamma_\mu$ is the effective quark–photon vertex immediately deduced from (7.93):

$$^*\Gamma_\mu(P, P') = \gamma_\mu - m_f^2 \int \frac{d\Omega}{4\pi} \frac{\hat{K}_\mu \hat{\not{K}}}{(P \cdot \hat{K})(P' \cdot \hat{K})} \tag{10.43}$$

with $\hat{K} = (-i, \hat{\mathbf{k}})$, $\omega = -q_4$, $\omega_p = -p_4$, $\omega_p' = \omega_p - \omega$.

The Dirac trace in (10.41) is split into three terms according to the number of HTL corrections to the vertices: these terms are denoted by Tr[0], Tr[1] and Tr[2] respectively:

$$\Pi_{\mu\mu} = \int \frac{d^4P}{(2\pi)^4} \Big(\text{Tr}[0] + \text{Tr}[1] + \text{Tr}[2] \Big) \tag{10.44}$$

We limit ourselves to writing the first two terms:

$$\text{Tr}[0] = -2 \sum_{r,s=\pm 1} (P_r \cdot P_s)\Delta_r\Delta_s \tag{10.45}$$

$$\text{Tr}[1] = 4m_f^2 \int \frac{d\Omega}{4\pi} \frac{1}{(P \cdot \hat{K})(P' \cdot \hat{K})}$$

$$\times \sum_{r,s=\pm 1} (P_r \cdot \hat{K})(P_s \cdot \hat{K})\Delta_r\Delta_s \tag{10.46}$$

When the analytic continuation to Minkowski space $i\omega \to q_0 + i\eta$ is performed, the functions with non-vanishing discontinuities are $\Delta_\pm(P)$ and $(P \cdot \hat{K})^{-1}\Delta_\pm(P)$; the associated spectral densities are called $\rho_\pm$ (see (6.141)) and $\sigma_\pm$ respectively:

$$\rho_\pm = -2 \text{ Im } \Delta_\pm(p_0 + i\eta, p) \tag{10.47}$$

$$\sigma_\pm = -2 \text{ Im } [(P \cdot \hat{K})^{-1}\Delta_\pm(P)]$$

$$= \text{P}\frac{1}{P \cdot \hat{K}}\rho_\pm - 2\pi\delta(P \cdot \hat{K})\alpha_\pm \tag{10.48}$$

$$-\Delta_\pm(p_0 + i\eta, p) = \alpha_\pm + \frac{i}{2}\rho_\pm \tag{10.49}$$

where $\alpha_\pm$ is the real part of $-\Delta_\pm$.

The imaginary part of Π_μ^μ is evaluated from (7.106); the analytically continued $\hat{K}$ is $(-1, \hat{\mathbf{k}})$, while the analytically continued $P_\pm$ is $(1, \pm\hat{\mathbf{p}})$;

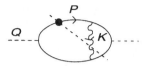

Fig. 10.8   Cut diagram displaying the collinear singularity.

moreover, one should not forget additional minus signs from $\delta_{\mu\nu} \to -g_{\mu\nu}$. The contribution from Tr[0] is computed as in section 8.2, but difficulties appear with the term Tr[1]. Since $\hat{K}$ is light-like, $(P \cdot \hat{K})$ may vanish only if P is space-like, and the same remark holds for $(P' \cdot \hat{K})$; note also that $(Q \cdot \hat{K}) \leq 0$. These kinematical remarks lead to

$$\mathbf{P}\frac{1}{P \cdot \hat{K}}\, \delta(P' \cdot \hat{K}) = \mathbf{P}\frac{1}{Q \cdot \hat{K}}\, \delta(P' \cdot \hat{K}) = \frac{1}{Q \cdot \hat{K}}\delta(P' \cdot \hat{K}) \qquad (10.50)$$

since, as $Q \cdot \hat{K} \leq 0$, there is no need for a principal value prescription. A non-integrable singularity due to massless quark exchange (fig. 10.8) develops when $(Q \cdot \hat{K}) = 0$, namely, when $\hat{\mathbf{k}} = -\hat{\mathbf{q}}$; the loop momentum $\hat{K}$ is collinear with the photon momentum Q. It is worth noting that the collinear divergence is related to the vanishing photon mass. If Q^2 were off-shell, the pole at $Q \cdot \hat{K} = 0$ would not be kinematically accessible for time-like Q^2 and it would be regularized by a principal value prescription for space-like Q^2. The collinear singularity is regularized by going to a space dimension $3 + 2\varepsilon$, the final result being

$$\operatorname{Im} \Pi^{\mu}_{\mu}|_{\text{reg}} = -\frac{2m_f^2}{\varepsilon q_0} \int [\mathrm{d}p]\delta(Q \cdot P)\left[\left(\rho_+\left(1 - \frac{p_0}{p}\right)\right.\right.$$
$$\left. + \rho_-\left(1 + \frac{p_0}{p}\right)\right)\left(\alpha'_+\left(1 - \frac{p'_0}{p'}\right) + \alpha'_-\left(1 + \frac{p'_0}{p'}\right)\right)$$
$$\left. + \text{sym.}\ (p_0, p \leftrightarrow p'_0, p')\right] \qquad (10.51)$$

where

$$\int [\mathrm{d}p] = \pi(1 - e^{\beta q_0})\int \frac{\mathrm{d}^3 p}{(2\pi)^3}\int_{-p}^{+p} \frac{\mathrm{d}p_0}{2\pi}\int_{-p'}^{+p'}\frac{\mathrm{d}p'_0}{2\pi}$$
$$\times \tilde{f}(p_0)\tilde{f}(p'_0)(2\pi)\delta(q_0 - p_0 - p'_0) \qquad (10.52)$$

The evaluation of Tr[2] in (10.44) is more cumbersome; the final result also exhibits a $1/\varepsilon$ collinear singularity. A part of it cancels that found in (10.51), but one is left with a $1/\varepsilon$ term which does not disappear. Thus resummation is unable to eliminate a logarithmic singularity in the soft photon production rate.

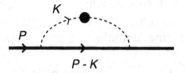

Fig. 10.9. Damping rate of a heavy fermion. The heavy dot indicates a resummed photon propagator.

10.4 Fermion damping rate

In section 7.3.3 we computed the plasmon damping rate and quoted the result for the fermion damping rate, the quasi-particles being at rest. In that case the resummation program works beautifully, because only the longitudinal component of the gauge boson propagator contributes, and long-range interactions are screened by the Debye mass. The situation deteriorates as soon as the quasi-particle moves because the transverse component of the gauge boson propagator begins to be relevant, and a logarithmic divergence arises due to the absence of magnetic screening. One possible way out is to appeal to the magnetic mass, but this does not work in the QED case, where the magnetic mass is strictly zero. Another possibility is to try a self-consistent approach by using the imaginary part of the fermion pole as an infrared regulator, while this imaginary part is precisely the quantity one is looking for; remember that this quantity is a priori of order $g^2 T$ in QCD, $e^2 T$ in QED.

In order to eliminate irrelevant details, we shall treat the case of a fast-moving heavy fermion in a QED plasma; $M \gg T$, $v = p/E \to 1$. One looks for complex poles in the fermion propagator; these are given by the equation

$$\det \{\gamma_0 p_0 - \gamma \cdot \mathbf{p} - M - \Sigma(p_0, \mathbf{p})\} = 0 \qquad (10.53)$$

and

$$p_0 = E - i\gamma \qquad (10.54)$$

This equation defines the damping rate γ (see fig. 10.9).

Given our kinematical conditions, $\mathrm{Re}\, p_0 \simeq E = (p^2 + M^2)^{1/2}$ and (10.53) yields a self-consistent equation for the damping rate γ (see exercise 10.2):

$$\gamma = -\mathrm{Im}\, p_0 \simeq -\frac{1}{4E} \mathrm{Im\, Tr} \left[(\not{P} + M)\Sigma(p_0, \mathbf{p}) \right]\Big|_{p_0 = E - i\gamma} \qquad (10.55)$$

In order to compute Σ we use a resummed photon propagator and limit ourselves to soft loop momenta, as this region gives the dominant contribution to γ, and to transverse photons which are responsible for the divergence. Note that if we set $p_0 = E$, instead of $p_0 = E - i\gamma$, namely, if we take p_0 *real* and on mass-shell, the RHS of (10.55) is half the interaction rate $\Gamma(E)$ of the heavy fermion, which was computed in section 8.1 (see (8.14)), and not the damping rate. The computations of section 8.1 were performed in the real time formalism. Let us now use imaginary time:

$$\Sigma \simeq -e^2 T \sum_{\omega_n} \int_{\text{soft}} \frac{d^3 q}{(2\pi)^3} {}^* \Delta_T(i\omega_n, \mathbf{q}) \left\{ \gamma_i S_F(P') \gamma_j P_{ij}^T \right\} \qquad (10.56)$$

Performing the frequency sums in order to derive $\text{Im}\,\Sigma$ should be standard routine by now (see (A.43)):

$$\text{Im}\,\Sigma(p_0, \mathbf{p}) = \frac{e^2 T}{E} \int_{\text{soft}} \frac{d^3 q}{(2\pi)^3} \int_{-q}^{+q} \frac{dq_0}{2\pi q_0} \rho_T(q_0, q)$$

$$\times (E\gamma_0 - (\mathbf{p}'.\hat{\mathbf{q}})(\gamma.\hat{\mathbf{q}}) - M)\,\text{Im}\frac{1}{p_0 - q_0 - E'} \qquad (10.57)$$

with $E' = [(\mathbf{p} - \mathbf{q})^2 + M^2]^{1/2}$, and we have used the approximation $f(q_0) \simeq T/q_0$. We easily recover the infrared divergence of section 8.1 when computing the interaction rate: we must take p_0 real, or more precisely $p_0 = E + i\eta$ in (10.58), in agreement with our continuation from imaginary to real time. Then

$$\text{Im}\,\frac{1}{p_0 - q_0 - E'} = -\pi\delta\left(p_0 - q_0 - \sqrt{(\mathbf{p} - \mathbf{q})^2 + M^2}\right)$$

$$\simeq -\pi\delta(-q_0 + qx) \qquad (10.58)$$

with $x = \hat{\mathbf{p}} \cdot \hat{\mathbf{q}}$. Making use of the sum rule (6.80):

$$\int_{-q}^{+q} \frac{dq_0}{2\pi q_0} \rho_T(q_0, q) \simeq \frac{1}{q^2} \qquad (10.59)$$

yields

$$\Gamma(p_0 \simeq E, p) \simeq \frac{e^2 T}{2\pi} \int^{eT} \frac{dq}{q} \qquad (10.60)$$

It is worth noting that the coefficient of the infrared divergent integral in (10.60) is gauge-fixing independent.

We now try a self-consistent approach by modifying the fermion propagator in (10.57) as follows:

$$\frac{1}{p_0 - q_0 - E' + i\eta} = \frac{1}{p_0' - E' + i\eta} \rightarrow \frac{1}{p_0' - (E' - i\gamma)} \qquad (10.61)$$

Then we find a convergent interaction rate (exercise 10.3). However the damping rate is still given by a divergent integral because Σ has to be evaluated at $p_0 = E - i\gamma$. This means that the narrow-width approximation, which is implicit in (10.61), cannot be valid.

The solution to the problem appears to depend in a crucial way on the analytic structure of the propagator. From first principles, the retarded propagator is an analytic function in the half-plane $\text{Im}\,p_0 > 0$. An expression such as

$$D_R(p_0) \simeq \frac{1}{p_0 - E + i\gamma} \qquad (10.62)$$

looks reasonable, but the point is that D_R should not have poles on the physical sheet. It would seem crucial to introduce branch cuts corresponding to thresholds for particle production, which send the pole at $p_0 = E - i\gamma$ into an unphysical sheet, as is the case for ordinary resonances. This idea has been studied in a toy model, where it can be shown that the narrow width approximation

$$\text{Im}\,\Sigma_R(p_0 = E, \mathbf{p}) \simeq \text{Im}\,\Sigma_R(p_0 = E - i\gamma, \mathbf{p}) \qquad (10.63)$$

remains valid. However, the inverse Fourier transform of the retarded propagator does not decay exponentially in time, so that γ may no longer be identified with a damping rate.

References and further reading

The validity of the KLN theorem at finite T was studied by Baier, Pire and Schiff (1988), Altherr, Aurenche and Becherrawy (1989) and Gabellini, Grandou and Poizat (1990), each group using a different infrared regularization. The KLN theorem in deep inelastic scattering has been examined by Cleymans and Dadič. The three-loop calculation in $g\varphi_6^3$ was performed by Grandou, Le Bellac and Poizat (1992), and the resummed calculation of section 10.1.2 is due to Le Bellac and Reynaud (1994). See also the article by Niégawa (1994) for a general discussion of the KLN theorem. Other work on the problem of infrared singularities may be found in Niégawa and Takashiba (1992), Weldon (1991, 1992), Landshoff and Taylor (1994). The results (10.32) and (10.33) were obtained by Kajantie and Kapusta (1985), while the problem with the partition function was discovered by Linde (1980); see also Braaten (1995). The review article by Gross, Pisarski and Jaffe (1981) is a useful reference on the problem of magnetic mass and dimensional reduction; for recent results, see Braaten and Nieto (1995). The difficulty with soft photon production was pointed out by Baier, Peigné and Schiff (1994) and by Aurenche, Becherrawy and Petitgirard (1994, unpublished). The problem of the fermion damping rate has been studied by many authors, starting with Pisarski (1989); the self-consistent approach was proposed by Lebedev and Smilga (1990, 1992), and the literature can be traced from the work by Peigné, Pilon and Schiff (1994). The role of branch cuts was examined by Baier and Kobes (1994) in a toy model.

Exercises

10.1 Show that for a heavy fermion at rest the interaction rate $\Gamma(M)$ is given by

$$\Gamma(M) = \frac{e^2 T}{4\pi}$$

Hint: start from (8.19) and use (8.1).

10.2 Write

$$\Sigma = a\gamma_0 + \gamma \cdot \mathbf{p} + c$$

and use the perturbative approximations $a, c \ll M$, $b \ll 1$ to derive (10.55).

10.3 (a) In the case $M \gg T$, $v \to 1$, show that the divergent contribution to the interaction rate is

$$\Gamma(E) = \frac{e^2 T}{\pi^2} \int \frac{dq}{q} \tan^{-1}\left(\frac{b^2}{q^2}\right)$$

where $b = \pi m^2 / 2$.

(b) Use (10.61) to show that the interaction rate may be written

$$\Gamma(E) \simeq \frac{2e^2 T}{\pi^3} \int \frac{dq}{q} \tan^{-1}\left(\frac{b^2}{q^2}\right) \tan^{-1}\left(\frac{q}{\gamma}\right)$$

Assuming that $\Gamma = 2\gamma$, show that

$$\Gamma(E) \simeq \frac{e^2 T}{2\pi} \ln \frac{1}{e}$$

(c) Show that the divergence persists if one computes the damping rate (we have assumed that γ is p-independent, but the same conclusion holds when the p-dependence is taken into account).

Appendix A
Formulary

A.1 Minkowski space

Metric tensor

$$g^{\mu\nu} = \operatorname{diag}(1,-1,-1,-1) \tag{A.1}$$

Scalar product

$$X \cdot Y = X_\mu Y^\mu = x^0 y^0 - \mathbf{x} \cdot \mathbf{y} \tag{A.2}$$

Four-divergence

$$\partial_\mu V^\mu = \partial_0 V^0 + \nabla \cdot \mathbf{V} \tag{A.3}$$

Fourier transform

$$f(P) = \int \mathrm{d}^4 X \ \mathrm{e}^{iP \cdot X} f(X) \qquad P_\mu = i\partial_\mu \tag{A.4}$$

Dirac equation

$$(i\gamma^\mu \partial_\mu - m)\psi(X) = (i\slashed{\partial} - m)\psi(X) = 0 \tag{A.5}$$

Gamma matrices

$$\{\gamma^\mu, \gamma^\nu\} = 2g^{\mu\nu} \qquad \gamma_5 = i\gamma^0\gamma^1\gamma^2\gamma^3 \tag{A.6}$$

Standard representation

$$\gamma^0 = \begin{pmatrix} 1 & 0 \\ 0 & -1 \end{pmatrix} \qquad \gamma^i = -\gamma_i = \begin{pmatrix} 0 & -\sigma_i \\ \sigma_i & 0 \end{pmatrix} \tag{A.7}$$

Dirac spinors

$$(\slashed{P} - m)u_s(\mathbf{p}) = 0 \qquad (\slashed{P} + m)v_s(\mathbf{p}) = 0 \tag{A.8}$$

Normalization

$$\bar{u}_s(\mathbf{p})u_s(\mathbf{p}) = -\bar{v}_s(\mathbf{p})v_s(\mathbf{p}) = 2m \tag{A.9}$$

A.2 Euclidean space

Fourier transform

$$f(P) = \int d^4 X \, e^{-iP \cdot X} f(X) \qquad P_\mu = -i\partial_\mu \qquad (\text{A.10})$$

Dirac equation

$$(i\gamma^\mu \partial_\mu - m)\psi(X) = (i\slashed{\partial} - m)\psi(X) = 0 \qquad (\text{A.11})$$

Gamma matrices

$$\{\gamma^\mu, \gamma^\nu\} = -2\delta_{\mu\nu} \qquad (\text{A.12})$$

Standard representation

$$\gamma_4 = i\gamma_0 = \begin{pmatrix} i & 0 \\ 0 & -i \end{pmatrix} \qquad \gamma_i = \begin{pmatrix} 0 & -\sigma_i \\ \sigma_i & 0 \end{pmatrix} \qquad (\text{A.13})$$

A.3   Feynman rules in Euclidean space (imaginary-time)

Spin zero boson propagator: $\omega_n = 2\pi n T$

$$D_F(Q) = \frac{1}{Q^2 + m^2} = \frac{1}{\omega_n^2 + q^2 + m^2} \qquad (\text{A.14})$$

Gauge boson propagator in a covariant gauge

$$D_{\mu\nu}^F(Q) = \frac{1}{Q^2}\left(\delta_{\mu\nu} - (1-\xi)\frac{Q_\mu Q_\nu}{Q^2}\right) \qquad (\text{A.15})$$

Gauge boson propagator in the Coulomb gauge

$$D_{\mu\nu}^F(Q) = \frac{\delta_{4\mu}\delta_{4\nu}}{q^2} + \frac{1}{Q^2} P_{\mu\nu}^T + \frac{\xi Q^2}{q^4}\frac{Q_\mu Q_\nu}{Q^2} \qquad (\text{A.16})$$

Transverse projector $P_{\mu\nu}^T$

$$P_{44}^T = P_{4i}^T = 0 \qquad P_{ij}^T = \delta_{ij} - \hat{q}_i \hat{q}_j \qquad (\text{A.17})$$

Fermion propagator: $\omega_n = \pi(2n+1)T$

$$S_F(P) = \frac{1}{\slashed{P} + m} = \frac{m - \slashed{P}}{P^2 + m^2} = \frac{m - \slashed{P}}{\omega_n^2 + p^2 + m^2} \qquad (\text{A.18})$$

$$\slashed{P} = \gamma_4 p_4 + \gamma \cdot \mathbf{p} = -\gamma_4 \omega_n + \gamma \cdot \mathbf{p} \qquad (\text{A.19})$$

Non-zero chemical potential

$$\omega_n \rightarrow \omega_n - i\mu \qquad (\text{A.20})$$

$g\varphi^4$-vertex

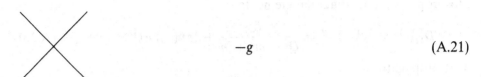

$$-g \qquad \text{(A.21)}$$

Three-gluon vertex

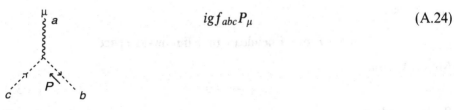

$$ig f_{abc} \big[\delta_{\mu\nu}(P-Q)_\rho$$
$$+ \delta_{\nu\rho}(Q-R)_\mu$$
$$+ \delta_{\rho\mu}(R-P)_\nu \big] \qquad \text{(A.22)}$$

Four-gluon vertex

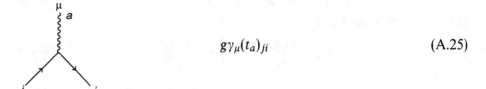

$$- g^2 \big[f_{eab} f_{ecd}(\delta_{\mu\rho}\delta_{\nu\sigma} - \delta_{\mu\sigma}\delta_{\nu\rho})$$
$$+ f_{eac} f_{edb}(\delta_{\mu\sigma}\delta_{\nu\rho} - \delta_{\mu\nu}\delta_{\sigma\rho})$$
$$+ f_{ead} f_{ebc}(\delta_{\mu\nu}\delta_{\sigma\rho} - \delta_{\mu\rho}\delta_{\sigma\nu}) \big] \qquad \text{(A.23)}$$

Gluon–ghost vertex

$$ig f_{abc} P_\mu \qquad \text{(A.24)}$$

Fermion–gluon vertex

$$g\gamma_\mu (t_a)_{ji} \qquad \text{(A.25)}$$

A.4 Feynman rules in Minkowski space (real-time)

Boson propagator: diagonal elements

$$D_{11}^F(Q) = (D_{22}^F(Q))^* = \frac{i}{Q^2 - m^2 + i\eta} + 2\pi n(q_0)\delta(Q^2 - m^2) \tag{A.26}$$

Cut propagators

$$D_F^>(Q) = 2\pi(\theta(q_0) + n(q_0))\delta(Q^2 - m^2)$$
$$= 2\pi\varepsilon(q_0)(1 + f(q_0))\delta(Q^2 - m^2) \tag{A.27}$$
$$D_F^<(Q) = 2\pi(\theta(-q_0) + n(q_0))\delta(Q^2 - m^2)$$
$$= 2\pi\varepsilon(q_0)f(q_0)\delta(Q^2 - m^2) \tag{A.28}$$

$D_F^> = D_{21}$ and $D_F^< = D_{12}$ if $\sigma = 0$

$$f(q_0) = \frac{1}{e^{\beta q_0} - 1} \qquad n(q_0) = \frac{1}{e^{\beta|q_0|} - 1} \tag{A.29}$$

Fermion propagators: diagonal elements

$$S_{11}^F(P) = (\not{P} + m)\tilde{S}_{11}^F(P) = (\not{P} + m)\left[\frac{i}{P^2 - m^2 + i\eta}\right.$$
$$\left. - 2\pi(\theta(-p_0) + \varepsilon(p_0)\tilde{f}(p_0 - \mu))\delta(P^2 - m^2)\right] \tag{A.30}$$
$$S_{22}^F(P) = (\not{P} + m)(\tilde{S}_{11}^F(P))^* \tag{A.31}$$

Cut propagators

$$S_F^>(P) = 2\pi\varepsilon(p_0)(1 - \tilde{f}(p_0 - \mu))(\not{P} + m)\delta(P^2 - m^2) \tag{A.32}$$
$$S_F^<(P) = -2\pi\varepsilon(p_0)\tilde{f}(p_0 - \mu)(\not{P} + m)\delta(P^2 - m^2) \tag{A.33}$$

$$\tilde{f}(p_0) = \frac{1}{e^{\beta p_0} + 1} \tag{A.34}$$

Vertices: multiply (A.22), (A.24) and (A.25) by $-i$, (A.23) by i and change $\delta_{\mu\nu} \to -g_{\mu\nu}$

A.5 From Euclidean to Minkowski space

Metric tensor

$$\delta_{\mu\nu} \to -g^{\mu\nu} \tag{A.35}$$

Scalar product

$$X \cdot Y = X_\mu Y_\mu \to -X \cdot Y = -X_\mu Y^\mu \tag{A.36}$$

Projector orthogonal to Q

$$\delta_{\mu\nu} - \frac{Q_\mu Q_\nu}{Q^2} \to -g^{\mu\nu} + \frac{Q^\mu Q^\nu}{Q^2} \tag{A.37}$$

Boson propagator

$$\frac{1}{Q^2 + m^2} \rightarrow \frac{i}{Q^2 - m^2} \tag{A.38}$$

Gamma matrices

$$\{\gamma_\mu, \gamma_\nu\} = -2\delta_{\mu\nu} \rightarrow \{\gamma^\mu, \gamma^\nu\} = 2g^{\mu\nu} \tag{A.39}$$

Fermion propagator

$$\frac{1}{\not{P} + m} = \frac{m - \not{P}}{P^2 + m^2} \rightarrow \frac{i}{\not{P} - m} = \frac{i(\not{P} + m)}{P^2 - m^2} \tag{A.40}$$

Fourth components

$$x_4 \rightarrow ix_0 \quad \gamma_4 \rightarrow i\gamma_0 \quad q_4 = -\omega \rightarrow iq_0 \tag{A.41}$$

A.6 Frequency sums

Boson–boson case ($\rho_l = 2\,\mathrm{Im}\,g_l$, $i\omega \rightarrow q_0 + i\eta$)

$$\mathrm{Im}\,T \sum_n g_1(i\omega_n)g_2(i(\omega - \omega_n)) = \pi(e^{\beta q_0} - 1) \int_{-\infty}^{\infty} \frac{dk_0}{2\pi}$$
$$\times \int_{-\infty}^{\infty} \frac{dk_0'}{2\pi} f(k_0)f(k_0')\delta(q_0 - k_0 - k_0')\rho_1(k_0)\rho_2(k_0') \tag{A.42}$$

Boson–fermion case ($i\omega \rightarrow p_0 + i\eta$)

$$\mathrm{Im}\,T \sum_n g_1(i\omega_n)g_2(i(\omega - \omega_n)) = \pi(e^{\beta p_0} + 1) \int_{-\infty}^{\infty} \frac{dk_0}{2\pi}$$
$$\times \int_{-\infty}^{\infty} \frac{dp_0'}{2\pi} f(k_0)\tilde{f}(p_0')\delta(p_0 - k_0 - p_0')\rho_1(k_0)\rho_2(p_0') \tag{A.43}$$

Fermion–antifermion case ($i\omega \rightarrow q_0 + i\eta$)

$$\mathrm{Im}\,T \sum_n g_1(i\omega_n)g_2(i(\omega_n - \omega)) = \pi(1 - e^{\beta q_0}) \int_{-\infty}^{\infty} \frac{dp_0}{2\pi}$$
$$\times \int_{-\infty}^{\infty} \frac{dp_0'}{2\pi} \tilde{f}(p_0)\tilde{f}(p_0')\delta(q_0 - p_0 - p_0')\rho_1(p_0)\rho_2(-p_0') \tag{A.44}$$

Appendix B
Operator formalism

In chapter 3 we derived the Feynman rules from a path integral formalism. It is instructive to rederive them by following the operator formalism, which relies on the interaction picture and Wick's theorem. This derivation will enable us to generalize the formalism to non-equilibrium situations. We consider for simplicity a neutral scalar field, without derivative interactions. Let us recall that when the full time-independent Hamiltonian H is written as

$$H = H_0 + V \tag{B.1}$$

where H_0 is the free Hamiltonian and V is the interaction, we may write the evolution operator $U_H(t, t_i)$ as

$$U_H(t, t_i) = U_{H_0}(t, t_i) U_I(t, t_i) \tag{B.2}$$

where $U_I(t, t_i)$, denoted simply by $U(t, t_i)$ in what follows, is the evolution operator in the interaction picture, which is assumed to coincide at time t_i with the Heisenberg picture. It obeys the equation

$$i \frac{dU(t, t_i)}{dt} = H_I(t) U(t, t_i) \tag{B.3}$$

where $H_I(t)$ is the interaction in the interaction picture (see (6.5)), or, equivalently,

$$U(t, t_i) = T \left(\exp \left[-i \int_{t_i}^{t} H_I(t') dt' \right] \right) \tag{B.4}$$

For arbitrary times t_1 and t_2 we have

$$U(t_2, t_1) = U(t_2, t_i) U(t_i, t_1) = U(t_2, t_i) U^{-1}(t_1, t_i) \tag{B5}$$

or

$$U(t_2, t_1) = e^{iH_0(t_2 - t_i)} e^{-iH(t_2 - t_1)} e^{-iH_0(t_1 - t_i)} \tag{B.6}$$

Assume that we wish to compute a two-point Green function:

$$G(x_1, x_2) = <\Psi | T\left(\hat{\varphi}_H(x_1)\hat{\varphi}_H(x_2)\right)|\Psi> \qquad (B.7)$$

where $|\Psi>$ is a state vector and $\hat{\varphi}_H(x)$ is the field operator in the Heisenberg picture. In the discussion which follows, we assume that $t_1 > t_2$, since this discussion can be immediately transposed to the case $t_1 < t_2$. The field operator $\hat{\varphi}_H(x)$ is related to the operator $\hat{\varphi}_{in}(x)$ of the interaction picture through

$$\hat{\varphi}_H(x) = U^{-1}(t, t_i)\hat{\varphi}_{in}(x)U(t, t_i) \qquad (B.8)$$

if the Heisenberg and interaction pictures coincide at some initial time $t_i < (t_1, t_2)$; we recall that the time evolution of $\hat{\varphi}_{in}(x)$ is governed by H_0, namely, that $\hat{\varphi}_{in}(x)$ evolves as a free field. We introduce a final time $t_f > (t_1, t_2)$ and write, using (B.8),

$$\begin{aligned}
G(x_1, x_2) &= <\Psi|U^{-1}(t_1, t_i)\hat{\varphi}_{in}(x_1)U(t_1, t_2)\hat{\varphi}_{in}(x_2)U(t_2, t_i)|\Psi> \\
&= <\Psi|U(t_i, t_f)U(t_f, t_1)\hat{\varphi}_{in}(x_1)U(t_1, t_2)\hat{\varphi}_{in}(x_2)U(t_2, t_i)|\Psi> \\
&= <\Psi|T\left(\hat{\varphi}_{in}(x_1)\hat{\varphi}_{in}(x_2)\exp\left[-i\int_C H_I(t)dt\right]\right)|\Psi> \qquad (B.9)
\end{aligned}$$

In (B.9), C is a contour going first from t_i to t_f, and then back from t_f to t_i: times on the second half of the contour are always 'later' than times on the first half, and on the second half there is in fact anti-time ordering. One usually takes $t_i \to -\infty$ and $t_f \to \infty$; then C is the so-called Schwinger–Keldysh contour. If we assume that the interaction is switched on and off adiabatically, we may take $|\Psi> = |\Psi_{in}(t \to -\infty)>$, where $|\Psi_{in}(t \to -\infty)>$ is chosen for example to be an eigenstate of the free Hamiltonian H_0.

Let us derive from (B.9) the Feynman rules at $T = 0$ first, and then the Feynman rules at thermal equilibrium. In the first case, one is interested in vacuum expectation values, so that $|\Psi> = |0>$, where $|0>$ is the vacuum state. Then $|\Psi_{in}(t \to -\infty)>$ is $|0_{in}>$, namely, the lowest eigenstate of the free Hamiltonian H_0. We then use the second line of (B.9) and remark that the action of $S = U(\infty, -\infty)$ on $|0_{in}>$ reduces to a phase factor. The Gell-Mann and Low formula follows from this, admittedly very crude, argument:

$$G(x_1, x_2) = \frac{<0_{in}|T\left(\hat{\varphi}_{in}(x_1)\hat{\varphi}_{in}(x_2)\exp\left[-i\int_{-\infty}^{\infty} H_I(t)dt\right]\right)|0_{in}>}{<0_{in}|T\left(\exp[-i\int_{-\infty}^{\infty} H_I(t)dt]\right)|0_{in}>} \qquad (B.10)$$

The perturbative expansion of the T-products in (B.10) is evaluated thanks to Wick's theorem, and this evaluation leads to the standard Feynman rules; the 'vacuum to vacuum graphs' arising from the denominator in

(B.10) cancel the corresponding graphs in the numerator, leaving us (in the case of the two-point function) with connected graphs only.

At non-zero temperature and at thermal equilibrium, we have to take a thermal average rather than an expectation value:

$$G(x_1, x_2) = \frac{1}{Z(H)} \text{Tr} \left(e^{-\beta H} T \left(\hat{\varphi}_{\text{in}}(x_1) \hat{\varphi}_{\text{in}}(x_2) \right. \right.$$

$$\left. \left. \times \exp\left[-i \int_C H_I(t) dt \right] \right) \right) \qquad \text{(B.11)}$$

Now $\exp(-\beta H)$ is formally an evolution operator in imaginary time, and (B.6) yields

$$U(t_i - i\beta, t_i) = e^{\beta H_0} e^{-\beta H} \qquad \text{(B.12)}$$

Then (B.11) may be cast into

$$G(x_1, x_2) = \frac{1}{Z(H)} \text{Tr} \left(e^{-\beta H_0} U(t_i - i\beta, t_i) \right.$$

$$\left. \times T \left(\hat{\varphi}_{\text{in}}(x_1) \hat{\varphi}_{\text{in}}(x_2) \exp\left[-i \int_C H_I(t) dt \right] \right) \right)$$

$$= \frac{\text{Tr} \left(e^{-\beta H_0} T \left(\hat{\varphi}_{\text{in}}(x_1) \hat{\varphi}_{\text{in}}(x_2) \exp\left[-i \int_{C+C_V} H_I(t) dt \right] \right) \right)}{\text{Tr} \left(e^{-\beta H_0} T \left(\exp\left[-i \int_{C+C_V} H_I(t) dt \right] \right) \right)} \qquad \text{(B.13)}$$

In (B.13), C_V is a vertical contour going from t_i to $t_i - i\beta$, times on C_V always being 'later' than times on C. The crucial point in (B.13) is that the density operator which is used to compute the thermal averages is now the free density operator: thermal averages are computed by using H_0. Note that there is no need to invoke adiabatic switching of the interaction. The contour $C + C_V$ is identical with the contour C of section 3.3 with the choice $\sigma = 0$ (fig. 3.6). As at $T = 0$, the denominator in (B.13) cancels the 'vacuum to vacuum graphs' which appear in the perturbative expansion of the numerator.

In order to derive Feynman's rules from (B.13), we need some equivalent of Wick's theorem. The standard operator form of the theorem is not convenient in this case, but fortunately one can derive the necessary result for thermal averages at equilibrium: this result will be called the 'thermal Wick theorem'. In order to keep notations simple, we first consider the case of one degree of freedom, namely, we prove the result for averages of position operators $\hat{q}(t)$, as in chapter 2, with a Hamiltonian H_0 which is the Hamiltonian of the harmonic oscillator with frequency ω. Let us take the T-product of four position operators with the time ordering $t_1 > t_2 > t_3 > t_4$. We write $\hat{q}(t)$ in terms of creation and annihilation operators $a = a_+, a^\dagger = a_-$

$$\hat{q}(t) = \frac{1}{\sqrt{2\omega}} \sum_{r=\pm} a_r e^{-ir\omega t} = \frac{1}{\sqrt{2\omega}} \sum_{r=\pm} a_r(t) \qquad (B.14)$$

The four-point thermal average reads

$$\langle \hat{q}(t_1)\hat{q}(t_2)\hat{q}(t_3)\hat{q}(t_4)\rangle = \frac{1}{4\omega^2} \sum_{r,s,t,u=\pm} \langle a_r(t_1)a_s(t_2)a_t(t_3)a_u(t_4)\rangle \qquad (B.15)$$

Now

$$\langle a_r a_s a_t a_u \rangle = [a_r, a_s]\langle a_t a_u \rangle + [a_r, a_t]\langle a_s a_u \rangle$$
$$+ [a_r, a_u]\langle a_s a_t \rangle + \langle a_s a_t a_u a_r \rangle \qquad (B.16)$$

and, from the cyclicity of the trace we have

$$\langle a_s a_t a_u a_r \rangle = e^{-r\beta\omega}\langle a_r a_s a_t a_u \rangle \qquad (B.17)$$

The preceding two equations yield

$$\langle a_r a_s a_t a_u \rangle = \frac{[a_r, a_s]}{1 - e^{-r\beta\omega}}\langle a_t a_u \rangle + \text{permutations} \qquad (B.18)$$

Suming over all indices and taking the following into account

$$\sum_{r,s} \frac{[a_r(t_1), a_s(t_2)]}{1 - e^{-r\beta\omega}} = (1 + n(\omega))e^{-i\omega(t_1 - t_2)} + n(\omega)e^{i\omega(t_1 - t_2)}$$
$$= 2\omega\langle \hat{q}(t_1)\hat{q}(t_2)\rangle \qquad (B.19)$$

leads to the 'thermal Wick theorem'

$$\langle T(\hat{q}(t_1)\hat{q}(t_2)\hat{q}(t_3)\hat{q}(t_4))\rangle$$
$$= \langle T(\hat{q}(t_1)\hat{q}(t_2))\rangle\langle T(\hat{q}(t_3)\hat{q}(t_4))\rangle + \text{permutations} \qquad (B.20)$$

A recursive proof allows us to generalize this result to $2N$-point functions. In the case of the scalar field, one easily proves the theorem by considering the mixed representation:

$$\hat{\varphi}(t, \mathbf{k}) = \frac{1}{\sqrt{2\omega_k}} \sum_{r=\pm} \sum_{\mathbf{k}} a_r(\mathbf{k})e^{-ir\omega_k t} \qquad (B.21)$$

It is a straightforward but instructive exercise to generalize these results to a non-zero chemical potential.

This thermal form of Wick's theorem allows us to express $2N$-point functions in terms of two-point functions, and is thus the analogue of the fundamental formula for Gaussian integration. This remark leads to an alternative derivation of the path integral formalism of section 3.3. It is clear that the choice of the contour $C + C_V$ leads to a doubling of the degrees of freedom, as in the path integral derivation. One also sees from (B.13) that neglecting the vertical part C_V of the contour amounts to taking thermal averages with the density matrix $Z^{-1}(H_0)\exp(-\beta H_0)$ rather than with the exact density matrix, while the dynamics of the field is still governed by the full Hamiltonian.

Off equilibrium, and for an homogeneous state, we may write the density matrix in the form

$$\rho = \sum_n p_n |n><n| \tag{B.22}$$

where the kets $|n>$ are eigenstates of H_0. Perturbation theory follows from

$$G(x_1, x_2) = \mathrm{Tr}\left(\rho\, T\left(\hat{\varphi}_{\mathrm{in}}(x_1)\hat{\varphi}_{\mathrm{in}}(x_2)\exp\left[-i\int_C H_I(t)\mathrm{d}t\right]\right)\right)$$

$$= \mathrm{Tr}\left(\rho\, S^{-1} T\left(\hat{\varphi}_{\mathrm{in}}(x_1)\hat{\varphi}_{\mathrm{in}}(x_2)\right.\right.$$

$$\left.\left. \times \exp\left[-i\int_{-\infty}^{\infty} H_I(t)\mathrm{d}t\right]\right)\right) \tag{B.23}$$

As in the equilibrium case we obtain a doubling of the degrees of freedom and a propagator in matrix form: contrary to the $T = 0$ case, the factor S^{-1} in (B.23) cannot be pulled out of the average value.

References

Adams, J.B., Ruderman, M.A. and Woo, C.H. (1963). *Phys. Rev.* **129**, 1383

Altarelli, G. (1992). QCD and experiments: status of α_s, in *QCD 20 Years Later,* eds. P.M. Zerwas and H.A. Kastrup (World Scientific, Singapore)

Altherr, T. (1990). *Phys. Lett.* **B238**, 360

Altherr, T. (1991). *Ann. Phys. (N.Y.)* **207**, 374

Altherr, T., Aurenche, P. and Becherrawy, T. (1989). *Nucl. Phys.* **B315**, 436

Altherr, T. and Kainulainen, K. (1991). *Phys. Lett.* **B262**, 79

Altherr, T., Petitgirard, E. and del Rio Gaztelurrutia, T. (1993). *Astropart. Phys.* **1**, 289

Altherr, T. and Salati, P. (1994). *Nucl. Phys.* **B421**, 662

Arnold, P. and Espinosa, O. (1993). *Phys. Rev.* **D47**, 3546

Ashida, N., Nakkagawa, H., Niégawa, A. and Yokota, H. (1992a). *Ann. Phys. (N.Y.)* **215**, 315

Ashida, N., Nakkagawa, H., Niégawa, A. and Yokota, H. (1992b). *Phys. Rev.* **D45**, 2046

Aurenche, P. and Becherrawy, T. (1992). *Nucl. Phys.* **B379**, 259

Baier, R. and Kobes, R. (1994). *Phys. Rev.* **D50**, 5944

Baier, R., Kunstatter, G. and Schiff, D. (1992). *Phys. Rev.* **D45**, R4381

Baier, R., Nakkagawa, H. Niégawa, A. and Redlich, K. (1992a). *Zeit. Phys.* **C53**, 433

Baier, R., Nakkagawa, H., Niégawa, A. and Redlich, K. (1992b). *Phys. Rev.* **D45**, 433

Baier, R., Peigné, S. and Schiff, D. (1994). *Zeit. Phys.* **C62**, 337

Baier, R., Pire, B. and Schiff, D. (1988). *Phys. Rev.* **D38**, 2814

Barton, G. (1990). *Ann. Phys. (N.Y.)* **200**, 271

Baym, G., Blaizot, J. P. and Svetitsky, B. (1992). *Phys. Rev.* **D46**, 4043

Baym, G., Friman, B.L.G., Blaizot, J.P., Soyeur, M. and Czyz, W. (1983). *Nucl. Phys.* **A407**, 541

Baym, G. and Mermin, J. (1961). *J. Math. Phys.* **2**, 232

Baym, G., Monien, H., Pethick, C.J. and Ravenhall, D.G. (1990). *Phys. Rev. Lett.* **64**, 1867

Bernard, C. (1974). *Phys. Rev.* **D9**, 3312

Bjorken, J.D. (1983). *Phys. Rev.* **D27**, 1983

Blaizot, J.P. (1992). Quantum field theory at finite temperature and density, in *High Density and High Temperature Physics,* eds. D.P. Min and M. Rho (Korean Physical Society, Seoul)

Blaizot, J.P. and Iancu, E. (1993a). *Nucl. Phys.* **B390**, 589

Blaizot, J.P. and Iancu, E. (1993b). *Phys. Rev. Lett.* **70**, 3376

Blaizot, J.P. and Iancu, E. (1994a). *Nucl. Phys.* **B417**, 608

Blaizot, J.P. and Iancu, E. (1994b). *Nucl. Phys.* **B421**, 565

Blaizot, J.P. and Iancu, E. (1994c). *Nucl. Phys.* **B434**, 662

Blaizot, J.P., Iancu, E. and Parwani (1995). *Phys. Rev.* **D52**, 2543

Blaizot, J.P. and Ollitrault, J.Y. (1990). Hydrodynamics of the quark gluon plasma, in *Quark Gluon Plasma,* ed. R.C. Hwa (World Scientific, Singapore)

Blaizot, J.P. and Ripka, G. (1986). *Quantum Theory of Finite Systems* (MIT Press, London)

Braaten, E. (1991). *Phys. Rev. Lett.* **66**, 1655

Braaten, E. (1995). *Phys. Rev. Lett.* **74**, 2164

Braaten, E. and Nieto, A. (1994). *Phys. Rev. Lett.* **73**, 2402

Braaten, E. and Nieto, A. (1995). *Phys. Rev. Lett.* **D51**, 6990

Braaten, E. and Pisarski, R.D. (1990a). *Nucl. Phys.* **B337**, 569

Braaten, E. and Pisarski, R.D. (1990b). *Nucl. Phys.* **B339**, 310

Braaten, E. and Pisarski, R.D. (1990c). *Phys. Rev. Lett.* **64**, 1338

Braaten, E. and Pisarski, R.D. (1990d). *Phys. Rev.* **D42**, 2156

Braaten, E. and Pisarski, R.D. (1992a). *Phys. Rev.* **D45**, 1827

Braaten, E. and Pisarski, R.D. (1992b). *Phys. Rev.* **D46**, 1829

Braaten, E., Pisarski, R.D. and Yuan, T. C. (1990). *Phys. Rev. Lett.* **64**, 2242

Braaten, E. and Segel, D. (1993). *Phys. Rev.* **D48**, 1478

Braaten, E. and Thoma, M. (1991). *Phys. Rev.* **D44**, 2625

Braaten, E. and Yuan, T. C. (1991). *Phys. Rev. Lett.* **66**, 2183

Brown, L.S. (1992). *Quantum Field Theory* (Cambridge University Press)

Cleymans, J. and Dadič, I. (1992). *Zeit. Phys.* **C42**, 133

Cleymans, J., Gavai, R. V. and Suhonen E. (1986). *Phys. Rep.* **130**, 217

Collins, J.C. (1983). *Renormalization* (Cambridge University Press)

Corianò, C. and Parwani, R. (1994). *Phys. Rev. Lett.* **73**, 2398

Creutz, M. (1983). *Quarks, Gluons and Lattices* (Cambridge University Press)

Dolan, L. and Jackiw, R. (1974). *Phys. Rev.* **D9**, 3320

Efraty, D. and Nair, V. P. (1992). *Phys. Rev. Lett.* **68**, 2891

Efraty, D. and Nair, V. P. (1993). *Phys. Rev.* **D47**, 5601

Elze, H. and Heinz, U. (1989). *Phys. Rep.* **183**, 81

Evans, T. (1992). *Nucl. Phys.* **B374**, 340

Evans, T.S. (1994). A new time contour for thermal field theories, in *Banff/Cap Workshop on Thermal Field Theory*, eds. F.C. Khanna, R. Kobes, G. Kunstatter and H. Umezawa (World Scientific, Singapore)

Fetter, A. and Walecka, J.D. (1971). *Quantum Theory of Many Particle Systems* (McGraw-Hill, New York)

Feynman, R.P. (1972). *Statistical Mechanics* (Benjamin, New York)

Fradkin, E.S. (1965). *Proc. Lebedev Inst.* **29**, 6

Frenkel, J. and Taylor, J. C. (1990). *Nucl. Phys.* **B334**, 199

Frenkel, J. and Taylor, J. C. (1992). *Nucl. Phys.* **B374**, 156

Gabellini, Y. Grandou, T. and Poizat, D. (1990). *Ann. Phys. (N.Y.)* **202**, 436

Gale, C. and Kapusta, J. I. (1991). *Nucl. Phys.* **B357**, 1991

Glimm, J. and Jaffe, A. (1987). *Quantum Physics* (Springer-Verlag, Berlin)

Grandou, T., Le Bellac, M. and Poizat, D. (1991). *Nucl. Phys.* **B358**, 408

Gross, D.J., Pisarski, R.D. and Yaffe, L.G. (1981). *Rev. Mod. Phys.* **53**, 43

Guérin, F. (1994). *Nucl. Phys.* **B432**, 281

Haber, H.E. and Weldon, H.A. (1982). *Phys. Rev.* **D25**, 502

Heiselberg, H. (1994). *Phys. Rev. Lett.* **72**, 3013

Heiselberg, H. and Pethick, C.J. (1993). *Phys. Rev.* **D48**, 2916

t'Hooft, G. and Veltman, M. (1974). Diagrammar, in *Particle Interactions at Very High Energies*, eds. D. Speiser, F. Halzen and J. Weyers (Plenum Press, New York)

Huang, K. (1963). *Statistical Mechanics* (John Wiley, New York)

Itzykson, C. and Zuber, J.B. (1980). *Quantum Field Theory* (McGraw-Hill, New York)

Jackiw, R. and Nair, V. P. (1993). *Phys. Rev.* **D48**, 4991

James, K. and Landshoff, P.V. (1990). *Phys. Lett.* **B251**, 167

Kadanoff, L.P. and Baym, G. (1962). *Quantum Statistical Mechanics* (W. A. Benjamin, Menlo Park CA)

Kajantie, K. and Kapusta, J. (1985). *Ann. Phys. (N.Y.)* **160**, 477

Kalashnikov, O.K. and Klimov, V. V. (1980). *Phys. Lett.* **B95**, 234

Kapusta, J.I. (1989). *Finite Temperature Field Theory* (Cambridge University Press)

Kapusta, J.I., Lichard, P. and Seibert, D. (1991). *Phys. Rev.* **D44**, 2774

Karsch, F. (1990). Simulating the quark gluon plasma on the lattice, in *Quark Gluon Plasma*, ed. R.C. Hwa (World Scientific, Singapore)

Karsch, F. (1992). Deconfinement and chiral symmetry restoration on the lattice, in *QCD 20 Years Later*, eds. P.M. Zerwas and H.A. Kastrup (World Scientific, Singapore)

Kelly, P.F., Liu, Q., Lucchesi, C. and Manuel, C. (1994). *Phys. Rev.* **D50**, 4209

Klimov, V.V. (1981). *Sov. J. Nucl. Phys.* **33**, 934

Kobes, R. (1990). *Phys. Rev.* **D42**, 562

Kobes, R. (1991). *Phys. Rev. Lett.* **67**, 1384

Kobes, R., Kunstatter, G. and Mak, K.W. (1992). *Phys. Rev.* **D45**, 4632

Kobes, R., Kunstatter, G. and Rebhan, A. (1990). *Phys. Rev. Lett.* **64**, 2992

Kobes, R., Kunstatter, G. and Rebhan, A. (1991). *Nucl. Phys.* **B355**, 1

Kobes, R. and Semenoff, G.W. (1986). *Nucl. Phys.* **B272**, 329

Kraemmer, U., Rebhan, A. and Schulz, H. (1995). *Ann. Phys. (N. Y.)* **238**, 286

Landshoff, P.V. and Rebhan, A. (1993a). *Nucl. Phys.* **B383**, 607

Landshoff, P.V. and Rebhan, A. (1993b). *Nucl. Phys.* **B410**, 23

Landshoff, P.V. and Taylor, J.C. (1994). *Nucl. Phys.* **B430**, 683

Landsman, N. P. and van Weert, Ch.G. (1987). *Phys. Rep.* **145**, 141

Le Bellac, M. (1992). *Quantum and Statistical Field Theory* (Oxford University Press)

Le Bellac, M. and Reynaud, P. (1994). *Nucl. Phys.* **B416**, 801

Lebedev, V.V. and Smilga A.V. (1990). *Ann. Phys. (N.Y.)* **202**, 229

Lebedev, V.V. and Smilga A.V. (1992). *Physica* **A181**, 187

Leibrandt, G. and Staley, M. (1994). Debye screening length in the imaginary time formalism in the temporal gauge, in *Banff/Cap Workshop on Thermal Field Theory*, eds. F.C. Khanna, R. Kobes, G. Kunstatter and H. Umezawa (World Scientific, Singapore)

Lifschitz, E. and Pitaevskii, L. (1981). *Physical Kinetics* (Pergamon Press, Oxford)

Linde, A.D. (1980). *Phys. Lett.* **B96**, 289

Ma, S.K. (1985). *Statistical Mechanics* (World Scientific, Singapore)

Mandl, F. (1988). *Statistical Physics* (John Wiley, New York).

McLerran, L.D. and Toimela T. (1985). *Phys. Rev.* **D49**, 1047

Mills, R. (1969). *Propagators for Many Particle Systems* (Gordon and Breach, New York)

Muta, T. (1987). *Foundations of Quantum Chromodynamics* (World Scientific, Singapore)

Negele, J.W. and Orland, H. (1988). *Quantum Many Particle Systems* (Addison-Wesley, New York)

Niégawa, A. (1989). *Phys. Rev.* **D40**, 1989

Niégawa, A. (1993). *Phys. Rev. Lett.* **71**, 3055

Niégawa, A. and Takashiba, K. (1992). *Nucl. Phys.* **B370**, 335

Parisi, G. (1988). *Statistical Field Theory* (Addison-Wesley, New York)

Pearson, A.C. (1994). Why the real time formalism does not factorize, in *Banff/Cap Workshop on Thermal Field Theory*, eds. F.C. Khanna, R. Kobes, G. Kunstatter and H. Umezawa (World Scientific, Singapore)

Peigné, S., Pilon, E. and Schiff, D. (1993). *Zeit. Phys.* **C60**, 455

Petitgirard, E. (1992). *Zeit. Phys.* **C54**, 673

Pisarski, R.D. (1989). *Phys. Rev. Lett.* **63**, 1129

Pisarski, R.D. and Wilczek, F. (1984). *Phys. Rev.* **D29**, 338

Ramond, P. (1980). *Field Theory: a Modern Primer* (Benjamin, Reading, Massachusetts)

Rebhan, A. (1992). *Phys. Rev.* **D46**, 4779

Rebhan, A. (1993). *Phys. Rev.* **D48**, R3967

Rebhan, A. (1994). *Nucl. Phys.* **B430**, 419

Reif, F. (1965). *Fundamentals of Statistical and Thermal Physics* (McGraw-Hill, New York)

Rivers, R.J. (1987). *Path Integral Methods in Quantum Field Theory* (Cambridge University Press)

Satz, H. (1990). Colour screening and quark deconfinement in nuclear collisions, in *Quark Gluon Plasma,* ed. R.C. Hwa (World Scientific, Singapore)

Satz, H. (1992). Probing the primordial state in high energy heavy ion collisions, in *QCD 20 Years Later,* eds. P.M. Zerwas and H.A. Kastrup (World Scientific, Singapore)

Schulman, L. (1981). *Techniques and Applications of Path Integration* (John Wiley, New York)

Schulz, H. (1994). *Nucl. Phys.* **B413**, 353

Selikhov, A. and Gyulassy, M. (1993). *Phys. Lett.* **B316**, 316

Shuryak, E. (1988). *The QCD Vacuum, Hadrons and the Superdense Matter* (World Scientific, Singapore)

Shuryak, E. (1992). *Phys. Rev. Lett.* **68**, 3270

Silin, V. (1960). *Sov. Phys. JETP* **11**, 1136

Svetitsky, B. and Yaffe, L.G. (1982). *Phys. Rev.* **D26**, 963

Taylor, J.C. and Wong, S.M.H. (1990). *Nucl. Phys.* **B346**, 115

Umezawa, H. (1994). Equilibrium and non equilibrium thermal physics, in *Banff/Cap Workshop on Thermal Field Theory,* eds. F.C. Khanna, R. Kobes, G. Kunstatter and H. Umezawa (World Scientific, Singapore)

Umezawa, H., Matsumoto, H. and Tachiki, M. (1982). *Thermofield Dynamics and Condensed States* (North Holland, Amsterdam)

van Eijk, M.A. and van Weert, Ch.G. (1992). *Phys. Lett.* **B278**, 305

van Weert, Ch.G. (1994). Aspects of thermal field theory, in *Banff/Cap Workshop on Thermal Field Theory,* eds. F.C. Khanna, R. Kobes, G. Kunstatter and H. Umezawa (World Scientific, Singapore)

Weldon, H.A. (1982a). *Phys. Rev.* **D26**, 1394

Weldon, H.A. (1982b). *Phys. Rev.* **D26**, 2789

Weldon, H.A. (1983). *Phys. Rev.* **D28**, 2007

Weldon, H.A. (1989). *Phys. Rev.* **D40**, 2410

Weldon, H.A. (1990). *Phys. Rev.* **D42**, 2384

Weldon, H.A. (1991a). *Phys. Rev. Lett.* **66**, 293

Weldon, H.A. (1991b). *Phys. Rev.* **D44**, 3955

Weldon, H.A. (1994). *Phys. Rev.* **D49**, 1579

Wilczek, F. (1992). QCD and asymptotic freedom: perspectives and prospects, in *QCD 20 Years Later,* eds. P.M. Zerwas and H.A. Kastrup (World Scientific, Singapore)

Wong, S. (1970). *Nuovo Cim.* **65A**, 689

Wong, S.M.H. (1992). *Zeit. Phys.* **C53**, 465

Zinn-Justin, J. (1989). *Quantum Field Theory and Critical Phenomena* (Oxford University Press)

Index